Lasers in Polymer Science and Technology: Applications

Volume I

Editors

Jean-Pierre Fouassier, Ph.D.

Professor
Laboratory of General Photochemistry
Ecole Nationale Superieure de Chimie
University of Haute-Alsace
Mulhouse, France

Jan F. Rabek, Ph.D.

Department of Polymer Technology
The Royal Institute of Technology
Stockholm, Sweden

CRC Press, Inc.
Boca Raton, Florida

Library of Congress Cataloging-in-Publication Data

Lasers in polymer science and technology:applications / editors, Jean
  -Pierre Fouassier and Jan F. Rabek.
   p. cm.
  Bibliography: p.
  Includes index.
  ISBN 0-8493-4844-7 (v. 1)
  1. Polymers--Analysis. 2. Laser spectroscopy. I. Fouassier,
Jean-Pierre, 1947- II. Rabek, J. F.
TP1140.L37 1990
668.9--dc20   89-9822
                                               CIP

Direct all inquiries to CRC Press, Inc., 2000 Corporate Blvd., N.W., Boca Raton, Florida, 33431.

© 1990 by CRC Press, Inc.

International Standard Book Number 0-8493-4844-7 (v. 1)
International Standard Book Number 0-8493-4845-5 (v. 2)
International Standard Book Number 0-8493-4846-3 (v. 3)
International Standard Book Number 0-8493-4847-1 (v. 4)

Library of Congress Number 89-9822
Printed in the United States

DEDICATION

To our wives, partners through life
Geneviève — Ewelina
and our children
Patrick, Laurence, and Yann — Dominika
for their patience and understanding.

PREFACE

Laser spectroscopy and laser technology have been growing ever since the first laser was developed in 1960 and cover now a wide range of applications. Among them, three groups came into prominence as regards polymer science and technology: molecular gas lasers (notably CO_2 lasers) in the IR region, gas, solid, and dye lasers in the visible and near IR region, and the relatively new group of UV excimer lasers. Lasers are unique sources of light. Many recent advances in science are dependent on the application of their uniqueness to specific problems. Lasers can produce the most spectrally pure light available, enabling atomic and molecular energy levels to be studied in greater detail than ever before. Certain types of laser can give rise to the shortest pulses of light available from any light source, thus providing a means for measuring some of the fastest processes in nature.

Measurements of luminescence (fluorescence and phosphorescence) provide some of the most sensitive and selective methods of spectroscopy. In addition, luminescence measurements provide important information about the properties of excited states, because the emitted light originates from electronically excited states. The measurement of luminescence intensities makes it possible to monitor the changes in concentration of the emitting chemical species as a function of time, whereas the wavelength distribution of the luminescence provides information on the nature and energy of the emitting species.

Such areas as laser luminescence spectroscopy, pico- and nanosecond absorption spectroscopy, CIDNP and CIDEP laser flash photolysis, holographic spectroscopy, and time-resolved diffuse reflectance laser spectroscopy, have evolved from esoteric research specialities into standard procedures, and in some cases routinely applied in a number of laboratories all over the world.

Application of Rayleigh, Brillouin, and Raman laser spectroscopy in polymer science gives information about local polymer chain motion, large-scale diffusion, relaxation behavior, phase transitions, and ordered states of macromolecules.

During the last decade the photochemistry and photophysics of polymers have grown into an important and pervasive branch of polymer science. Great strides have been made in the theory of photoreactions, energy transfer processes, the utilization of photoreactions in polymerization, grafting, curing, degradation, and stabilization of polymers. The progress of powerful laser techniques has not been limited to spectroscopical studies in polymer matrix, colloids, dyed fabrics, photoinitiators, photosensitizers, photoresists, materials for solar energy conversion, or biological molecules and macromolecules; it has also found a number of practical and even industrial applications.

One of the most important applications of lasers is the use of a high intensity beam for material processing in polymers. In these materials, the laser beam can be employed for drilling, cutting, and welding. Lasers can produce holes at very high speeds and dimensions, unobtainable by other processing methods.

Lasers can be successfully used to study surface processes and surface modification of polymeric materials, such as molecular beam scattering, oxidation, etching, annealing, phase transitions, surface mobility, and thin films and vapor phase deposition.

UV laser radiation causes the breakup and spontaneous removal of material from the surface of organic polymers (ablative photodecomposition). The surface of the solid is etched away to a depth of a few tenths of a micron, and the products are expelled at supersonic velocity. This method has found practical applications in photolithography, optics, electronics, and the aerospace industries.

The newest process includes stereolithography, which involves building three-dimensional plastic prototypes (models) from computer-aided designs. Stereolithography is actually a combination of four technologies: photochemistry, computer-aided, laser light, and laser-image formation. The device (which consists in a mechanically scanned, computer driven three-dimensional solid pattern generator) builds parts by creating, under the laser exposure, cross sections of the part out of a liquid photopolymer, then "fusing" the sections together until a complete model is formed.

Another new development is technology of micromachines such as gears, turbines, and motors which are 100 to 200 μm in diameter which can be used in a space technology, microrobots, or missile-guidance systems. These micromachines are made by a process of etching patterns on silicon chips. Beside making such micromachines, microscopic tools on a catheter, inserted through a blood vessel, would enable surgeons to do "closed heart" surgery. Developing of micromachine technology would not be possible without photopolymers and UV lasers.

The editors went to great lengths in order to secure the cooperation of the most outstanding specialists to complete this monography. A number of invited authorities were not able to accept our invitation, due to other commitments, but all authors who presented their contributions "poured their hearts out" in this endeavour. We would like to thank them for their efforts and cooperation. This monography strongly favors the inclusion of experimental details, apparatus, and techniques, thus allowing the neophyte to learn the "tricks of the trade" from the experts. This is an effort to show, in compact form, the bulk of information available on applications of lasers to polymer science and technology. The editors are pleased to submit to the readers the state-of-the art in this field.

J.-P. Fouassier and J. R. Rabek

THE EDITORS

Dr. Jean-Pierre Fouassier is head of the Laboratoire de Photochimie Générale, Ecole Nationale Supérieure de Chimie de Mulhouse, and Centre National de la Recherche Scientifique, and Professor of Physical Chemistry at the University of Haute Alsace.

Prof. J. P. Fouassier graduated in 1970 from The National School of Chemistry, Mulhouse, with an Engineer degree and obtained his Ph.D. in 1975 at the University of Strasbourg. After doing postdoctoral work at the Institüt für Makromolekulare Chemie, Freiburg, (West Germany), he was appointed as lecturer. It was in 1980 that he assumed his present position.

Prof. J. P. Fouassier is a member of the Société Française de Chimie, the Groupe Français des Polymères, the European Photochemistry Association, the ACS Polymer Division, and Radtech Europe.

Prof. J. P. Fouassier has been the recipient of research grants from the Centre National de la Recherche Scientifique, the Ministère de la Recherche, the Association Nationale pour la Valorisation et l'Aide à la Recherche, and French and European private industries. He has published more than 100 research papers. His current major research interests include time-resolved laser spectroscopies, excited state processes in photoinitiators and photosensitizers, laser-induced photopolymerization reactions, development of photosensitive systems for holographic recording, and light radiation curing.

Dr. Jan F. Rabek is Professor of Polymer Chemistry in the Department of Polymer Technology, The Royal Institute of Technology, Stockholm, working in the field of polymer photochemistry and photophysics since 1960. His research interests lie in the photodegradation, photooxidation, and photostabilization of polymers, singlet oxygen photooxidation, spectroscopy of molecular complexes in polymers, and recently photoconducting polymers and polymeric photosensors.

Dr. Rabek obtained his D.Sc. in Polymer Technology at the Department of Polymer Technology, Technical University, Wroclaw, Poland (1965) and his Ph.D. in Polymer Photochemistry at the Department of Chemistry, Sileasian Technical University, Gliwice, Poland (1968). He has published more than 120 research papers, review papers, and books on the photochemistry of polymers.

CONTRIBUTORS

Richard D. Burkhart, Ph.D.
Professor
Department of Chemistry
University of Nevada
Reno, Nevada

Michel Clerc, D.Sc.
Research Scientist
CEA
DESICP/DPC
Gif-Sur-Yvette, France

M. Fofana, Ph.D.
Research Fellow
PCSM Laboratory
Ecole Superieure de Physique et
 Chimie Industrielle
Paris, France

Curtis W. Frank, Ph.D.
Professor
Department of Chemical Engineering
Stanford University
Stanford, California

Tomiki Ikeda, Ph.D
Research Associate
Research Laboratory of Resources
 Utilization
Tokyo Institute of Technology
Yokohama, Japan

K. A. McLauchlan, Ph.D.
Physical Chemistry Laboratory
Oxford University
Oxford, England

Jean-Claude Mialocq, D.Sc.
Research Scientist
Department of Physical Chemistry
CEN Saclay
Gif-Sur-Yvette, France

L. Monnerie, Ph.D.
Professor
PCSM Laboratory
Ecole Superieure de Physique et
 Chimie Industrielle
Paris, France

Donald B. O'Connor
Research Assistant
Department of Chemistry
University of California
Riverside, California

David Phillips, Ph.D.
Department of Chemistry
Imperial College of Science,
 Technology and Medicine
London, England

Garry Rumbles, Ph.D.
Department of Chemistry
Imperial College of Science,
 Technology and Medicine
London, England

Gary W. Scott, Ph.D.
Professor
Department of Chemistry
University of California
Riverside, California

William C. Tao
Department of Chemical Engineering
Stanford University
Stanford, California

Shigeo Tazuke, Ph.D., D.Eng.
Professor
Research Laboratory of Resources
 Utilization
Tokyo Institute of Technology
Yokohama, Japan

V. Veissier, Ph.D.
Research Fellow
PCSM Laboratory
Ecole Superieure de Physique et
 Chimie Industrielle
Paris, France

J. L. Viovy, Ph.D.
Research Associate
PCSM Laboratory
Ecole Superieure de Physique et
  Chimie Industrielle
Paris, France

Mitchell A. Winnik, Ph.D.
Professor
Department of Chemistry
University of Toronto
Toronto, Canada

SERIES TABLE OF CONTENTS

Volume I

Lasers for Photochemistry

Photophysical and Photochemical Primary Processes in Polymers

Time-Resolved Laser Spectroscopy Nanosecond and Picosecond Techniques, Principles, Devices, and Applications

Photophysical Studies of Triplet Exciton Processes in Solid Polymers

Photophysical Studies of Miscible and Immiscible Amorphous Polymer Blends

Elucidation of Polymer Colloid Morphology through Time-Resolved Fluorescence Measurements

Studies of the Depolarization of Polymer Luminescence: Lasers or Not Lasers?

Laser Studies of Energy Transfer in Polymers Containing Aromatic Chromophores

Flash Photolysis Electron Spin Resonance and Electron Polarization

Volume II

Magnetic Field Effects and Laser Flash Photoloysis — ESR of Radical Reactions in Polymers and Model Compounds

Application of the CIDNP Detected Laser Flash Photolysis in Studies of Photoinitiators

Studies of Radicals and Biradicals in Polymers and Model Compounds by Nanosecond Laser Flash Photolysis and Transient Absorption Spectroscopy

Stepwise Multiphoton Processes and their Applications in Polymer Chemistry

Application of Laser Flash Photolysis to the Study of Photopolymerization Reactions in Nonaqueous Systems

Laser Spectroscopy of Excited State Processes in Water-Soluble Photoinitiators of Polymerization

Laser Spectroscopy of Photoresistant Materials

Primary Photophysical and Photochemical Processes of Dyes in Polymer Solutions and Films

Diffuse Reflectance Laser Flash Photolysis of Dyed Fabrics and Polymers

Time-Resolved Total Internal Reflection Fluorescence Spectroscopy for Dynamic Studies on Surface

Application of Lasers to Transient Absorption Spectroscopy and Nonlinear Photochemical Behavior of Polymer Systems

Volume III

Laser-Induced Photopolymerization: A Mechanistic Approach

Potential Applications of Lasers in Photocuring, Photomodification, and Photocrosslinking of Polymers

Three-Dimensional Machining by Laser Photopolymerization

Polymers for High-Power Laser Applications

Ablative Photodecomposition of Polymers by UV Laser Radiation

Holographic Spectroscopy and Holographic Information Recording in Polymer Matrices

Applications of Holographic Grating Techniques to the Study of Diffussion Processes in Polymers

Laser Photochemical Spectral Hole Burning: Applications in Polymer Science and Optical Information Storage

Volume IV

Laser Mass Spectrometry: Application to Polymer Analysis

Laser Optical Studies of Polymer Organization

Application of Lasers in the Scattering for the Study of Solid Polymers
Laser Spectroscopy in Life Sciences
Emission and Laser Raman Spectroscopy of Nucleic Acid Complexes
Picosecond Laser Spectroscopy and Optically Detected Magnetic Resonance on Model
  Photosynthetic Systems in Biopolymers

TABLE OF CONTENTS

Chapter 1
Lasers for Photochemistry 1
Michel Clerc and Jean-Claude Mialocq

Chapter 2
Photophysical and Photochemical Primary Processes in Polymers 53
Shigeo Tazuke and Tomiki Ikeda

Chapter 3
Time-Resolved Laser Spectroscopy Nanosecond and Picosecond Techniques,
Principles, Devices, and Applications 91
David Phillips and Garry Rumbles

Chapter 4
Photophysical Studies of Triplet Exciton Processes in Solid Polymers 147
Richard D. Burkhart

Chapter 5
Photophysical Studies of Miscible and Immiscible Amorphous Polymer Blends 161
William C. Tao and Curtis W. Frank

Chapter 6
Elucidation of Polymer Colloid Morphology through Time-Resolved Fluorescence
Measurements 197
Mitchell A. Winnik

Chapter 7
Studies of the Depolarization of Polymer Luminescence: Lasers or Not Lasers? 211
J. L. Viovy, L. Monnerie, V. Veissier, and M. Fofana

Chapter 8
Laser Studies of Energy Transfer in Polymers Containing Aromatic
Chromophores 237
Donald B. O'Connor and Gary W. Scott

Chapter 9
Flash Photolysis Electron Spin Resonance and Electron Polarization 259
K. A. McLauchlan

Index 305

Chapter 1

LASERS FOR PHOTOCHEMISTRY

Michel Clerc and Jean-Claude Mialocq

TABLE OF CONTENTS

I.   Introduction 2

II.   Fundamental Physical Principles 3

III.   Stimulated Emission and Amplification 4

IV.   Pumping and Population Inversion 7
       A.   Pumping Methods 8
       B.   Longitudinal Cavity Modes 8
       C.   Transverse Modes 9

V.   Solid-State Lasers 10
       A.   General Remarks 10
       B.   Ruby Lasers 11
       C.   Neodymium YAG Laser 12
       D.   Semiconductor Lasers 15

VI.   Carbon Dioxide Laser 16
       A.   Pulsed Tea CO_2 Lasers 18
       B.   Hybrid Tea CO_2 Laser 18
       C.   Axial Flow Laser 19
       D.   Sealed CO_2 Lasers 19
       E.   Ultraviolet Excimer Lasers 21
       F.   Metal Vapor Laser 24

VII.   Ultrashort Laser Pulses 26
       A.   Mode Locking 26
       B.   Q-Switched and Mode-Locked Solid-State Lasers 27
       C.   Continuous Wave Mode-Locked Laser Sources 28
       D.   Picosecond Rare Gas Halide Lasers 29

VIII.   Frequency Conversion 29

IX.   Dye Lasers 31
       A.   Introduction 31
       B.   Photophysical Properties of Laser Dyes 32
       C.   Stimulated Emission 34
       D.   Solvent Effects and the Influence of Additives 34
       E.   Continuous Wave Dye Lasers 36
       F.   Pulsed Dye Lasers 36
       G.   Flashlamp-Pumped Dye Lasers 37
       H.   Passive Mode Locking of Flashlamp-Pumped Dye Lasers 38

I.   Laser-Pumped Picosecond Dye Lasers.................................. 38
J.   Picosecond and Femtosecond Continuous Wave Dye Lasers 40
K.   Pulsed Dye Amplifiers 42

X.   Free Electron Lasers 43

Acknowledgments.. 45

References.. 45

I. INTRODUCTION

The acronym LASER (light amplification by stimulated emission of radiation) first appeared 30 years ago in a paper by Townes and Schawlow.[1] Stimulated emission had been predicted by Einstein as early as 1917. Lasers are now entering the industrial applications period but they continue to be valuable laboratory instruments. Several publications[2,3] have been devoted to the principle of laser operation, with the fundamental physics treated at very different levels of scientific knowledge. There is a much greater lack of data on the use of lasers, such as the investment costs and operating costs (power consumption and maintenance). It is still too soon to obtain precise practical information for conventional equipment. Lasers are still being extensively developed and new techniques are appearing such as free electron lasers (FELs). In the course of the last 4 years the civilian market for lasers has equaled and then overtaken the military market. This situation allows us to characterize a few types of lasers (among the numerous commercial varieties) which appear to have promise for significant industrial development. The carbon dioxide lasers and the Nd:YAG lasers have already found a place in production units for machining, cutting, soldering, or the thermal treatment of materials. The UV excimer lasers are starting to find applications in this field and have an assured future in photochemistry. The high spatial coherence of lasers makes them irreplaceable in applications where precise geometry in the deposition of energy is of prime importance.

This is the case in the production technique in high implantation density semiconductor circuits. Lasers make possible direct engraving, annealing, and the localized deposition of doping materials by laser chemical vapor deposition. The large investments necessary in the manufacture of integrated circuits do not allow other than a conservative development based on well-tested material. This is why the use of lasers which seems so promising in laboratories is slow in appearing in this industry. On the other hand, the development of medical and surgical applications is very rapid and the working conditions contribute a great deal by their requirements to the reliability of the material.

In chemistry there was previously much discussion of the comparative cost of photons and of usual sources of energy (electricity and oil). The conclusions were a little falsified by the low yields of lasers (10^{-3} and even 10^{-4} in the visible) until the appearance of excimer lasers with yields of the order of 10^{-2}.

The yield is defined as the ratio of the luminous energy measured at the exit of the laser to the electrical energy consumed. This purely energetic reasoning does not leave much chance in photochemistry except in the fields where the conventional techniques required extremely high energy consumption. This is the case, for example, in the separation of uranium isotopes where 2 MeV/separated atom of uranium-235 is used in thermal diffusion. The photochemical and radical polymerization reactions with quantum yields much greater than the inverse of the electrical yield of lasers could also be profitable. Finally it seems that the most promising applications are those where the laser contributes to the manufacture of products with a high intrinsic value. Often the energy cost in the price of these products is not the dominant one. It is also necessary that the specific characteristics of lasers such as spectral selectivity, polarization, and the small divergence are all, or in part, involved in the process considered.

In addition to the lasers already mentioned, metal vapor lasers (copper or gold) are reaching average powers of the order of 100 W, in the visible spectrum wavelengths (570 and 510 nm for copper) and with electrical yields near 1%. The champion of lasers with high electrical yields is the semiconductor laser. This laser was for a long time considered as being limited to the IR and as having very low power; now the average power is > 1W. The semiconductor laser has a yield >10% and a reliability that is measured in thousands of hours.

The wavelengths of semiconductor lasers now extend into the red visible spectrum and the powers obtained in pulsed operation allow us in the short term to envisage frequency doubling owing to the progress now being made in nonlinear crystals.

In this chapter we have chosen to sacrifice a little of the usual treatment of the fundamental physical principles of lasers, which are very well presented in a large number of books.[2-4] We shall devote more effort to the technological description of the lasers that are most likely to have industrial applications in photochemistry. Believing that a simple diagram is worth more than a photograph, we shall base our comments on laser technology on simplified drawings where the most important elements are clearly shown.

II. FUNDAMENTAL PHYSICAL PRINCIPLES

Atoms and molecules have the property of distributing their internal energy in a discontinuous series of energy states.

This distribution of energy takes place at thermodynamic equilibrium according to Boltzmann's law, where the ratios of the energy differences of the quantum levels considered E_i, E_j, and the average thermal energy kT appear in the exponential term. The distribution of the populations N_i/N_j in the states E_i, E_j is given by:

$$N_i/N_j = (g_i/g_j)\exp[-(E_i - E_j)/kT]$$

(g_i and g_j are the multiplicities of the levels i and j)

For an energy difference $[E_i - E_j]$ the photons emitted or absorbed have a frequency such that $[E_i - E_j] = h\nu$. h is Planck's constant ($6.62 \cdot 10^{-34}$ J.s) and k Boltzmann's constant ($1.38 \cdot 10^{-23}$ J. molecule^{-1} .K^{-1}). The temperature is measured in degrees Kelvin : K. The fluorescences of atoms that are subjected to certain electrical or chemical excitations give rise to light spectra which show discrete lines characteristic of the elements. The energy states of the atoms are defined by a set of quantum numbers which describe the state of excitation of an electronic orbit of an electron of the atom. For molecules there are other discrete quantized levels, which are due to vibration and rotation of the atoms which make up the molecule. In addition, for each set of electronically excited atomic states there correspond one or several electronically excited states of the molecule formed by these atoms.

Whatever the type of quantization considered (electronic, vibrational, rotational, or the

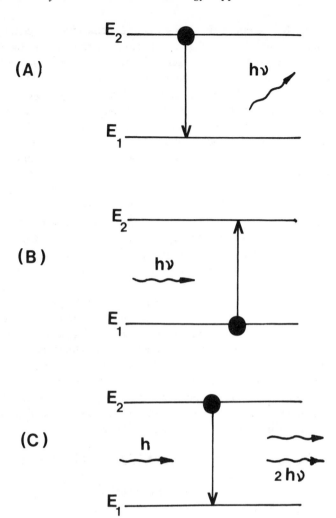

FIGURE 1.   (A) Spontaneous emission, (B) absorption, and (C) stimulated emission.

resultant of the three), the transition between two different energy states can be accompanied by the emission or the absorption of a photon whose energy is given by Planck's law, $E = h\nu$.

The energy of a transition can also be dissipated nonradiatively, for example, by collision or by internal energy conversion in the case of a molecular structure with several degrees of freedom.

The spontaneous emission from a state E_2 to a state E_1 (Figure 1A) is characterized by a coefficient "A" expressed in s^{-1}. A is the inverse of the lifetime τ of the excited state. The absorption (Figure 1B) is characterized by Einstein's second coefficient "B_{12}".

III. STIMULATED EMISSION AND AMPLIFICATION

Let us consider the interaction of a set of atoms or molecules which can be in the states (1) or (2), with energies E_1 or E_2 such than $[E_1 - E_2] = h\nu$.

Taking B_{21}, the stimulated emission coefficient, and B_{12}, the absorption coefficient, and with

$$\rho(\nu) = 8\pi \frac{n^3\nu^2}{c^3} \times \frac{h\nu}{[\exp h\nu/kT - 1]} \tag{1}$$

the radiation field of a black body with a density $\rho(\nu)$, one can write

$$(W'_{21})_i = B_{21}\rho(\nu) \tag{2}$$

$$\text{and} \quad (W'_{12})_i = B_{12}\rho(\nu) \tag{3}$$

which are, respectively, the probabilities of stimulated emission and absorption. Adding the contribution of spontaneous emission to the probability $(W'_{21})_i$ one obtains

$$W'_{21} = B_{21}\rho(\nu) + A \tag{4}$$

At thermal equilibrium the average populations of the two states are constant and the number of transitions from (2) to (1) is equal to the number of transitions from (1) to (2)

$$N_2W'_{21} = N_1W'_{12} \tag{5}$$

where N_1 and N_2 are the population densities of levels 1 and 2. Combining Equations 2, 3, and 5 and using Equation 1 for $\rho(\nu)$, one obtains

$$N_2\left[B_{21} \cdot \frac{8\pi n^3 h\nu^3}{c^3(\exp h\nu/kT - 1)} + A \right] = N_1 \cdot B_{12} \cdot \frac{8\pi n^3 h\nu^3}{c^3(\exp h\nu/kT - 1)} \tag{6}$$

At thermal equilibrium one has

$$N_2/N_1 = g_2/g_1 \exp(-h\nu/kT) \tag{7}$$

g_1 and g_2 being the multiplicities of levels (1) and (2). For example, for an atom with a quantum number J, the multiplicity is equal to $(2J + 1)$.

Equations 6 and 7 can only be simultaneously satisfied if one has

$$B_{12} = B_{21} \cdot g_2/g_1 \tag{8}$$

and at the same time

$$A/B_{21} = 8\pi n^3 h\nu^3/c^3 \tag{9}$$

(or as a function of the wavelength $\lambda = c/\nu$):

$$A/B_{21} = 8\pi n^3 h/\lambda^3 \tag{9}$$

Finally, one can write the number $(W'_{21})_i$ of transitions induced per second and per atom under the influence of the energy density $\rho(\nu)$ of black body radiation

$$(W'_{21})_i = \frac{A\lambda^2}{8\pi n^2 h\nu} \cdot \frac{c}{n} \cdot \rho(\nu) \tag{10}$$

The radiation spectrum of a black body is very wide in comparison to the spectral width of the absorption or emission of the transition $[E_2 - E_1]$ at the frequency ν. In a laser, one must consider the interaction with radiation that is much more monochromatic and with a small spectral width at the frequency ν. The efficiency of the interaction depends on the difference between the frequency of the stimulating radiation and the resonance frequency E/h of the transition.

If the number of transitions induced per second by the monochromatic radiation is $(W'_{21})_i$, *then (W_{21}) must be proportional to a function g(ν) centered at the transition frequency ν = E/h.*

W_{21} *and W'_{21} must approach each other when the spectral width becomes very large.* These two conditions are satisfied in writing:

$$(W_{21})_i = \frac{C^3}{8\pi n^3 h\nu^3} \cdot A \cdot \rho_\nu \cdot g(\nu) \tag{11}$$

To avoid any confusion in this equation ρ_ν is the *total energy density* centered around the frequency and contained in a spectral width that is very narrow in comparison to that of the black body, whereas in Equations 1 to 3, $\rho(\nu)$ refers to a spectral width.

ρ_ν is expressed in joules per unit volume and g(ν) is in joules per unit volume per hertz. Let us note moreover that the function g(ν) is normalized;

$$\int_{-\infty}^{+\infty} g(\nu)d\nu = 1 \tag{12}$$

Taking the radiation intensity centered at the frequency ν as I_ν so that $I_\nu = (c/n) \cdot \rho_\nu$ one obtains on the basis of [11] the equation expressed as a function of the wavelength $\lambda = c/\nu$.

$$(W_{21}) = \frac{\lambda^2}{8\pi n^2 h\nu} \cdot A \cdot I_\nu \cdot g(\nu) \tag{13}$$

The variation of the power ΔP per unit volume is written

$$\Delta P/V = [N_2(W_{21}) - N_1(W_{12})]h\nu \tag{14}$$

Using equations [2], [3], [8] and [14] we have

$$\Delta P/V = \left(N_2 - \frac{g_2}{g_1} \cdot N_1\right) \frac{\lambda^2}{8\pi n^2} \cdot A \cdot g(\nu) \cdot I(\nu) \tag{15}$$

This power is added to the incident wave which increases in the z direction:

$$dI(\nu,z)/dz = \frac{\Delta P}{V} = \gamma(\nu) \cdot I(\nu,z) \tag{16}$$

$\gamma(\nu)$ is the exponential coefficient of amplification:

$$\gamma(\nu) = \frac{\lambda^2}{8\pi n^2} \cdot A \cdot g(\nu)\left[N_2 - \frac{g_2}{g_1} N_1\right] \tag{17}$$

The amplification for $\gamma > 0$ is written:

$$I(v,z) = I_0(v)\exp[\gamma(v) \cdot z] \tag{18}$$

IV. PUMPING AND POPULATION INVERSION

Boltzmann's equation shows that at thermal equilibrium at room temperature the majority of the species are in the ground state. For example, at 300 K the average energy is kT = 0.025 eV. For a laser emitting in the 600-nm range, in a transition $E_2 \rightarrow E_1$ which would be, for example, to the ground state, one would have $hv = 2.06$ eV. In this case the exponential term in the Boltzmann's equation would be equal to exp $[-2.06/0.025] = 1.64 \cdot 10^{-36}$.

For this reason one avoids choosing the pair of levels E_2 and E_1 with E_1 being the ground state. This is, however, as we shall see later, the case with the ruby laser where practically the lowest level of the Cr^{3+} ion is split by the field of the crystal (this splitting is however much less than kT).

A diagram of pumping involving four levels is shown in Figure 2B. This possesses the advantage of having a level E_1 at the termination of the laser emission which is situated at an energy well above the thermal energy kT. The level E_1 is therefore much less populated than the ground state level E_0 (Figure 2B) which greatly facilitates the pumping aimed at obtaining a population inversion between E_1 and E_2. In this four-level mechanism, pumping takes place by the intermediary of level E_3. The nonradiative transfer process from E_3 to E_2 brings about the desired population inversion between E_2 and E_1. When the stimulated emission from E_2 to E_1 is ended, the return to the ground state takes place by radiative or nonradiative relaxation of the E_1 state. This relaxation is sometimes very slow and results in blocking the stimulated emission. In some cases, such as the helium-neon laser as well as the copper vapor laser, the relaxation of the E_1 state occurs mainly by collisions of the metastable atoms with the walls. For the copper vapor laser relaxation also takes place by superelastic collisions with the electrons. The third pumping method (Figure 2C) is quite similar to a four-level system, but in addition it makes use of the possibility of the transfer of resonant energy between two different species. This is the case in the $(CO_2 + N_2)$ laser where there is energy transfer between the vibrationally excited nitrogen and a level of the CO_2 in nearly perfect resonance, from which results the well-known CO_2 laser emissions around 9 and 10 μm.

Pumping and population inversion are described by a system of kinetic equations. For example, for a four-level system (Figure 2B) one writes

$$\frac{dN_1}{dt} = R_{21}N_2 + R_{31}N_3 - R_{10}N_1 \tag{19}$$

$$\frac{dN_2}{dt} = R_{32}N_3 - R_{21}N_2 \tag{20}$$

$$\frac{dN_3}{dt} = PN_0 - R_3N_3 \tag{21}$$

$$\frac{-dN_0}{dt} = \frac{dN_1}{dt} + \frac{dN_2}{dt} + \frac{dN_3}{dt} \tag{22}$$

In these equations N_1, N_2, and N_3, designate the populations expressed in species per cubic centimeter. $-R_{xy}$ are the rate constants in s^{-1}.

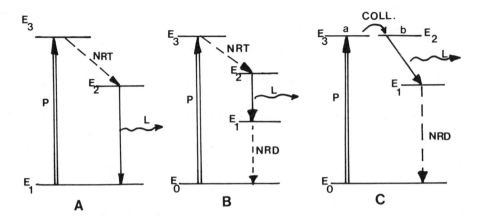

FIGURE 2.   Pumping and inversion. (A) Three-level scheme, (B) four-level scheme, and (C) four levels with resonant collisional transfer, NRT: nonradiative transfer and NRD: nonradiative decay.

$-P$ is the rate constant of pumping of level E_3, and R_3 the constant of the depopulation of this level by various mechanisms, but mainly the transfer of energy from E_3 to E_2.

Population inversion occurs when $N_2 > g_2 N_1 / g_1$. This simplified writing of the system of kinetic equations assumes that the excited levels have very long lifetimes, which allows us to neglect the spontaneous emission process. In fact, in some of the cases which will be discussed in this chapter, the radiative lifetimes are very short (10^{-8} s) and the spontaneous emission ($A > 10^8$ s^{-1}) is in competition with the pumping. This is the case, for example, in excimer lasers.

A. PUMPING METHODS

The pumping level E_3, which is shown in the three schematics of Figure 2, is often made up of a set of excited levels which are pumped simultaneously. In general, these levels are very strongly coupled which allows us to treat them kinetically as a single pumping level E_3.

Therefore, the pumping does not need to be carried out by an external source as selective as the schematics in Figure 2 could let one think. In the case of a gaseous medium, a continuous or pulsed electric discharge is generally used. For liquid solutions (dyes) or solids (glass or crystal lasers) incoherent polychromatic sources such as arcs or flashes are used.

As a reminder, since they are pumping methods which are not used for the lasers which interest us in this chapter, we shall cite chemical pumping and pumping by an accelerated particle beam.

The best pumping yield is obtained when the external source is precisely resonant with the absorption bands of the amplifying medium. This is the case in the laser pumping used for dye lasers or for some solid amplifying media (color center lasers). The lasers most often used for pumping are the ionized rare gas lasers, the pulsed Nd:YAG lasers used in their second harmonic at 532 nm, the copper vapor lasers, and the UV excimer lasers. These are the principal lasers described in this chapter.

B. LONGITUDINAL CAVITY MODES

The width of the amplifying band varies from some hundreds of megahertz for the atomic lines to several hundred gigahertz for the dyes. The resonant cavity formed by the pair of mirrors "O" and "R" (Figure 3) is a Fabry-Perot etalon with a free spectral range equal to c/2L. For example, for a 50-cm long cavity (corrected for refraction indices of the different media) one has c/2L = 0.3 GHz. There can be, therefore, several modes spaced at c/2L which oscillate simultaneously in the spectral range of the gain band (Figure 3).

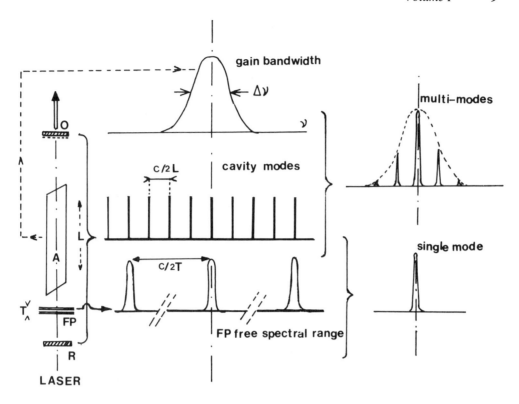

FIGURE 3.   Laser cavity properties (longitudinal modes). A = Amplifying medium, R = maximum reflectance mirror, O = output coupler (partially reflecting mirror), FP = Fabry-Perot, T = thickness of the FP, and L = cavity length from mirror to mirror. (From Boulnois, J. C., *Le Laser,* Edition Lavoisier Paris, 1984. With permission.)

In these conditions the laser emission spectrum is called "longitudinal multimode". If a spectrally selective element is introduced in the cavity, for example, a Fabry-Perot etalon with a small width "T" and a free spectral range c/2T, the combined transmission of the assembly allows the isolation of a single longitudinal mode. For an average wavelength of 500 nm, $N = L/\lambda = 10^6$. Even in choosing very stable materials and controlling the temperature very precisely it is impossible to have the length "L" constant too near to 10^{-6}.

It is generally necessary to control the system by using, for example, mirrors mounted on piezoelectric blocks subjected to a signal which compares the laser wavelength to a reference wavelength such as, for example, an atomic line.

In the case of pulsed operation when several modes are emitted, a strong modulation of the time envelope of the pulse is observed. When the laser is operating in "longitudinal single mode" the time evolution of the pulse is smooth.

C. TRANSVERSE MODES

Until now we have considered only the resonance conditions of the cavity in the direction of the optical axis of the resonator. In fact the diameters of the amplifying medium and the mirrors are large enough to allow the previously defined axial frequencies to have directions that are slightly oblique with respect to the optical axis. When no special precautions are taken to limit the oscillation to single axially symmetric mode TEM_{00} (transverse electromagnetic mode), for example, in placing a diaphragm in the cavity, one obtains beams showing patterns distributed in the plane perpendicular to the beam. These patterns may be symmetric with respect to the axis of the cavity or with respect to planes passing through this axis. Transverse multimode operation is rarely wanted (at least for photochemical applications). In general it is sufficient to reduce the diameter of the cavity with a diaphragm so as to have the limiting diffraction conditions: $\Theta = k \cdot \lambda/d$, with $k = 1.22$ for a plane wave and $k = 2/\pi$ for a gaussian beam. Θ is the limiting angle and d the diameter of the diaphragm. The problem of the choice of cavities combining plane, concave, or convex mirrors is quite complicated and will be found exhaustively treated by Ronchi[5] or by Young[6] in articles on laser optics. In some cases where the transverse dimensions of the amplifying medium are significant and where one wants to extract the maximum energy, one uses combinations of concave and convex mirrors which give an annular cross-section beam in a near field. Most often one wants a gaussian distribution of the intensity around the $\vec{z}$ axis of the resonator, this distribution corresponds to $I(r) = I_0 \exp[-2r^2/W^2(z)]$, where $W(z)$ is the radius of the circle for which the intensity is equal to I_0/e^2, and r is the radius of the transverse cross-section perpendicular to the axis in the case of transverse single mode operation of the cavity. The reason for this is that the gaussian distribution is kept in the course of the beam transportation through the multiple optical systems used to deliver the beam.

V. SOLID-STATE LASERS

A. GENERAL REMARKS

With the exception of semiconductor lasers, which will be described in another section, the amplifying medium in solid-state lasers is a glass or a crystal. The optically active transitions occur between discrete electronic states of metal ions. The most commonly used ions are those of chromium for the ruby laser and those of neodymium for the Nd:YAG laser.

The glasses are silicates or phosphates, and the crystals are, for example, alumina for the ruby laser or a garnet of yttrium and aluminum for the YAG. There are a large variety of crystalline substrates for the transition metal ions. The amplifier is usually a cylindrical rod with faces treated with an antireflecting coating at the wavelength of the oscillation. The faces are ground at an angle of a few degrees with respect to the axis so that the rod does not act as a subcavity.

The flash lamps parallel to the axis of the cylinder (Figure 5), are usually filled with xenon or krypton in order to obtain the maximum overlap between the range of the emission of the flash lamps and the absorption spectrum (in particular in the case of Nd:YAG). This spectral overlap is very limited which is one of the reasons for the low efficiency of solid-state lasers. Considerable progress in the pumping of lasers has been recently made by replacing flash lamps by semiconductor diode lasers.

The first solid-state Nd:YAG lasers pumped by the radiation from diode lasers date from 1986, their continuous wave (CW) power at 1.064 μm is still limited to a fraction of a watt. Another recent advance, which is especially interesting for Nd:YAG, is to no longer use cylindrical rods but instead slabs, which opens the possibility of solid-state lasers emitting several kilowatts of average power.

In the design of solid-state lasers there are problems of thermal equilibrium between the energy deposited by the flashes, the useful energy extracted from the medium in the form of coherent radiation, and the waste thermal energy removed by the cooling system.

The choice of a medium (glass or crystal) is made as a function of parameters, among which the thermal conductivity plays an important role. The success of the Nd:YAG laser is due in part to its good thermal conductivity. The other optional parameters concern the optics; for example, the substrate plays a role in the spectral broadening of the lines of the ions. The coefficients related to the nonlinear properties of the substrate are also critical, as for certain values of the power density catastrophic autofocusing may occur.

For the photochemical applications which interest us here the glasses (silicates or phos-

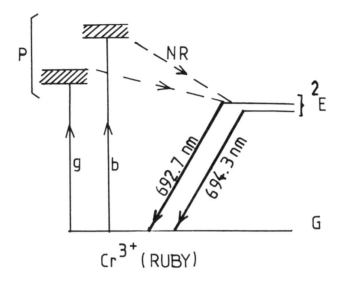

FIGURE 4. Cr^{3+} ion energy levels in the Ruby Laser, g and b = green and blue absorption bands, P = set of pumping levels, and NR = non-radiative energy transfer.

phates) are ruled out as they are not suited to a repetition rate above 1 Hz. The YAG as well as alexandrite and the other crystals have been developed commercially.[30,31]

B. RUBY LASERS

The crystal is an alumina substrate doped with 0.05% Cr_2O_3. The operation of the laser involves three quantum levels with the states $^4F_1$ and $^4F_2$ of Cr^{3+}, which correspond to ruby absorption bands in the blue and the green. A very rapid nonradiative transfer occurs between the ^4F states and the ^2E state.

The stimulated emission arises from the ^2E state which has a population inversion allowing the emission of a doublet at 694.3 and 692.7 nm. The first of these at 694.3 is the most important and the optics of the cavity of a ruby laser are in general optimized for this wavelength. The ^2E doublet is due to the splitting of the excited state by the crystalline field of the ruby (Figure 4).

This laser was invented by Maiman[7] in 1960 and a large number of pumping and cavity switching systems applicable to solid-state lasers have been used with the ruby laser (Q switching, mode locking, etc.). The cavity of the oscillator of a solid-state laser is shown schematically in Figure 5 along with the optical components that are conventionally used.

It has appeared very interesting from the start to allow the oscillation in the cavity to take place only when the maximum of excited states have been optically pumped in the rod. The ruby laser has allowed the development of fast blocking and deblocking of the cavity based on its Q factor. This Q-switching technique allows the production of pulses of several hundreds of millijoules in a few nanoseconds. This has been obtained (in historical order) by rotating mirrors and by saturable filters, consisting of dyes which absorb at 694.3 nm used either in solution in a liquid or in glasses (saturable glasses). Finally, the present techniques most often used are based on the Pockels effect, the Kerr effect, or on an acoustooptic modulation.

The passive or active methods of mode locking to obtain trains of picosecond pulses, which are described in detail elsewhere (see Section VII), are applicable to ruby lasers.

Although the ruby laser has a very high pumping threshold, it has been shown in the laboratory that it can be pumped and made to oscillate giving a continuous emission. There

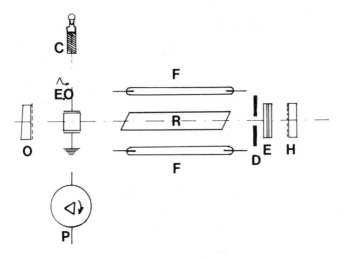

FIGURE 5.   Q-switched solid state laser. Optical shutters: C: saturated dye cell, E.O.: electro-optic shutter (Pockels), P: rotating prism, R: laser rod, D: diaphragm, E: Fabry-Perot etalon, H: high reflectance mirror, O: partially reflecting output coupler, and F: flashes.

has been no commercial development of this mode of operation. The thermal conductivity of crystalline alumina is good, higher than that of glasses and even that of YAG. The crystal is birefrigent.

Ruby lasers can function only at a fairly low rate; for example, at 30-s intervals for pulses of several hundreds of millijoules or at a fraction of a hertz for pulses that do not exceed a few millijoules.

In mode-locked operation trains of 20 to 30 pulses with a duration of 3 to 4 ps are obtained.

It can be tuned with a much reduced output in the range of 330 GHz but this is rarely made use of.

Although there are no possible photochemical applications, the ruby laser continues to be commercially produced for it provides a wide (0.1- to 1-m) range of coherence lengths and small divergences, well suited to metrology and holography.

C. NEODYMIUM YAG LASER

In a 1985 evaluation this laser alone accounted for 22% of the market of lasers for civilian applications. The YAG is the competitor for the CO_2 gas laser in the field of mechanical applications.

The active ion is Nd^{+3}; this laser is thus called Nd:YAG or, by shortening, simply YAG, the initials of yttrium aluminum garnet ($Y_3Al_5O_{12}$). The diagram of the energy levels of the Nd^{+3} ion is given in Figure 6. The stimulated emission at 1.064 μm takes place between the $^4F_{3/2}$ and the $^4I_{11/2}$ levels, but there are other possibilities of laser emission at 0.9 and 1.35 μm which are more rarely used. The laser effect of Nd^{+3} ions in a YAG matrix has been observed by Geusic[8] in 1964. An exhaustive review relative to Nd:YAG lasers has been very recently published by Zverev et al.[9]

The effective cross section for stimulated emission of the Nd:YAG is 35 times greater than that of the ruby. As a result the pumping threshold is much lower, allowing the Nd:YAG laser to perform as well with a continuous output as in Q-switched operation with a very high repetition rate. To give an order of magnitude, one can now obtain 100 mJ, and 10- to 20-ns pulses with a repetition rate of several tens of hertz out of 2- to 5-cm³ rods.

In CW operation several to 100 watts are obtained, depending on whether the beam is single mode or transverse multimode.

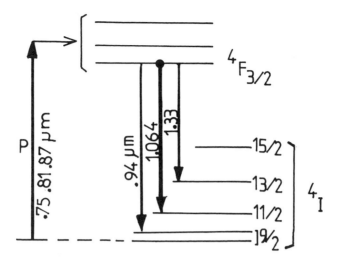

FIGURE 6. Nd³⁺ ion in a YAG crystal (yttrium aluminium garnet). Main emission at 1.064 μm and weak emissions at 0.94, and 1.3 μm. P = principal absorptions at 0.75, 0.81, and 0.87 μm.

Most often, either for photochemistry or for the pumping of dye lasers, the fundamental wavelength 1.064 μm is not used, but the second (0.532 μm) or the third (0.354 μm) harmonic. The second harmonic is obtained with a yield of about 30%, at the optimum of the beam at 1.064 μm, and with a specific power depending on the nonlinear doubling crystal selected (KDP, KTP, LiNbO₃, etc.).

The third harmonic is obtained by parametric mixing of the fundamental wavelength at 1.064 μm and of the second harmonic at 0.532 μm. The use of the fourth harmonic at 0.266 μm has become much less frequent since the near UV region is now directly accessible (with much more energy available per pulse) using excimer lasers.

The second harmonic at 0.532 μm corresponds to photons with an energy of 2.3 eV, equivalent to 55 kcal/mol⁻¹. This order of magnitude corresponds to numerous atomic and molecular absorptions, thus justifying the interest in this wavelength in photochemistry. Also the energy of these photons is high enough for the pumping of dye lasers which emit from the yellow to the near IR. This green wavelength causes less photochemical damage than is the case of pumping by UV excimer lasers, as, for example, XeCl at 308 nm (≃ 4 eV or 95 kcal/mol).

The form of the resonant cavity of a Nd:YAG laser is fundamentally the same as that used for the ruby laser. The quality of the pulse generators for electrooptical systems of the Pockels type makes them preferable to rotating mirrors of saturable dyes, which are shown only as a reminder in Figure 6. As the gain of the Nd:YAG is much higher than that of the ruby and also because the absorption bands of Nd³⁺ are in the near IR from 0.75 to 0.87 μm, the flashes are different from those of the ruby and the pumping dissipates much less waste thermal energy. The rod diameter is 6, 7, or 9 mm and the length 5 to 7 cm, and it is pumped by one or two flash lamps arranged in a reflector with an elliptical cross section. The rod is placed at one of the foci of the ellipse and the flash at the other. Of course, with two flash lamps two elliptical sections with a common focus are used and so forth.

The aim here is to have uniform illumination of the rod around the optical axis of the

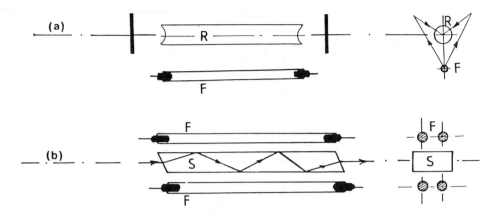

FIGURE 7.   (a) Cylindrical laser rod R, showing (exaggerated) thermal barrel deformation of refractive indexes and (b) slab showing propagation by internal reflection and flash pumping on both sides.

cavity and over the length of the rod. This is not perfectly attained especially when high average power (>100 W) is desired.

In effect, it is not ideal to optically pump the rod radially from the exterior to the interior, while trying to remove the calories in the inverse direction. This results in a radial gradient of index produced by the thermal gradient in the rod and schematized as a barrel deformation in Figure 7.

This results in an effect equivalent to that of a lens convergent or divergent by a fraction of a diopter, which can lead to undesirable focusing in the frequency doubling crystals or in the amplifiers (if these are used).

The use of slabs, prismatic plates of YAG crystals, allows us to solve some of these technological problems. Figure 7 shows the progression by internal reflection of the average ray of a laser beam in a slab.

The use of slabs allows us, in theory, to obtain more uniform pumping across the thickness of the plate and to have better removal of calories by the cooling water. To our knowledge there are no commercial units based on the use of slabs, but tests show that some kilowatts of average power in the emission at 1.064 μm can thus be obtained.

This important technological advance strengthens the position of YAG lasers on the market of high average power lasers (mainly that of CO_2 lasers, excimers, and copper vapor lasers).

The problem with these is the production of large crystalline plates. As a matter of fact, YAG garnet (as well as homologous garnets such as YLF) is obtained by growing crystals from molten salts (Czochralski's method).

It is very difficult to obtain monocrystalline rods with a diameter of more than 5 cm and a length of more than 10 cm, and in these the slabs must be machined from the most homogeneous region, at the price of the loss of much of the crystal.

The flash pumping yield varies from 0.2 to 1% depending on whether the beam is single mode (diffraction limited) or transverse multimode. The flash lamps give radiation equivalents to that of a black body at 6500 K and the use of xenon or krypton allows an improvement in the pumping yield,[10] but the majority of the electrical energy stored in the condensers is lost in the form of heat in the cooling circuit for the laser heads. The most recent advance made in the pumping of solid-state lasers has been the use of gallium arsenide (GaAs) semiconductor solid-state lasers, which will be discussed in Section V.D. As a matter of fact, semiconductor lasers emit precisely in the region from 0.75 to 0.87 μm, where the Nd^{3+} ion absorption bands are located.

The semiconductor lasers are characterized by an excellent electric yield, of the order of 30% at the threshold. A single junction can now emit 0.5 W, and it is possible to choose diodes that coincide perfectly with the Nd^{3+} absorption bands. Thus continuous emissions from the Nd:YAG at 1.064 μm, at more than 100 mW, with a yield of more than 10% can be produced.

The diode pumping of Nd:YAG is carried out at precisely 808.5 nm; there are instruments which emit 35 or 70 mW at 1.064 μm and 12 or 25 mW at 1.319 μm (depending on the number of pumping diodes and the optical processes chosen). A diode pumped laser having similar outputs at 1.047 and 1.313 μm uses a YLF matrix (Li Y F_4).

The geometry is very simple: the Nd:YAG or YLF rod is a small cylinder with a diameter and a length of a few millimeters, which is pumped along its axis by a 100 to 200 mW Ga.Al.As diode laser.

The fundamental emission at 1.064 μm can be doubled to 532 nm; one thus obtains from 4- to 11-mW in the green, the doubling crystal used is potassium titanyl phosphate (KTP).

Progress in the field of diode laser pumping is very rapid and in development laboratories work is being carried out using multistripe diode arrays where up to several tens of channels can emit 20 W in 150-μs pulses at 50 Hz.[11] Much can be hoped for in the use of diodes in the pumping of solid-state lasers; for example, the lifetime of a tungsten-halogen lamp is of the order of 100 h, compared with the lifetime or diodes which can be more than 5000 h. This lifetime, however, is strongly dependent on the operating temperature.

Tunable solid-state lasers are now coming at a state of development allowing their direct application to a wide field of applications including photochemistry. The advantages of solid-state lasers are pointed out by Moulton;[30] the Alexandrite laser (Cr:Be Al_2O_4) is tunable from 700 to 800 nm, and offers many similarities with the Nd:YAG from robustness standpoint.[31]

Ti:Al_2O_3, which is tunable from 700 to 1000 nm, can be pumped by flashes, CW lasers, or by highly repetitive copper vapor lasers. By some aspects, this crystal behaves as a dye laser, in a wavelength range where dye lasers are generally difficult to operate.

D. SEMICONDUCTOR LASERS

Until very recently diode lasers or semiconductor lasers would not have found a place in a chapter devoted to lasers for photochemistry. It is perhaps still a little premature, but the orders of magnitude of power and yield are becoming so attractive that it is important to be informed in this field. One is beginning to see, as a matter of fact, the commercialization in the U.S., Japan, and Germany of diode lasers that can emit a fraction of a watt per channel and tens of watts from multiple arrays. We saw above that the most immediate application is the pumping of solid-state lasers such as Nd:YAG and Nd:YLF. The shorter wavelengths that can be emitted by semiconductor lasers are near to the red band, 0.7 to 0.8 μm. In photochemistry, strictly defined, there is as yet no example of the direct use of diode lasers, but on the other hand in atomic physics there have been experiments where carefully selected GaAlAs diodes were used in investigations with resonance radiation from cesium at 0.85 μm.[13]

For a long time a diode laser has been pictured as a junction between two semiconductors, one p-doped and the other n-doped (according to whether electrons or holes are dominant).

The first lasers that appeared in 1962 were of this type, called homojunction.[11,12] These laser diodes (LDs) could not operate at room temperature for the junction current required to obtain the desired inversion between the valence and the conduction bands would have been destructive. It is difficult to summarize in a few words the considerable progress and the manifold technological advances that have been made from 1962 to the present.

From our point of view, this development will be summarized as follows: the methods of epitaxy of crystalline layers, first in liquids, then with atomic jets, and finally MOCVD

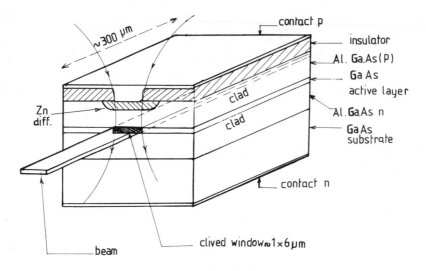

FIGURE 8.   Schmatic view of a semiconductor laser, consisting of a heterostructure of GaAs barriers.

(metal organic chemical vapor deposition) have in turn brought about the improvements in the three dimensional building up of layers of variously doped semiconductors. The channelizing of the pumping current in the volume of the active zone of the laser has been progressively obtained by the use of the heterojunction technique. The stacked layers in a very complex mixed structure (Figure 8) have a thickness of a fraction of a micron. The emitting channel of the laser is typically 5 μm wide and 300 μm long. The electrons are confined in the active layer of GaAs by lateral potential barriers of p-type GaAlAs which make up the heterojunction. The electrons are thus confined in the active layer where they recombine with the holes to produce photons. The band of active layers plays the role of a light guide. Because of the small dimensions, particularly of the thickness, of the cross section of the channel, the radiation emitted is strongly diffracted.

In the spectral range from 0.7 to 2 μm, the materials are AlGaInP, AlGaAsP, AlGaAs, InGaAsP, and InGaAs.

These new heterojunctions operate at room temperature and the luminous energy that can be extracted from a stripe is of the order of 100 to 200 mW in GaAs. A limit is fixed by the optical damage that the exit faces of the channel can tolerate. These faces are most often produced by cleavage of the crystal. The limit is some megawatts per squared centimeter in CW operation. This development of semiconductor lasers to higher and higher powers is described in a recent article.[11]

For the moment the spectral range covered by the heterojunctions of GaAs or quaternary alloys such as Al.Ga.In.P is limited to 0.7 μm on the short wavelength side. The major applications are those in telecommunications where there is more interest in the IR side towards 1.3 μm, and those of the videodisc or laser printer type where the search for short wavelengths is not a major objective. Nevertheless, there already have been laboratory scale experiments on the generation of harmonics by frequency doubling in optical guides of lithium niobate. The radiation widths of semiconductor lasers are very small (10 MHz) and the cavity modes are widely spaced (>500 GHz) as the typical length of a cavity is about 300 μm.

VI. CARBON DIOXIDE LASER

The technological success of the carbon dioxide (CO_2) lasers is due to their relatively

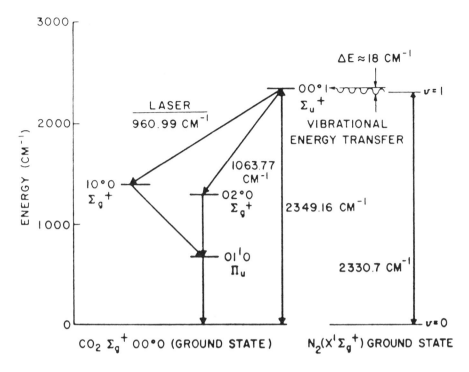

FIGURE 9. CO_2 Laser. Quantum energy levels showing the energy transfer from N_2 to CO_2.

simple construction, at least for lasers with <1 kW average power. The photophysics of the medium has not essentially changed since the initial work of Patel.[14]

Stimulated emission is obtained between vibrational quantum levels and is in the IR around 9 and 10 μm. The CO_2 molecule is triatomic and linear. The two oxygen atoms are symmetrically located with respect to the central carbon atom. This molecule therefore has $(3N - 5) = 4$ degrees of freedom, to which correspond three fundamental vibrational modes: ν_1, ν_2, and ν_3, one of which (ν_2) is doubly degenerate. The quantum level of excitation in a mode is given in the order ν_1, ν_2, ν_3, according to the nomenclature (000) for the ground state, (001) for ν_3, v = 1, etc. Figure 9 shows the energy diagram for pumping and emission in a CO_2 laser.

The CO_2 laser uses a mixture of helium (He), nitrogen (N_2), and carbon dioxide (CO_2) in proportions such as 4/1/1 or 8/1/1 according to the operating mode and the type of excitation chosen. The helium allows, for example, the possibility of varying the electrical impedance of the discharge. The nitrogen is responsible for the rather long tail (1 μs) which follows the quite short (100 ns) main emission in the case of a transverse discharge pulsed laser.

The CO_2 laser operates in pulsed or continuous mode. The molecular nitrogen, N_2, is excited to the vibrational level v = 1 of the ground state. This level, populated directly by the electrons of an electric discharge, and by cascade from higher vibrational levels, is the energy storage reservoir level of the laser. The remarkable property of the N_2 v = 1 vibrational level is that it is in nearly perfect resonance ($\Delta\epsilon = 18$ cm^{-1}) with the level ν_3 (001) of CO_2 which gives rise to the two emissions distributed around 9 and 10 μm. These emissions correspond to the transitions (001) to (100) around 960.99 cm^{-1} (10 μm), and (001) to (020) at 1063.77 cm^{-1} (9 μm). The word around means that the two vibration bands distributed among the P and R branches ($\Delta J = \pm 1$) are made up of more than 100 rotational lines. In the most common case, naturally occurring isotopic CO_2, $^{12}C^{16}O_2$, there are 100 lines from 9.1 to 10.7 μm, the tabulation of which has recently been brought up to date.[15] The CO_2 laser can operate with molecules containing other isotopes, ^{13}C and ^{18}O.

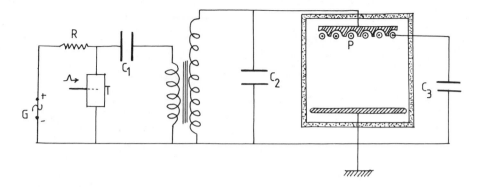

FIGURE 10.   Electrical circuitry and discharge channel for a CO_2 TEA laser. G = generator, R = charging resistor, T = thyratron or spark gap, C_1, C_2, C_3 = capacitors, and P = preionization electrode array.[17]

In the case of $^{12}C^{16}O^{18}O$, one has twice as many rotational lines for the degeneracy because the symmetry of the CO_2 molecule no longer exists.[16] In practice, the use of isotopic species other than $^{12}C^{16}O_2$ is only conceivable in a sealed laser and in using catalytic regeneration of the decomposed CO_2.

The average separation between rotational lines is of the order of 1.5 cm^{-1}, this gap can be filled by using pulsed lasers operating under high pressure (8 to 10 atm) by profiting from the broadening of the laser lines, which is of the order of 0.193 cm^{-1}.atm^{-1}.[16] However these high-pressure lasers operate only at a low repetition rate (a few hertz) as they make use of very high voltage generators.

All the excitation and transfer cross sections of the reactions in the mechanism of Figure 9 are well known. For example, the excitation of N_2 by 1- to 5-eV electrons has a cross section of 4.10^{-16} cm^2. The cross sections for the direct excitation of the CO_2 levels are from 3 to 6.10^{-16} cm^2; these values are relatively high.

The CO_2 laser is thus a four-level laser which is characterized by a storage level with a long lifetime.

A. PULSED TEA CO_2 LASERS

In general the commercial pulsed CO_2 lasers operate with transverse electrical pumping as shown schematically in Figure 10. The mixture of He, N_2, and CO_2 is used at a pressure close to atmospheric. These operating conditions are the origin of the initials TEA (transverse electric atmospheric) which currently designate this type of CO_2 laser. The TEAs can easily provide pulses of the order of 0.1 to 1 J for the rotational lines with the highest gain.

The pulse produced by a TEA laser has an initial short-lived (70- to 100-ns) component, followed by a rather long tail, of the order of a microsecond. This second component can be partially reduced by decreasing the proportion of nitrogen. The proportions of the mixture of He, N_2, and CO_2 are typically 4:1:1. The operating conditions are well known and understood and are well detailed in the literature.[17,18]

Figure 10 shows that the electric circuit of a TEA laser is very simple. The capacitor C_2 (charged through a step-up voltage transformer from a capacitor C_1) is connected to the electrodes by a circuit with low self-inductance. The discharge is prepared by an array of preionization electrodes. The method of producing the preionization and the arangement as well as the nature of the preionization electrodes vary considerably from one manufacturer to another. All the TEA lasers are characterized by an excellent electrical yield, which reaches 10% of the energy stored in the capacitor banks.

B. HYBRID TEA CO_2 LASER

The combination, in the same resonant cavity of a TEA laser and a CW gain section

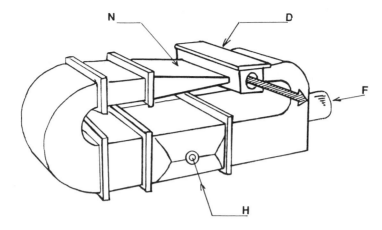

FIGURE 11. Gas circulation system for a high repetition rate TEA CO_2 laser. H = heat exchanger, D = discharge channel, and N = aerodynamic nozzle.[17]

form what is called a hybrid laser. This system, whose characteristics are well described by Lachambre,[18] allows one to have longitudinal single mode pulsed operation. The time envelope of the pulse is thus free from oscillations and is "smooth" up to a repetition rate of 350 Hz.[18]

In order to increase their repetition rate the TEA systems have been enclosed in aerodynamic loops where the gas mixture flows transversely. Figure 11 shows schematically this type of structure, which includes a compressor, a heat exchanger, and a nozzle near the electrodes. Since 1971, Michon[19] has had a laser of this type operating at 2 kW average power and a repetition rate of 100 Hz. later Baranov[20] reached 10 kW at 1000 Hz.

The limitations of this type of laser are now technological. For example, from the point of view of reliability of operation it is wished to extend this to 1000 h or more. Transverse rapid flow is the preferred solution for "multikilowatt" lasers, but up to 1 kW the technologically simpler axial flow solution is the one most often used commercially.

C. AXIAL FLOW LASER

The structure of these lasers resembles that of the majority of CW longitudinal lasers (Figure 12). The gases are recycled through a heat exchanger, using a Roots type compressor, in order to minimize the helium consumption. This consumption is approximately 100 l (STP) of the mixture per hour and per kilowatt of laser emission.

The power emitted is directly proportional to the length of the discharge which has been possible under optimum flow conditions to extract 400 W.m^{-1}. This type of laser is currently chosen for powers of 1 to 2 kW. The advantages of continuous axial flux lasers over TEA lasers is that the electrical components are less severely stressed and have a longer life. Nevertheless, only the CO_2 TEA lasers have >2 kW average power.

For photochemical applications CO_2 lasers are equipped with a diffraction grating at the Littrow angle. Folding of the discharge tube allows the construction of CO_2 lasers with reasonable dimensions. The discharge tube is water cooled and the Brewster angle windows are made of zinc selenide (ZnSe). The solid copper mirrors are internally water cooled.

D. SEALED CO_2 LASERS

From the outset, the possibility of the operation of the CO_2 laser in a sealed tube has been studied with a great deal of interest.

As we have seen previously, sealed operation would allow the use of isotopic varieties of CO_2 which would considerably increase the spectral tuning range.[16]

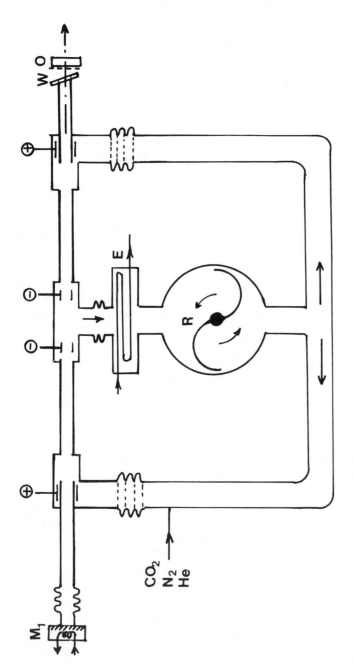

FIGURE 12.   Schematic representation of a longitudinal flow CO_2 laser. R = roots pump, E = heat exchanger, W = brewster angle window (NaCl or ZnSe), O = output coupling mirror, and M_1 = high reflectance water cooled mirror.

After 20 years of work in this direction, the technology of sealed CO_2 lasers provides instruments with a power of 10 to 100 W having mean times between failures (MTBFs) of several thousand hours.

The excitation method which gives the best results is RF (radio frequency) excitation, as it eliminates the problem of electrode oxidation and the decomposition of the CO_2 which is the weak point of ordinary continuous discharges.

The laser channel has a diameter of 2 to 3 mm and a length of 300 to 400 mm. In most cases these lasers emit spontaneously at 10 μm on the highest gain lines (P_{18} or P_{20}), and some manufacturers offer the option of a diffraction grating for the selection of rotational lines from 9 to 11 μm.

For the moment, the photochemical applications are not the main objective of this type of sealed lasers. They have been developed for military applications and their civilian uses are rather in surgery or in material transformation.

The RF excitation can also be pulsed at repetion rates of the order of kilohertz. The technique, which is in full commercial development, is described in a review article by Newman and Hart.[21] The perspective of lasers of this type emitting an average power of 1000 W seems to be quite realistic.

E. ULTRAVIOLET EXCIMER LASERS

The rare gas atoms have completed electron configurations. The collision of two of these atoms in their ground state is therefore completely repulsive, neglecting the weak long-distance interactions such as those of Van der Waals. When at least one of the atoms is in an excited electronic state there can be a bound state of the corresponding excited diatomic molecule. The collision of rare gas atoms produces homonuclear excited dimers (Ar_2, Kr_2, Xe_2, etc.) and in addition mixtures of rare gases and halogens lead to heteronuclear molecules, stable in the excited state, or weakly bound in the ground state (ArF, KrF, XeF, XeCl).

For the first case, excited dimers, the acronym EXCIMER has been adopted. For the second case, excited complexes, the acronym is EXCIPLEX. Unfortunately present usage has eliminated this distinction and now each is indifferently called EXCIMER.

The energy potential curves of rare gas dimers are shown in schematic form in Figure 13. This figure is excessively simplified to clearly show the essential characteristic of excimer lasers, that is, having a completely repulsive terminal level. Even in the case where a weak binding energy of the Van der Waals type exists in the ground state, this energy is very much less than the thermal energy kT of the electric discharge used to excite the laser. This figure also shows symbolically the possibility of an absorption between two excited states. In fact, the potential energy diagrams are more complicated and, for example, in He_2* there are more than 23 electronic states from 0 to 35000 cm^{-1} (energy close to that of the He_2^+ ion). The good yield of excimer lasers is based largely on this property of total population inversion of the excited state with respect to a totally repulsive ground state.

Figure 14 shows the case of a well studied exciplex, XeCl, which produces an emission in the near UV at 308 nm. The excited state XeCl* responsible for the laser emission is populated by a whole series of reactions involving electrons, atomic and molecular ions, and metastable atomic and molecular states.

The different paths for the formation of XeCl* are shown schematically in block diagram (Figure 14). In this case the halogen donor is hydrochloric acid, HCl. The best yield is obtained when some neon is present in the mixture, as Ne^+, Ne^*, and the dimer Ne_2* play an important role in the formation of Xe*.

The reaction mechanisms and the rate constants are well known for numerous excimers.[22] Here we will retain only one essential and common characteristic of these reaction mechanisms, which leads to quite severe restraints from the point of view of the pumping of these lasers. The excited states have very short lifetimes (1 to 10 ns). The pumping must

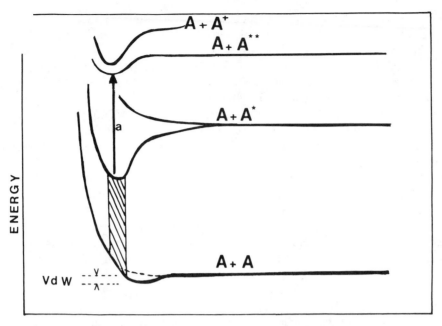

FIGURE 13.   Schematic (not to scale) representation of potential energy curves of rare gas excimers.
VdW = possible Van der Waals bonding.

therefore be carried out in a very short time. Apart from experimental pumping systems using electron guns, the method most developed commercially is the transverse electric discharge. This discharge system is technically very similar to that of the TEA CO_2 laser (Figure 10), but the electric circuits are designed to produce electric discharges typically 100 times shorter than in the case of CO_2 lasers.

In order that the discharge takes place with a rise time as short as possible (<1 ns), part of the energy stored in the condensers is used to preionize the gas by a series of auxiliary electrodes.

In the systems where a great deal of energy (from 1 J to 1 kJ) is wanted, at the expense of the pulse repetition rate, preionization is carried out by beams of electrons or X-rays. For lasers that are to provide high average power (up to 100 W) and repetition rates from 50 to 250 Hz, auxiliary preionization electrodes are presently used.

At the beginning of the commercialization of these lasers the existence of a very intense electric discharge in the presence of corrosive gases gave rise to some concern, but these conditions are now well under control.

The progress has come about by a better choice of materials as well as the use of purification systems for the gas mixture. Another critical point of these lasers is the electric discharge circuit. The electric voltages used are of the order of 30 kV, and the switching of the energy is done by the use of thyratrons, which are gas (H_2 or D_2) tubes where the triggering is controlled by a grid. The role of these thyratrons is to assure a very short discharge (1 ns). Originally these tubes were developed for radar equipment, but since then specially designed models have been developed for lasers (CO_2, copper vapor, and excimer lasers). Some models[23] withstand peak currents of 15000 A with voltages of 35 kV.

A recent improvement in discharge circuits consists in using compression stages of the lectric pulse by saturable inductances. This system allows the use of slower discharges in the thyratron, which is thus subjected to less severe operating conditions. The lifetime of this tube, which represents 5% of the capital cost of the laser, now reaches 10^9 pulses. As

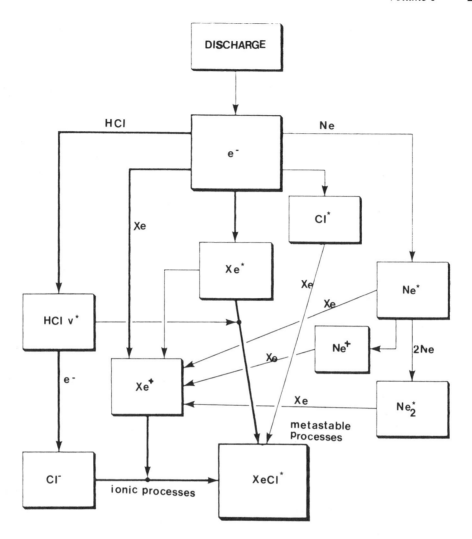

FIGURE 14. Block diagram of the main energy transfer steps in a XeCl exciplex laser. The mixture consists of HCl + Ne + Xe. (From Clerc, M., Rigny, P., and de Witte, O., *Le Laser*, Editions Lavoisier Paris, 1984. With permission.)

excimer lasers presently operate at 200 or 250 Hz it can be seen that a billion pulses are reached in 1000 h.

Excimer lasers are potentially interesting for several applications, such as photochemistry, material transformation processes (machine engraving), as well as for microlithography.

The wavelengths and typical yields of excimer lasers in current use are the following:

	ArF	KrF	XeCl	XeF
(nm)	193	248	308	351
y.%	2	3.5	2.5	2

As we have seen, excimer lasers are lasers with a high gain ($\sim10\%$ cm^{-1}) and a short storage time. Owing to the precision of their thyratron discharge system, it is nevertheless possible to connect them in an oscillator + power amplifier (OPA) series. This configuration allows up to 100 W average power. In low-repetition rate operation, one can thus obtain 1-J pulses starting with two excimer laser modules connected in OPA. Another advantage is that the beams thus obtained are less divergent (<0.4 mrad) with very homogeneous intensity profiles.

F. METAL VAPOR LASER

The first metal vapor laser was invented by Gould et al.[25] in 1966. A review article devoted to this subject was written by Piper in 1978.[26] At this time CVLs pumped by electric pulses at rates from 5 to 13 kHz reached average powers of >40 W.

Since then, owing to the stimulation given by the programs of laser enrichment of uranium isotopes, the average powers that can be extracted from a CVL tube have passed the 100-W level, but the power available at the commercial level is still limited to 60 or 70 W.

If one could give only one essential characteristic of metal vapor lasers, it would be the possibility of operating at very high repetition rates (5 to 10 kHz) with 10- to 40-ns pulse duration.

The electrical yield of these lasers is in general of the order of 0.5%, but in a recent article Isaev et al.[29] reported raising this yield to 1% by using a double exciting pulse.

The emissions of CVL are at 510.5 nm (green) and 578.2 nm (yellow). The green line represents rougly 70% of the power emitted, but the ratio of the green and yellow lines can vary slightly with the choice of discharge conditions. The lead emission is at 723 nm, with operating rates of 1 to 6 kHz, and average powers of the order of 1 W.

The barium emission is at 1.5 μm, with an average power of 12 W. The manganese is at 534 nm and 1.3 μm with an average power of 2 W at 5 kHz. Finally, the gold emission is at 628 nm, a wavelength of interest for medical applications for it corresponds to the spectral transparence band of organic tissues.

However, it is the CVL which has the greatest interest, for the laser lines at 510.5 and 578.2 nm are particularly well suited to the pumping of dyes which have the highest quantum yield. For many applications operation at a repetition rate as high as 10 kHz is comparable to continuous operation, which makes the CVL a potential competitor of argon ion lasers for the pumping of laboratory dye lasers.

At first view, the internal structure of a CVL seems rather simple and the technology less sophisticated than that of argon ion laser tubes. In actual fact the design of a CVL tube involves difficult thermal problems and a rigorous choice of discharge conditions. The operating temperature of CVLs is typically 1500°C. A diagram of the excited states of the CVL is shown in Figure 15.

The electrons in the electric discharge excite the excited electronic states of copper, which by cascading populate the $^2P_{3/2}$ and $^2P_{1/2}$ levels situated at approximately 3.8 eV. The emissions $^2P_{1/2} \rightarrow {}^2D_{3/2}$ and $^2P_{1/2} \rightarrow {}^2D_{5/2}$ correspond, respectively, to the green and yellow laser lines.

The ^2D states are metastable with respect to the fundamental state $^2S_{1/2}$. The relaxation of these metastable states is therefore a very important kinetic component of the CVLs.

For small diameter lasers the wall plays an important role in the relaxation of the ^2D metastable states. Work done in the U.S.S.R. and Israel has shown that if neon is used as a buffer gas instead of helium, then part of the volume relaxation of the ^2D states is due to the neon.[27,28] The use of neon as well as the optimization of other parameters of the electric discharge and the heat exchange has allowed an enhancement of the average power of CVL tubes by increasing their diameter to 70 mm. The temperature required to vaporize copper, which is of the order of 1450 to 1500°C, is reached and maintained by the electric discharge.

The very high purity alumina tube is surrounded by thermal barriers and coaxial insulating layers (Figure 16). The copper, also very pure, is in the alumina tube in the form of melted droplets. The buffer gas is maintained at a pressure of 10 to 100 torr. Only an intentionally schematic diagram of the components of a CVL tube is given in Figure 16. For this laser,

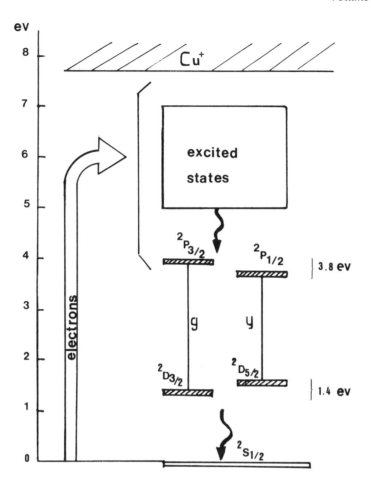

FIGURE 15.   Copper vapor laser energy levels.

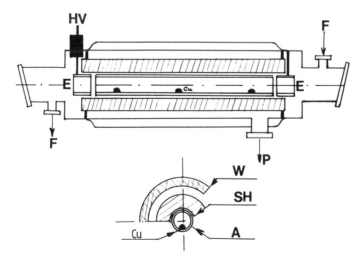

FIGURE 16.   Schematic representation of the main internal components of a copper vapor laser. HV = high voltage (5 to 6 kV), E = hollow cylindrical electrodes, F = slow flow of He or Ne, Cu = copper droplet melted at 1500°C, A = high purity alumina, SH = thermal shield, and W = double jacketed water cooling system.

which is making a commercially important breakthrough, different manufacturers have used very different solutions. Some carry out the sealing by joints involving the alumina tube which means that these must operate under very severe conditions of temperature. With others the outer wall is vacuum tight and the assembly is swept by the buffer gas, which on the other hand has the task of slowly degassing the thermal insulators.

Finally, the population inversion is obtained by very short rise time electrical discharges with 5- to 10-kHz repetition rates. The electrical circuit uses thyratrons under severe operating conditions. Important advances have recently been made in this field by reducing the peak current in the thyratrons owing to a technique of compression of the electrical pulse by the use of saturable inductances. Some thyratrons reach 10^{10} discharges, equal to 300 h at 10 kHz.

The CVL is not a sealed tube laser; there is continuous sweeping by an inert gas (He or Ne). The metal vapor is thus slowly removed and must be replaced. In the best case there can be 400 h of operation before this refurbishment, but in practice it is most often done after 50 h of continuous operation.

VII. ULTRASHORT LASER PULSES

A. MODE LOCKING

As seen previously a large number of modes may exist in the Fabry-Perot cavity of a laser (Figure 3). The angular frequency separation between each mode is $\Delta\omega = \pi c/l$, where c is the speed of light and l, the cavity length. For small l values, $\Delta\omega$ is larger and the number of oscillating modes for which the gain is greater than the losses decreases. While the phase difference between modes is random in a conventional laser, the modes are forced to oscillate together in a mode-locked laser. The phase and the amplitude of each mode are defined in the frequency domain and in the time domain by the two complex functions V(t) and V(ω) related by the Fourier transform:

$$V(t) = \frac{1}{\sqrt{2\pi}} \int_{-\infty}^{+\infty} \tilde{\tilde{V}}(\omega)\exp[-i\omega t]d\omega$$

$$\tilde{\tilde{V}}(\omega) = \frac{1}{\sqrt{2\pi}} \int_{-\infty}^{+\infty} V(t)\exp[i\omega t]dt$$

V(t) and $\tilde{V}(\omega)$ are given by

$$V(t) = A(t)\exp[i(\varphi(t) - \omega_0 t)]$$

$$\tilde{V}(\omega) = a(\omega)\exp[i\varphi(\omega)]$$

In the case of mode locking, $\varphi(\omega) = 0$. The phase relationship produced results in completely constructive interference between all the modes and a pulse of light travels back and forth between the cavity mirrors. Each time it is incident upon the semitransparent output mirror, a pulse of light is delivered and the time interval between pulses is equal to the round trip time T = 2.l/c, typically around 10 ns. The pulse width Δt is given by

$$\Delta t = \frac{2\pi K}{\Delta\Omega}$$

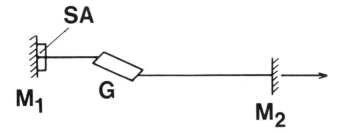

FIGURE 17. Passive mode locking. M1, 100% mirror; SA, saturable absorber; G, gain medium; and M2, output mirror.

TABLE 1
Gain Media and Saturable Absorbers

Gain media	Wavelength (μm)	Saturable absorber	Ref.
Nd^{+3} glass rod	1.06	Eastman 9740	33, 35
		Eastman 9860	35
		Heptamethine pyrylium dye $\neq$ 5	41
		N 3274	42
Ruby	0.6943	Cryptocyanine	35, 43
		DDI[a]	35, 43
		dicyanine	35
Alexandrite	0.72—0.78	Cryptocyanine	44

[a] DDI = 1,1'-diethyl-2-2'dicarbocyanine iodide.

where K is a constant which depends on the shape of the spectral distribution and $\Delta\Omega$ is the spectral bandwidth. For a gaussian profile, K = 0.441. The pulse duration thus decreases as the pulse bandwidth increases.

Locking of He-Ne modes was demonstrated by Hargrove et al.[32] in 1964, but since then mode locking in solid-state lasers such as neodymium glass, neodymium YAG, ruby, and in dye lasers has enabled the generation of much shorter pulses ($1 < \Delta t < 30$ ps).[33-40]

There are several ways to accomplish mode locking, involving either passive or active methods. In passive mode locking, which has been the more widely used method, a saturable absorber is inserted in the laser cavity (Figure 17). The saturable absorber is generally a dye solution which attenuates low-intensity optical fluctuations. The absorber is however bleached when a random noise pulse is sufficiently large. The leading edge is absorbed and the central portion is transmitted. The absorber relaxation time must be short enough ($\tau < T = 2.l/c$) to attenuate later portions of the optical noise. The transmitted light is amplified in the active gain medium. After several round trips in the cavity, a perfect pulse shaping is obtained.

B. Q-SWITCHED AND MODE LOCKED SOLID-STATE LASERS

Gain media and some of the more reliable saturable absorbers are listed in Table 1. One problem of the method is the damaging of the dielectric mirror in contact with the dye solution.[35] A Galilean telescope is often included in the laser cavity to enlarge the beam diameter thus reducing the laser power density (power per unit cross-sectional area of the beam).[35]

For neodymium glass or YAG lasers, the so-called no. 5 dye, which has a relaxation time of 2.7 ps in 1-2-dichloroethane, appears to be the most photochemically stable and produce the shortest pulses.[41] For the generation of picosecond ruby laser pulses, 1,1'-

diethyl-2,2'-dicarbocyanine iodide (DDI) is the best mode-locking dye and the pulse duration is ~12 ps.[43]

Active mode locking which was first achieved in a 0.6328-μm HeNe laser[32] is commercially available for Nd-YAG lasers. The combination of active and passive mode locking results in stable mode-locked pulse trains. Stabilization of the amplitude and frequency of the modes is obtained by modulating the laser at a synchronous frequency equal to the reciprocal round trip time.[45,46] The modulator is a fused quartz block which, when excited at a longitudinal acoustic resonance, acts as a diffraction grating and produces diffraction orders oscillating at twice the driving frequency.

Actually the most reliable picosecond solid-state lasers are the Nd-YAG and Nd-glass lasers. Active mode locking improves the reproducibility of Nd-YAG laser trains but the pulse duration increases from 27 ps using passive mode locking and Eastman 9740 as a saturable absorber to 35- to 40-ps in the combined mode locking. Using dye no. 5 results in a ≤25-ps pulse duration at the expense of a higher threshold.

Even shorter pulses ~5 ps are obtained with passively mode-locked Nd glass lasers,[38] but the characteristics of the output pulses are influenced by the intensity-dependent index of refraction change,[37,38] which results in a phase change and a chirp. To generate chirp-free picosecond pulses, a single pulse should be extracted from the leading part of the pulse train and then amplified.[38,39] Stable active-passive mode locking of a Nd-phosphate glass laser has been shown using Eastman no. 5 dye at a pulse repetition rate above 1 Hz.[47]

These lasers are particularly appropriate for picosecond spectroscopic studies which necessitate a large energy exciting pulse, for example, a few tens of millijoules to generate a continuum of white light for absorption studies.[40] A drawback is the relatively low repetition rate, 10 Hz for picosecond Nd-YAG lasers.

C. CONTINUOUS WAVE MODE-LOCKED LASER SOURCES

Acoustooptic mode-locked argon ion and krypton ion lasers are commercially available. The pulse duration is ~200 ps. They are used in the synchronous pumping of dye lasers.[48] The mode-locker drive frequency should be stabilized to ~±100 Hz/40 MHz for successful pumping. The argon ion 514.5-nm line is currently used but the 351-nm line (140 mW) and all three UV laser lines (400 mW) have also been used to pump a CW stilbene 3 dye laser (420 to 470 nm).[49]

More recently, CW mode-locked Nd-YAG lasers have also become commercially available, with a pulse duration of ~100 ps at 1064 nm and an average output power of 10 W at ~80 MHz repetition rate (Figure 18). Second harmonic generation leads to 60- to 70-ps pulse duration at 532 nm and 1 W average output power. However, there are crucial disadvantages in using these lasers to obtain 532 nm. They require an expensive $KTiOPO_4$ crystal for harmonic generation with a high conversion efficiency and despite the high damage threshold of this crystal, it is easily damaged by the unwanted hazardous Q-switching of the Nd-YAG laser. It is also possible to simultaneously Q-switch and mode lock CW Nd-YAG lasers and produce trains of picosecond mode-locked pulses at rates up to 1 KHz. The energy of the largest mode-locked pulse so obtained is ~80 μJ at 1064 nm.

Recent advances have been obtained by optical pulse compression which utilizes self-phase modulation (SPM) to chirp the pulse in a 100- to 2000-m long single mode optical fiber and compression in a grating pair (Figure 19). Using this method, 1064-nm input pulses of 80-ps duration have been compressed to 1.8-ps duration[50] and 532-nm input pulses of 32-ps duration to 0.41 ps.[51] Without a grating pair delay line, spectrally chirped 1064-nm pulses amplified in pulsed Nd-YAG stages to energies >10 mJ, have been compressed from 80 to 12 ps.[52] Amplified pulses have been recompressed in a grating-pair delay line, because, as pointed out by Strickland and Mourou,[53] amplifying the stretched pulse after the optical fiber rather than the compressed pulse allows higher energies before the occurrence of self focusing. Even a 1.3-J energy pulse, with a duration of 2.3 ps, has been recently obtained at 1064 nm.[54]

D. PICOSECOND RARE GAS HALIDE LASERS

Passive mode locking of a XeCl laser has been achieved[55] with BBQ (4,4'''di(2-butyl-octoxy-1)-*p*-quaterphenyl) and BPBD (2-(4'-terbutyl-phenyl)5-(4''-biphenylyl) 1,3,4-oxadi-azole) but a gain duration of 150 ns was needed to obtain an almost 100% modulated train of 12 pulses with BBQ and 9 pulses with BPBD, and the pulse duration was only ~2 ns. The generation of picosecond excimer laser pulses up to 40 mJ energy has been obtained at 308 nm with a XeCl amplifier[56] and at 248.5 nm with a KrF amplifier.[57,58] Such an experimental setup is now commercially available and makes use of an XeCl pumped, quenched *p*-terphenyl dye laser which delivers one pulse of ~1mJ energy and 0.5-ns rise time for the pumping of a distributed-feedback dye laser (see Section IX.I) containing coumarin 307 (λ_L = 497 nm)[56,57] or rhodamin B(λ_L = 616 nm).[56] After suitable amplification and frequency doubling, the pulse width is ~10 ps.[57] Through injection in a double pass excimer amplifier, high energy picosecond pulses are thus obtained up to 25 Hz. Recently, the method has been refined even further decreasing the pulse duration to ~230 fs at 308 nm[59] and ~80 fs at 248.5 nm[58] (Figure 20), and the 15-mJ energy is high enough to start new subpicosecond absorption studies with UV excitation of molecules.

VIII. FREQUENCY CONVERSION

It is often desirable to convert the output frequency of the master laser to another one more suited for a particular application. Nonlinear crystals are used for harmonic generation and parametric generation. Stimulated Raman scattering in gases, atomic vapor, and liquids is also an efficient method.

In the second harmonic generation in a nonlinear crystal, two photons of frequency v are converted to a single photon of frequency $2v$.[60] In the expression for the dependence of optical polarization P on the applied optical field E

$$P = XE(1 + a_1E + a_2E^2 + ...)$$

the second term Xa_1E^2 gives rise to the second harmonic.[60] This effect is easily explained if one considers the electric field $E = E_0 \sin \omega t$ and the quadratic term which provides a contribution p to the polarization in the crystal.[60]

$$p = Xa_1E_0^2\sin^2\omega t = Xa_1(E_0^2/2)(1 - \cos2\omega t)$$

The cos $2\omega t$ dependence of the polarization is responsible for the second harmonic generation. Severe restrictions on the harmonic generation are imposed by the optical dispersion of the crystal. Refractive indices of the currently used crystals potassium dihydrogen phosphate (KDP) and ammonium dihydrogen phosphate (ADP) have been measured between 0.214 and 1.529 μm.[61] Phase matching conditions require the choice of a direction of propagation which is defined by the angle Θ between the optical axis of the crystal and the direction of the incident beam. When doubling the fundamental frequency of a tunable dye laser, the second harmonic generation can be obtained through temperature tuning or angle tuning. At high laser powers, the optical damage threshold of the crystal becomes an important criterion. For example, severe optical damage can be caused to $LiNbO_3$ and $LiIO_3$ crystals. KDP, deuterated KD*P, and cesium-dihydrogen arsenate or CDA are preferred for Q-switched lasers.[62,63] For second harmonic generation, down to 217 nm, potassium pentaborate (KB_5O_8, 4 H_2O) showed excellent transmission,[64] and a new material β-BaB_2O_4

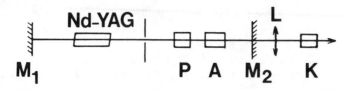

FIGURE 18. CW mode-locked Nd-YAG lasers. M1, 100% mirror; D, diaphragm; P, polarizer; A, acoustooptic mode locker; M2, output mirror; L, lens; and K, KTP crystal.

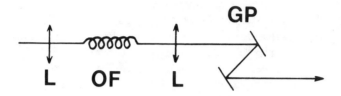

FIGURE 19. Optical pulse compressor. L, lens; OF, optical fiber; and GP, grating pair.

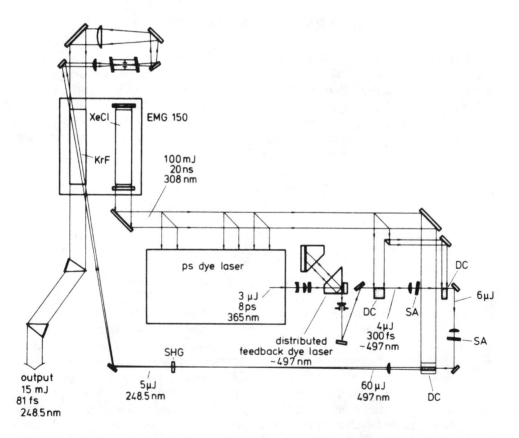

FIGURE 20. Experimental arrangement. DC: dye cell, SA: saturable absorber, and SHG: second harmonic generator. (From Szatmari, S., Schäfer, F. P., and Muller-Horsche, E., *Opt. Commun.*, 63, 305, 1987. With permission.)

which is nonhydroscopic, chemically inert, and mechanically stable has been recently shown to be very attractive for high-power generation at 2048 Å.[65,66] Because the threshold power of the harmonic generation (~100 MW) in BaB_2O_4 is significantly higher than in KDP (70 MW), its efficiency is not better.[66] Moreover, it has a low acceptance angle but it has a higher damage threshold, a lower-temperature sensitivity, and its resistance to thermal fracture is about an order of magnitude greater than that of any other nonlinear material for which data exists.[66] Potassium titanyl phosphate (KTP) is very efficient for doubling the frequency of CW-Nd:YAG lasers due to its high nonlinear coefficient and large acceptance angle.[67]

Frequency tripling involves mixing of the fundamental frequency with the second harmonic. Tripling the YAG frequency is common and very efficient.[68]

Stimulated Raman scattering enables the availability of specific frequencies at high peak powers. A list of materials found to exhibit stimulated Raman effect includes liquids such as benzene, carbon tetrachloride, etc. and gases such as hydrogen, methane, etc.[69] Commercial systems making use of the Raman frequency of hydrogen $\nu_R = 4155$ cm^{-1} allow the transformation of a tunable dye laser radiation ν_L into a tunable red-shifted radiation $\nu_{S1} = \nu_L - \nu_R$. Methane ($\nu_R = 2917$ cm^{-1}) is also a good Raman medium.[70]

Parametric amplification is a less-known method which results from optical mixing.[71,72] The interaction of two fields with different frequencies ω_1 and ω_2 gives rise to emitted light containing the frequencies $\omega_1 + \omega_2$ and $\omega_1 - \omega_2$. Three waves must be phase matched, ω_1, ω_2, and $\omega_3 = \omega_1 \pm \omega_2$. The optical parametric oscillator (OPO) consists of a parametric amplifier placed in an optical resonator to obtain much higher powers. Tuning is obtained by rotating the crystal. Intense tunable picosecond pulses have been generated in the IR with a conversion efficiency exceeding 1%,[73,74] which can be used to populate well-defined vibrational states of polyatomic molecules and to study vibrational relaxation.[75] Another important application is the so-called "frequency upconversion" to detect IR light at a frequency for which a fast or sensitive detector is not available. The light at frequency ω_1 is mixed in a crystal with an intense light at frequency ω_2 and the detection is made at frequency $\omega_3 = \omega_1 + \omega_2$. This method has been combined with multichannel optical detection for wide spectral coverage.[76]

IX. DYE LASERS

A. INTRODUCTION

After the initial demonstration of a ruby laser by Maiman in 1960, the first dye laser action was not obtained until 1966 when Sorokin and Lankard[77] observed laser emission from a solution of an organic compound, namely chloroaluminium phtalocyanine. Soon afterwards, rhodamine 6G, which is the most widely used laser dye, was found in 1967. Apart from solid-state lasers incorporating Cr^{3+}[78] or Ti^{3+}[79] and tunable from 700 to 920 nm, the tunability of solid-state lasers is severely restricted. Similarly di- and triatomic molecular lasers (HF, DF, CO, CO_2) are only tunable at discrete energy levels. Organic dye lasers are easily tunable from the UV (300 nm) to the IR (1800 nm) as a consequence of the large number of laser dyes[80,81] and of their broadened electronic levels which are strongly sensitive to solute-solvent interactions. By spectrally narrowing by using dispersive elements in the laser cavity, the laser emission bandwith can be reduced to a few megahertz and locked to atomic or molecular lines. While much better performances are currently obtained in metrology laboratories, we shall, for the purposes of photochemistry, focus on the performances easily obtained in using commercial lasers.

According to the terminology of dyers and colorists, dyes are colored substances able to impart color to other substances. This definition has been extended to organic molecules having a strong absorption and containing a system of conjugated bonds. For good laser

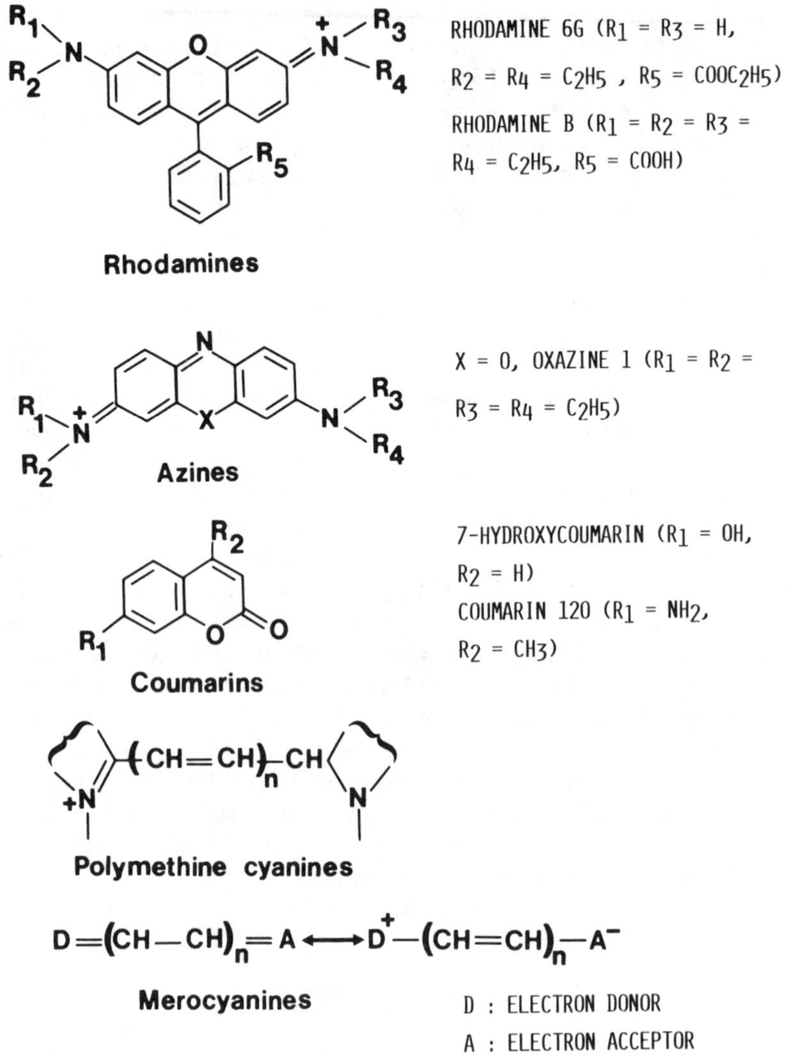

RHODAMINE 6G ($R_1 = R_3 = H$,
$R_2 = R_4 = C_2H_5$, $R_5 = COOC_2H_5$)

RHODAMINE B ($R_1 = R_2 = R_3 =$
$R_4 = C_2H_5$, $R_5 = COOH$)

Rhodamines

$X = 0$, OXAZINE 1 ($R_1 = R_2 =$
$R_3 = R_4 = C_2H_5$)

Azines

7-HYDROXYCOUMARIN ($R_1 = OH$,
$R_2 = H$)
COUMARIN 120 ($R_1 = NH_2$,
$R_2 = CH_3$)

Coumarins

Polymethine cyanines

$$D = (CH-CH)_n = A \longleftrightarrow D^+ - (CH=CH)_n - A^-$$

Merocyanines

D : ELECTRON DONOR
A : ELECTRON ACCEPTOR

FIGURE 21.   Principal classes of laser dyes.

dyes, very special requirements must be met with respect to their molecular structure and photophysical and photochemical properties. An extensive list of 546 laser dyes has been given up to 1984 by Maeda[81] but the number of the common laser dyes which are efficient enough to cover the 250- to 1800-nm spectral domain,[82] using the fundamental frequency or the second harmonic, is about 50. For example, one can pump the very efficient and photochemically stable rhodamine dyes at 532 nm (second harmonic of Nd-YAG lasers), 510.6 nm (CVL), 501.7, or 514.5 nm (CW argon ion laser), and the second harmonic frequency can be obtained with a good efficiency.

For the many newcomers to the field of dye lasers, the reference book of Schäfer, *Dye Lasers* (1977),[77] and the review article of Nair[80] are advisable.

B. PHOTOPHYSICAL PROPERTIES OF LASER DYES

Some formulas representing the principal classes of efficient laser dyes are given in Figure 21. These dyes may have different commercial names depending on their origin.[83-85]

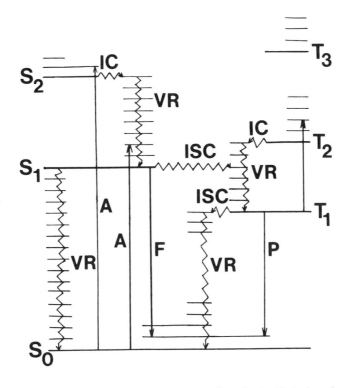

FIGURE 22.   Diagram of the energy levels of a molecule. The horizontal nonradiative transition processes are internal conversion (IC) between same multiplicity states and intersystem crossing (ISC) between states of different multiplicity. The vertical transition processes include absorption (A), fluorescence (F), phosphorescence (P), and vibrational relaxation (VR).

The absorption spectrum of dyes is attributed to transitions from vibronic levels of the electronic ground state (or singlet S_0) to electronic excited singlet states $S_1, S_2 . . .$ (Figure 22). To a first approximation the position of the first absorption band is determined by the length of the conjugated chain and the number of π electrons moving along the chain, according to the free electron gas model of Kuhn.[86] After the initial Franck-Condon transition (10^{-16} s), relaxation to the lower vibronic level of S_1 is rapid, within a few picoseconds, and the radiative process $S_1 \rightarrow S_0 + h\nu$ is responsible for the spontaneous emission (fluorescence) and for the stimulated emission. The complex molecular structures give rise to a large number of vibrations and rotations of the molecular skeleton which are coupled to the electronic transition and broadened by solute-solvent interactions. The density of energy levels per spectral unit (cm^{-1}) is so high that the absorption and the fluorescence bands of a dye can be considered as continuous. The energy pumped into an electronic excited state will be available for laser emission over the whole fluorescence band with a spectral bandwidth, depending on the dispersive elements of the laser cavity.

The radiative lifetime τ_r of the S_1 state is related to the Einstein coefficient $A = 1/\tau_R$ for spontaneous emission and can be calculated using the Strickler-Berg equation.[87]

$$\frac{1}{\tau_r} = 2.88 \times 10^{-9} \, n^2 \, \frac{\int I(\tilde{\nu})d\tilde{\nu}}{\int \tilde{\nu}^{-3} I(\tilde{\nu})d\tilde{\nu}} \int \frac{\epsilon(\tilde{\nu})d\tilde{\nu}}{\tilde{\nu}}$$

where n is the refraction index of the solution, $F(\tilde{\nu})$ the fluorescence intensity at the wavenumber $\tilde{\nu}$ (in cm^{-1}), and $\epsilon(\tilde{\nu})$ the molar decadic extinction coefficient.

The real lifetime of the first singlet excited state which is shorter than the radiative lifetime due to the competitive nonradiative processes is given by

$$\tau = \tau_r \times \phi_F$$

where ϕ_F is the fluorescence quantum yield, independent of the excitation wavelength.

The nonradiative deactivation processes include

- Internal conversion $S_1 \rightarrow S_0$, which depends on the dye molecular structure (structural rigidity) and the solvent properties.
- Intersystem crossing (isc), $S_1 \rightarrow T_1$, to the long-lived triplet state T_1, which traps the dye molecules preventing their stimulated emission. It obeys the loop rule of Drexhage[88]: the orbital magnetic moment of the π-electrons making a loop couples with the spin of the electrons, thereby enhancing the rate of isc. ISC can be also enhanced if the dye is substituted with heavy elements (Br, I, S, or Se).
- Isomerization (polymethine cyanines).
- Charge transfer interactions: the quenching ability of anions decreases in the order iodide, thiocyanate, bromide, chloride, and perchlorate.[88]
- Energy transfer.
- Excited state reactions: quenching owing to collisions of the excited dye molecules with ground state molecules as evidenced in concentrated rhodamine 6G solutions, protolytic reactions as in coumarin derivatives, and ring-opening reactions.

C. STIMULATED EMISSION

Light amplification is carried out in the fluorescence band of the dye solution but absorption transitions of excited molecules ($S_n \leftarrow S_1$, $T_n \leftarrow T_1$) constitute losses for the pumping and the emission. The population of the triplet state can be kept low enough by rapidly pumping the dye to the laser threshold in a time τ_0 shorter than the time constant τ_{isc} of the $T_1 \leftarrow S_1$ isc. process. Typical values are $\tau_0 < \tau_{isc} \sim 100$ ns. This condition is fulfilled under giant pulse laser pumping but continuous laser pumping and flashlamp pumping necessitate low isc quantum yield or the use of triplet quenchers. The cross section for stimulated emission is related to the Einstein coefficient B.[77] It must be realized that the maximum values of the cross sections for absorption $\sigma_a (\bar{\nu})$ and for stimulted emission $\sigma_e (\bar{\nu})$ are often equal due to the mirror symmetry of the absorption and fluorescence bands of many dyes. It can be easily shown using the oscillation condition equation that the laser wavelength λ_L increases with the dye concentration and the active length of the laser cuvette.[77] If broad band reflectors are used, a wide tunability range is obtained but dispersive elements may be needed for applications in high resolution spectroscopy or laser isotope separation.

For the generation of ultrashort laser pulses (a few femtoseconds), dye lasers are particularly suitable since the pulse duration is related to the spectral bandwidth of the gain curve.

D. SOLVENT EFFECTS AND THE INFLUENCE OF ADDITIVES

The choice of the solvent is very important for the solubility and solvatochromism (Figure 23)[89] properties of the dye and for thermal effects in the laser cuvette which can lead to deterioration of the operation of the laser.[90-97] The dye solubility and the tendency to form dimers or aggregates even at low concentration are strongly dependent on the solvent. The laser wavelength is largely dependent on the refractive index of the solvent. In fact water is the best solvent in view of its low dn/dt.[98,99] Some values of the refractive index n_D ($\lambda = 589.26$ nm) are listed in Table 2 for water, methanol, ethanol, and determinations of the temperature dependence of refractive index of various liquids are given in Table 3.

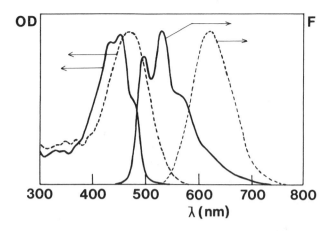

FIGURE 23. DCM absorption and fluorescence spectra in various solvents;[88] OD, optical density and F, fluorescence intensity in arbitrary units; —— isooctane and ---- methanol.

TABLE 2
Refractive Index n_D ($\lambda = 589.26$ nm) in Water, Methanol, and Ethanol as a Function of Temperature

Temp (°C)	Water (100)	Methanol		Ethanol	
10	—	1.33225	(101)	1.3656	(102)
20	1.33299	1.32840	(102)	1.36048	(100)
		1.32844	(101)		
30	1.33192	1.32457	(102)	1.35639	(100)
		1.32459	(101)		
40	1.33051	1.3207	(100)	1.35522	(100)
		1.32071	(101)		
50	1.32894	1.3169	(100)	1.34800	(100)

TABLE 3
Temperature Dependence of the Refractive Index of Various Solvents at 632.8 nm

Substances	$-\dfrac{dn}{dT} \times 10^4$ (K^{-1})	Ref.
H_2O	0.989[a]	98
	0.965	99
Methanol	3.941[b]	103
CCl_4	5.70 [a]	98
	6.118[b]	103
C_6H_6	6.21 [a]	98
	6.523[b]	103
C_6H_{12}	5.38 [a]	98
	5.561[b]	103

[a] At 298.15°K, 1 atm.
[b] At 30°C.

The dn/dt value is -39×10^{-5} in methanol between 20 and 40°C[103] and only -9.7×10^{-5} in water.[98,99] Therefore, to overcome the tendency to dimerization and aggregate formation in aqueous solution, which can induce losses in the laser cavity, additives can be used as surfactants: ammonyx LO or *N,N*-dimethyldodecylamine-*N*-oxide[88,93,96,104-106] Triton® X-100,[107] sodium dodecylsulfate,[108] and decon 90.[96] The disaggregation properties of thiourea have also a beneficial influence on the lasing threshold and the output power.[109] Additives that increase the viscosity or "thickeners" (polyvinyl alcohol, etc.) are also used.[95]

The other important additives are triplet quenchers such as O_2,[110-112] cyclooctatetraene (COT),[104,113-115] and diphenylbutadiene.[116-118] However, in aerated dye solutions, the formation of excited singlet oxygen is in some cases an important drawback due to its reaction with ground state dye molecules leading to their decomposition.[119] Another compound 1,4-diazabicyclo.2.2.2 octane (DABCO) has been recently used by Treba and Koch[20] with coumarin dyes, as a singlet oxygen quencher and a dye triplet quencher. The solvent acidity must be controlled in some cases to displace the acid-base equilibria in the ground state, as well as the fluorescent excited state,[88,108] to extend the tunability, and sometimes improve the photostability.[108]

For industrial applications of dye lasers to uranium enrichment, the study of the economics has to take into account the low but significant cost of laser dyes and solvents chemistry since their photodegradation is appreciable under high repetition rate laser pumping.

E. CONTINUOUS WAVE DYE LASERS

The operation of a CW dye laser[80,121] is quite similar to that of a pulsed dye laser. However, significant losses due to the accumulation of the dye triplet state can decrease the laser efficiency or even inhibit the laser operation. Long-term heating of the active medium by the pumping laser may produce optical inhomogeneities.[93-95,104] Pump sources for CW dye lasers are gas lasers. The most powerful is the argon ion laser and argon lines at 514.5; 488 nm are often used for excitation while the less powerful lines 351 and 364 nm can be used for coumarin dyes. The lifetime of argon laser tubes is limited to between 1000 and 10000 h and they are expensive. Krypton lasers which emit over the range 350 to 800 nm (strongest emission at 647.1 nm) are also used.

The pumping light is focused on a flat jet of dye solution, close to the output of a nozzle. The focal point is a few tens of microns, the dye concentration is about 10^{-3} mol.dm^{-3} and for hydrodynamic reasons the solvent is usually ethylene glycol. High flow velocities up to a few 10 m/s enable small transit times of 10^{-7} s across the pumped region, thereby, reducing the triplet state population. An efficiency of 20% can be obtained. An output of 5.6 W from a CW rhodamine 6G dye laser has been obtained under 24 W-all lines pumping from an argon laser.[104] Monochromatic emission and tunability are obtained using dispersive optics such as a diffraction grating, a prism,[122] Fabry-Perot etalons, birefringent filters, or a Michelson-type interferometer[105] within the cavity. The cavity length ranges from 60 to 200 cm and the space between the longitudinal modes is therefore a few gigahertz. Using a movable mirror on a piezoelectric translator and an atomic line as spectral reference, the laser wavelength can be stabilized within a few tens of kilohertz.

The dye lasers can also be pumped using CW mode-locked laser sources to produce ultrashort pulses on a continuous basis (see Section IX.J).

F. PULSED DYE LASERS

Dyes can show laser action under longitudinal or transverse nanosecond laser pulse excitation.[123] It is often advantageous to set a cylindrical lens in front of the dye cell. Many dyes have been successfully pumped using the second harmonic output (1 to 10 MW/cm²) of a giant-pulse ruby laser and the second or third harmonic of a neodymium laser. The output of the nitrogen laser (3371 Å) despite the low energy available (1 to 10 mJ) is also

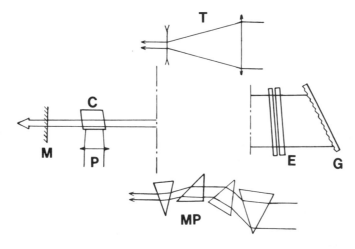

FIGURE 24.   Tunable dye laser. M, output mirror; C, dye cell; P, pump
laser; T, inverted telescope; MP, multiple prism; E, Fabry-Perot etalon;
and G, grating.

a convenient pump but excimer lasers (XeCl, 308 nm) tend to be preferred in view of their high repetition rate (~500 Hz) and high energy pulses (100 to 200 mJ). However with CVLs a still higher pulse repetition frequency of 5 KHz at 510.6 nm is currently obtained.[124-126] Moreover, pumping in the visible leads to a minor dye degradation than pumping in the UV.

Tuning of the dye laser was demonstrated by Soffer and McFarland[127] with a diffraction grating, but tunable narrow band emission with a bandwidth less than 300 MHz or 0.004 Å was first obtained by Hänsch in 1972[128] using a high dispersion echelle grating in a Littrow mount as end reflector, an inverted telescope for collimation and beam expansion, and a tilted Fabry-Perot etalon (Figure 24). The cavity length (40 cm) is in the order of the laser light coherence length to avoid discrete axial mode structure. Using a Fabry Perot interferometer outside the laser cavity, Hänsch could reduce the laser bandwith to 7 MHz full-width half maximum (~8×10^{-5} Å), thereby increasing the pulse duration to 30 ns FWHM. A prism beam expander can also be used in place of the telescope to obtain similar linewidths.[129] Narrow band emission is necessary for applications to uranium laser isotope separation and recently a bandwidth of 100 MHz and a power of 300 mW were obtained from a single mode dye laser oscillator and a high power dye amplifier pumped by a CVL at a pulse repetition frequency of 5 kHz.[125]

G. FLASHLAMP-PUMPED DYE LASERS

Excellent reviews deal with flashlamp pumped dye lasers.[80,91,123,130-134] Flashlamp systems with rise times less than 1 μs have been developed to overcome the effects of intersystem crossing to the triplet state. Some devices utilize a coaxial flashlamp mounted on a low inductance disk capacitor with a hole in its center to accommodate the flashlamp and the dye cell.[130] Other devices make use of a conventional linear xenon flashlamp driven by a low-inductance capacitor.[133] The optical resonator features flat mirrors consisting of multiple dielectric coatings with broadband spectral reflectivities to use several different dyes. The output mirror should be highly reflecting if a poor laser dye is used. Otherwise a 50% transmission mirror is advisable.

The solution must be flowed for repetitive pulsing because of the strong variation of the solvent refractive index with temperature. Air bubbles and liquid turbulence should be avoided.

The overall efficiency of a flashlamp-pumped dye laser does not exceed 1%: for 50 J of electrical input energy, the laser pulse energy is typically 200 mJ.[130]

The wavelength of the dye laser increases with dye concentration. Tuning of a flashlamp-pumped dye laser can be achieved using a diffraction grating,[130] a Fabry-Perot etalon,[135] or a multiple-prism tuner.[136]

A flashlamp-pumped dye laser system repetitively pulsed at 20 pps delivered a power output of 200 mW.[137] Coaxial lamps have a short lifetime and are not suited for high repetition rates. Moreover, the close coupling between discharge and dye cell leads to index gradients induced by heat transfer. Preionized linear flashlamps were also used at repetition rates up to 10 Hz with a lamp lifetime exceeding 10^5 flashes to deliver laser pulses of 200 mJ energy and 800 ns duration.[138]

H. PASSIVE MODE LOCKING OF FLASHLAMP-PUMPED DYE LASERS

Mode locking in a flashlamp-pumped rhodamine 6G laser using a 3,3′-diethyloxadi-carbocyanine iodide (DODCI) solution as saturable absorber was first obtained by Schmidt and Schäfer.[139] A train of picosecond pulses of a few hundred millijoules of energy and of 1- to 2-μs duration can be generated using a coaxial linear air flashlamp in a cylindrical elliptical pumping reflector.[140] Pulse durations of ~5 ps are reliably produced but the buildup of picosecond pulse generation is relatively slow. The very first pulses can be ~100 ps duration and the shortest pulses are obtained after 45 round trips in the cavity.[141] For a more complete description of flashlamp-pumped mode-locked dye lasers, the reader should refer to previously published reviews.[142-144] These lasers were the first and are still low-priced tunable picosecond sources. The frequency range covered using the laser dyes and corresponding saturable absorbers listed in Table 4 extends from 475 to 805 nm. The laser threshold and the full modulation of the pulse train depend largely on the excited state lifetime of the mode-locking dye which can be varied using solvents of different viscosities.[146,147,150] Despite the low price and extended tunability of flashlamp-pumped dye lasers, they are not greatly used because of the low repetition rate due to thermal effects in the long amplifying dye cell and the need to frequently change the laser dye solution (~ 2 d). For these reasons, CW mode-locked dye lasers tend to be preferred. However, they are still used for single shot fluorescence lifetime measurements using a streak camera device.

I. LASER-PUMPED PICOSECOND DYE LASERS

Mode locking of a dye laser by pumping with another mode-locked laser was achieved for the first time by Glenn et al.[154] The length of the dye laser cavity was equal to or a submultiple of the length of the cavity of the pump laser but ultrashort cavity lengths (50 μm) can also be used.[155] However, in this case the oscillating axial modes and the spectral tuning cannot be controlled easily. Pulse trains of a mode locked Nd-YAG can be used to pump IR dye lasers efficiently. The photochemically stable IR dye no. 26 (4-[7-(2-phenyl-4H-1-benzothiopyran-4-ylidene)-4-chloro-3,5-trimethylene-1,3,5-heptatrienyl]-2-phenyl-1-benzothiopyrylium perchlorate) gave 5.5-ps duration laser pulses between 1.15 and 1.24 μm.[156] This wavelength range is of considerable interest because of the high transmission in optical fibers. Tunability has been extended to 1.53 μm with dye no. 5 and a 2% energy conversion is made possible despite the very short ground state recovery time of this dye.[157] Stimulated emission of IR dyes between 1.4 and 1.8 μm was also obtained using single pulse excitation from a Nd-glass laser and the pulse duration was 6 ps.[82]

For the UV-visible region, dye lasers pumped by Q-switched solid lasers,[97] nitrogen lasers,[158-161] and excimer lasers[162] deliver picosecond pulses based on a self-Q-switching effect and distributed feedback[163] provided by the gain modulation within the active medium itself. The coherent light of the pump laser is split into two beams which interfere in a dye solution, producing a periodic spatial modulation of the gain and the refractive index of the

TABLE 4
Picosecond Pulse Generation in Flashlamp-Pumped Dye Lasers

Laser dye	Laser dye solvent	Saturable absorber	Absorber solvent	Tuning range (nm)	Ref.
Coumarin 102 (2.2 × 10^{-4} M)	CH_3OH	DOCI	EG-CH_3OH (90—10)	475—490	146
Coumarin 314 (2.5 × 10^{-4} M) Coumarin 6 (10^{-5} M)	CH_3OH	PIC	Glycerol 87%	515—525	147
Fluorol 7GA Coumarin 6 (10^{-4} M)	CH_3OH	DQOCI		550—580	148
Rhodamine 6G	C_2H_5OH	DODCI	C_2H_5OH	584—625	140, 142, 149
Rhodamine 6G (1.5 × 10^{-4} M)	5% Ammonyx LO	DODCI	C_2H_5OH	595—625	106
Rhodamine 6G	CH_3OH	Pinacyanol	Glycerol	585—605	150
Rhodamine B		DQTCI		605—639	149
		DODCI		615—645	
	CH_3OH	Pinacyanol	Glycerol	611—634	150
Rhodamine 101	C_2H_5OH	DQTCI	C_2H_5OH	630—693	151
DCM	C_2H_5OH	DQTCI DTDCI		630—670	151
Cresyl violet + Rhodamine 6G	C_2H_5OH	DTDCI		652—704	
		DDCI		644—680	149, 152
		DOTCI		644—680	
DOTCI	DMSO	HITC	DMSO	795—805	153

Note: DOCI = 3,3′ diethyloxacarbocyanine iodide, DODCI = 3,3′ diethyloxadicarbocyanine iodide, DQTCI = 1,3′-diethyl-4,2′-quinolylthiacarbocyanine iodide, DOTCI = 3,3′-diethyloxatricarbocyanine iodide, HITCI = 1,1′,3,3,3′,3′-hexamethylindotricarbocyanine iodide, PIC = pseudoisocyanine iodide, and DMSO = dimethylsulfoxide.

medium.[97] The emissions of the excited dye molecules undergo gains in the inverted regions along the induced grating. The generated laser wavelength λ_L is related to the refractive index of the solution n and the separation of the interference fringes Λ,

$$\lambda_L = 2 \cdot n\Lambda$$

Λ is given by

$$\lambda_p = 2\sin\Theta \cdot \Lambda$$

where λ_p is the pump wavelength and Θ the incidence angle on the dye cell. The term describing the temporal characteristics of the distributed feedback dye laser (DFDL) is τ_C, average lifetime of a photon in the DFDL.[163] The self-Q-switching behavior of the DFDL is linked to the proportionality of τ_C to the square of the population of the upper laser level.[163] Relaxation oscillations are related to the variation of τ_C as a function of time and a short DFDL output pulse is emitted, or if the pump pulse intensity is still above the laser threshold several such pulses are emitted. Applications of the DFDL to the generation of high power picosecond UV pulses have been described in Section VII.D.

Mode locking of dye lasers pumped by a long pulse (150 ns) XeCl laser has also been achieved around 497 and 580 nm, resulting in pulse durations of 5.5 and 7 ps respectively.[164] However, the short cavity length needed to obtain several tens of round trips during the 150-ns pump duration does not allow easy pulse extraction.

Picosecond tunable dye lasers pumped by the second harmonic of a single pulse extracted

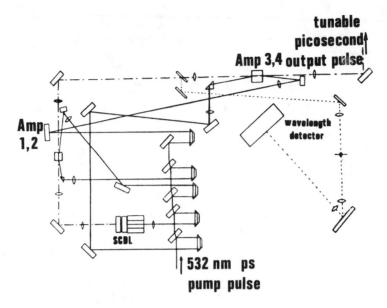

FIGURE 25.   Generation of high energy tunable picosecond pulses. (From Kortz, H. P., Pax, P., and Aubert, R., *Proceedings of SPIE*, 533, 32, 1985. With permission.)

from the train of pulses of an active-passively mode-locked Nd:YAG laser, making use of a short cavity and delivering up to 3 mJ at 10 Hz in a <10-ps pulse[165] (Figure 25), are commercially available.

J. PICOSECOND AND FEMTOSECOND CONTINUOUS WAVE DYE LASERS

The mode locking of CW dye lasers has been previously reviewed.[166]

Three different methods result in stable continuous emission of tunable mode-locked pulses:

1. Active mode locking by acoustic loss modulation
2. Pumping with mode-locked pulses from a CW laser source
3. Passive mode locking by a saturable absorber

Although the shortest pulses were first obtained using the passive mode-locking technique, synchronous pumping appears to be the most versatile method due to the large tunability of the laser pulses and the wide application to time-resolved spectroscopy. Fluorescence decay measurements can be carried out using a synchroscan streak camera that operates repetitively in synchronism with the mode-locked CW laser,[167] or a single photoelectron counting apparatus which provides a superior dynamic range.[168] In the latter case, cavity dumping is needed to lower the pulse repetition rate below ~4 MHz. Mode-locked argon ion and krypton ion lasers were first used for synchronous pumping, but the development of mode-locked CW Nd:YAG lasers with shorter pulse duration (<100 ps) (see Section VII.C) has enabled the synchronously pumping of dye lasers at 1.06 μm[169,170] or at 532 nm, with the second harmonic generated in a KTP crystal. These mode-locked Nd:YAG lasers can also be Q-switched at a 500-Hz repetition rate to deliver pulse trains with higher energy pulses, in the order of 50 μJ at 532-nm and 55-ps duration, which in turn pump a dye laser oscillator.[171]

A refinement has been recently introduced in synchronously pumped dye lasers by compressing either the IR output of the mode-locked Nd-YAG[172] or its second harmonic.[173,174] Using these methods, dye laser pulse widths < 220 fs were obtained.[172,173] By adding a cavity dumper, pulses of 500-fs duration tunable from 570 to 630 nm were generated.[172]

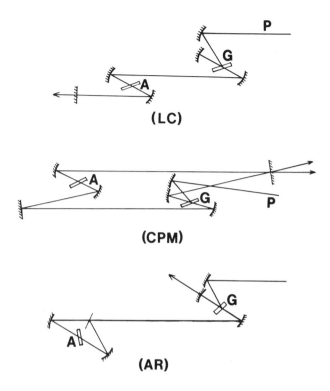

FIGURE 26. Passive mode-locking configurations: LC, linear cavity; CPM, colliding pulse mode-locking; AR, antiresonant ring; P, pump beam; G, gain medium; and A, saturable absorber.

Passively mode-locked CW dye lasers,[175] which were the first lasers to produce sub-picosecond pulses,[176] were originally pumped by the continuous output of an argon-ion laser. The active medium (rhodamine 6G) and the saturable absorber (DODCI) can be mixed together in a single flowing dye stream[176,178] or used in two free-flowing dye streams[177] (Figure 26). To obtain shorter pulses, a careful optimization of the cavity bandwith is needed.[179] Furthermore, the use of the colliding-pulse mode locking (CPM)[180] in which the counterpropagating pulses collide in a thin saturable-absorber dye jet (Figure 26) enabled the generation of pulses as short as 27 fs.[181]

Temporal compression of the pulses[177,182] via the grating-pair technique of Treacy resulted in yet shorter and shorter pulses: 16,[183] 8,[184] and 6 fs,[185] i.e., three optical periods. These extremely short pulses suffered from the nontunability of the CPM laser, and in the same way, frequency-tunable pulses from a synchronously pumped mode-locked dye laser have been compressed from 5.4 ps to 16 fs.[186] However, the lack of suitable active-passive dye combinations to extend the tunability of passively mode-locked CW dye lasers has been recently overcome and a list of new combinations is given in Table 5.

When femtosecond pulses synchronized with another laser or amplifiers are needed, hybrid mode locking is preferable to the extracavity fiber and grating pulse compression technique which is more expensive and complicated. In the hybrid mode-locked CW dye lasers, several experimental arrangements have been used:

1. Mode-locking dye cell in contact with the 100% reflectivity mirror[96]
2. Gain dye and absorber dye mixed in the same jet stream[197-199]
3. Linear cavity containing a gain dye jet and terminated in an antiresonant ring in which a saturable absorber jet is precisely positioned at the center of this ring[200,201] (Figure 26)

TABLE 5
Subpicosecond Pulse Generation in Passively Mode-Locked CW Dye Lasers

Laser dye	Saturable absorber	Tuning range (nm)	Pulse duration (fs)	Ref.
Coumarin 102	DOCI	492—506	580	187
Rhodamine 110	HICI	581	80	188
	HICI	553—570	150	189
Rhodamine 6G	DASTBI	570—600	520	190
	DODCI	590—610	—	175
Rhodamine B	DQTCI	616—658	220	191
DCM	DQTCI	655—673	680	192
Rhodamine 6G	DQTCI or DCCI	652—694	<500	193
Sulforhodamine 101		652—694	<500	193
Rhodamine 700	DOTCI	727—740	850	194
	HITCI	762—778	850	194
DCM	DDI	742—761	<400	195
Rhodamine 700		742—761	<400	195

Note: DOCI = 3,3'-diethyloxacarbocyanine iodide, HICI = 1,1',3,3,3',3'-hexamethylindocarbo-
cyanine iodide, DASTBI = 2-(p-dimethyl-aminostyryl)-benzthiazolylethyl iodide, DODCI =
3,3'-diethyloxadicarbocyanine iodide, DQTCI = 1,3'-diethyl-4,2'-quinolylthiacarbocyanine iodide, DCCI = 1,1'-diethyl-2,4'-carbocyanine iodide, DOTCI = 3,3'-diethyloxatricarbocyanine
iodide, HITCI = 1,1',3,3,3',3'-hexamethylindotricarbocyanine iodide, and DDI = 1,1'-die-
thyl-2,2'-dicarbocyanine iodide.

4. A dual-jet linear resonator[202]
5. A CPM ring laser[203]

A great advantage of hybrid mode locking is that the cavity length adjustment is much
less critical than in the pure synchronous mode locking.[196] The antiresonant ring preserves
the CPM effect on the pulse duration of dye lasers which enabled the generation of the
shortest pulses in passively mode-locked dye lasers. The antiresonant ring has provided 60-
to 70-fs pulse duration,[200,201] the CPM ring cavity 65-fs pulse duration,[203] and the dual jet
linear resonator even shorter pulses, 55 fs.[202] A list of hybridly mode-locked CW dye lasers
is provided in Table 6.

K. PULSED DYE AMPLIFIERS

Broadband light amplification in laser dyes pumped by high power solid lasers or by
flashlamp is often needed for a large number of scientific and technical applications.

In the case of flashlamp-pumped dye amplifiers, Flamant and Meyer[209] defined the
amplification gain G_1 as the ratio of the amplifier outputs when pumped (I') or not (I) and
the absolute energy gain G_2 the ratio of the pulse energy before (I_o) and after the amplifiers
(I'). G_1 could be as high as 700 but G_2 was only 6. Amplification of picosecond dye laser
pulses into two flashlamp-pumped amplifiers has been also considered.[106] Typically, a 70-
µJ picosecond pulse could be amplified to 1 mJ in a first amplifier, to only 4 mJ in the
second and 7 mJ in the third. This is due to high small-signal gain and saturation at low
energy densities, 3 mJ/cm². The use of a N_2 laser transversely pumped amplifier enabled
amplification of a single pulse from 2.5 µJ to ~100 µJ.[210] The use of a flashlamp pumped
dye laser amplifier is indeed wasteful because the flashlamp pump time (~ 1 µs) is much
longer than the storage time of the dye amplifier (~ 5 ns).

The gain characteristics of transversely pumped dye-laser systems have been analyzed
as a function of input signal intensity and pumping rate and compared with experimental
results.[211-213] Iteration of the amplification is also possible with N_2 lasers[214,215] and excimer

TABLE 6
Subpicosecond Pulse Generation in Hybridly Mode-Locked CW Dye Lasers

Laser dye	Saturable absorber	Tuning range (nm)	Pulse duration (fs)	Ref.
Disodium fluorescein	Rhodamine B	535—575	450	198
	DODCI	561	580	198
Rhodamine 110	Rhodamine B	545—585	250	198
Rhodamine 6G	DODCI	595—610	700	196
	DODCI	577—615	350	198
	DQOCI-DODCI	~583	69	202
	DODCI	—	64	201
Rhodamine B	DDBCI	604—632	430	204
	Oxazine 720	~650	187	205
Rhodamine 101	DQTCI	~675	59	205
Sulforhodamine 101	DQTCI	~675	55	202, 205
Rhodamine 700	DOTCI	710—718	470	206
Oxazine 750	HDITC-P	730—835	800	197
Oxazine 725	HITCI	750—780	600	207
Rhodamine 700	HITCI	770—781	550	206
Styryl 9	IR 140	840	550	208
Dye styryl 9	IR 140	850	65	203

Note: Refer to Table 5 for definitions of abbreviations.

lasers.[162,216] Even a high repetition rate of 5 kHz is possible with CVL.[124-126,217,218] The amplified output is superimposed on a background of amplified spontaneous emission (ASE) which can be removed by focusing the laser beam into a saturable absorber,[216] and by spatial and spectral filtering.[216,217] Efficient amplification ($\times 4000$) of femtosecond pulses has been successful in a gain jet, 1 mm thick, consisting of a dye dissolved in a mixture of ethyleneglycol-glycerol to enhance the viscosity and achieve a laminar flow.[219]

A particular type of amplification is the spectral narrowing of a dye laser output obtained by the injection of monochromatic radiation in a dye cell placed in a resonator. The emission from a N_2-pumped dye laser has been narrowed from 400 to 0.0016 Å with injection of 5145 Å argon laser light.[220] Injection of tunable light from a CW dye laser into a pulsed dye laser is of great importance for high resolution spectroscopy. Locking of a flashlamp-pumped dye laser has provided 50-MHz line width and the narrow band output saturated at 18 mJ for a 2-mW injection power.[221] The method has been extended recently over a 200 nm tuning range between 572 and 720 nm, at a band width of 6 MHz and a repetition rate of 10 Hz.[222] The band width of the amplified pulse can be limited by the duration of the amplifier pumping pulse; Lavi et al.[217] obtained 30 MHz in using a 30-ns CVL and 60 MHz in using a 15-ns Nd:YAG laser.

X. FREE ELECTRON LASERS

As we showed at the beginning of this chapter, the operation of the majority of lasers is described by interactions involving atomic or molecular quantum states.

The semiconductor laser which operates by population inversion between the valence and the conduction bands differs only slightly from classical mechanisms with three or four levels. In the case of the FEL there is a complete departure from this representation.

In 1977, Madey and his group at Stanford University[223] succeeded in producing the first coherent emission from a FEL at 3.417 μm, with a radiation width of 8 nm and an average power of 0.36 W. The previous year the same group demonstrated the amplification of an "ordinary" laser at 10 μm.[224]

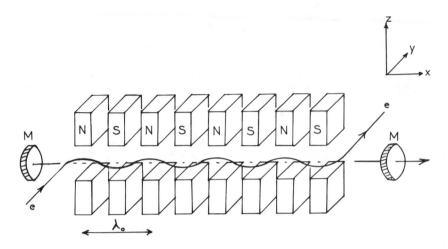

FIGURE 27.   Magnetic wiggler and optical cavity of a free electron laser showing the average electron trajectory. λ_0 is the period of the undulator.

Because the FEL is based on the interaction of an electron that is not in an atomic or molecular orbital, it is considered here as free. In fact a strictly free electron, in the absence of an electromagnetic field, can neither emit nor absorb light. We will thus consider, as Renieri[225] pointed out, that free signifies only that the electron is not bound to the quantum state of an atomic nucleus.

It is known that electrons at relativistic velocities, centrifugally accelerated by a magnetic field, emit continuous polychromatic radiation called synchroton radiation. This synchroton radiation is used, as such, in physics and photochemistry experiments.

As is shown in Figure 27 let us suppose that relativistic electron crosses an alternating magnetic structure with a period λ_0.

This alternating magnetic structure is, for example, made up of a series of alternate magnetic dipoles over a length of one or several meters. At each passage in a magnetic dipole, the trajectory of the electron is curved and an electromagnetic ray is emitted according to the tangent in the angle $\Theta = E/m_0c^2$. In the alternating magnetic structure of a wiggler the trajectory followed by the electron is sinusoidal and is in a plane perpendicular to the magnetic dipoles.

The construction of the magnetic wiggler is such that it has a period λ_0 of the order of centimeters. The time taken by the electron to travel λ_0 is $t = \lambda_0/v$. During this time the first photon emitted at the first curve of the trajectory has travelled a distance $c.t = c.\lambda_0/v$. At the end of a period of the wiggler the difference in the path is therefore $\lambda_0.(c/v - 1)$.

If this difference is equal to λ the emitted wavelength originating from the continuous spectral range of the synchrotron radiation, then constructive resonance at this wavelength may occur. This very simplified approach allows us however to show that if one tries to produce coherent emission with a short wavelength it is necessary to have a wiggler with a short period λ_0, or to have very high velocity electrons.

Knowing that $\gamma = (1 - v^2/c^2)^{-1/2}$, it can be seen that for relativistic electrons $\lambda \simeq \lambda_0/2\gamma^2$.

As a nonlimiting example, the relation $W = eV = m_0c^2\gamma^{-1}$ shows that for 147-MeV electrons ($\gamma = 290$) and with a wiggler $\lambda_0 = 7.8$ cm as that used at the University of Orsay on the ACO ring, λ is calculated to be 464 nm. In actual fact, the experiments carried out at this installation in 1982[226] produced an emission at 488 nm, for the formula for calculating λ as a function of γ should take a correcting term $K = e\, B_0\lambda_0/2\pi\, m\, c$ into account, so that

$$\lambda = \frac{\lambda_0}{2\gamma^2} \left(1 + \frac{k^2}{2} \right)$$

where Θ is the angle between the axis of the electron trajectory and the axis of propagation of the plane wave. The larger the number of periods of the wiggler the smaller will be the spectral width of the radiation at the exit. From this, if the electrons are in a synchrotron ring, it will be necessary to have the straight sections where the wigglers are placed as long as possible. One understands why the most ambitious projects, such as those being carried out at Stanford University, use linear accelerators, allowing several meters of wigglers. The use of linear accelerators also makes it possible to employ the HF superconducting cavity technique, and the possible recycling of the electrons. It is thus hoped, in a short time, to reduce the investment costs of FEL and to obtain hundreds of watts or kilowatts of emission tunable from 0.35 to 6 μm.

At first, the possibility of an operating FEL was shown by using synchrotron rings or linear accelerators already existing, but not always very well suited to the FEL. One now sees the development of projects with superconducting cavity linear accelerators which will make it possible, in the near future, to undertake a thorough evaluation of FELs from the point of view of maintenance and reliability.

The luminous pulses from FELs are made up of macropulses, typically 10^{-4} s, and of very short (a few picoseconds) micropulses. The interval between the macropulses in the case of superconducting accelerators such as that under construction at Stanford, is typically 84.6 ns. The peak power of the micropulses should be of the order of 1 to 4 MW.

ACKNOWLEDGMENTS

The authors wish to thank Maurice Michon and Denis Doizi for their assistance in correcting the manuscript and suggesting many improvements.

REFERENCES

1. **Schawlow, A. L. and Townes, C. H.,** Laser theory, *Phys. Rev.,* 112, 1940, 1958.
2. **Svelto, O.,** *Principles of Laser,* 2nd ed., Plenum Press, New York, 1982.
3. **Yariv, A.,** *Quantum Electronics,* 2nd ed., John Wiley & Sons, New York, 1975.
4. **Shimoda, K.,** *Introduction to Laser Physics* (Springer Series in Optical Sciences), Springer-Verlag, Berlin, 1986.
5. **Ronchi, L.,** Optical resonators, in *Laser Handbook,* Vol. 1, Arecchi, F. T. and Schulz-Dubois, E. O., Eds., North-Holland, Amsterdam, 1972, 151.
6. **Young, M.,** *Optics and Lasers,* (Springer Series in Optical Sciences), Springer-Verlag, Berlin, 1977.
7. **Maiman, T. H.,** Stimulated optical radiation in ruby masers, *Nature,* 187, 493, 1960.
8. **Geusic, J. E., Marcos, H. M., and Van Virtet, L. G.,** Laser oscillations in Nd doped yttrium aluminium, yttrium gallium and gadolinium garnets, *Appl. Phys. Lett.,* 4, 182, 1964.
9. **Zverev, G. M., Golyaev, Yu.D., Shalaev, E. A., and Shokin, A. A.,** Neodynium activated yttrium-aluminium garnet (YAG:Nd) Lasers, *J. Sov. Laser Res.,* 8(3), 189, 1987.
10. **Smith, B.,** Lamps for pumping solid-state lasers: performance and optimization, *Laser Focus/Electro-optics,* 22, 58, 1986.
11. **Harnagel, G., Welsch, D., Cross, P., and Scifres, D.,** High power laser arrays: a progress report, *Laser Appl.,* June 1986, 135.
12. **Botez, D.,** Recent developments in high power single element fundamental mode diode lasers, *Laser Focus/Electro-optics,* 23(3), 68, 1987.
13. **Watts, R. N. and Wieman, C. E.,** Manipulating atomic velocities using diode lasers, *Optics Lett.,* 11(5), 291, 1986.
14. **Patel, C. K. N.,** Continuous-wave laser action on vibrational rotational transitions of CO_2, *Phys. Rev.,* A, 136, 1187, 1964.

15. **Bradley, L. C., Soohoo, K. L., and Freed, C.,** Absolute frequencies of lasing transitions in nine isotopic species, *IEEE J., Quantum Electron.,* QE-22, 2, 234, 1986.

16. **Gibson, R. B., Boyer, K., and Javan, A.,** Mixed isotope multiatmosphere CO_2 laser, *IEEE J. Quantum Electron.,* QE-15, 11, 1224, 1979.

17. **Michon, M., Dumanchin, R., and Lavarini, B.,** High power per unit volume CO_2 lasers in continuous and pulsed operation, *IEEE J. Quantum Electron.,* QE-6, 1, 1970.

18. **Lachambre, J., Lavigne, P., Verreault, M., and Otis, G.,** Frequency and amplitude characteristics of a high repetition rate hybrid TEA-CO_2 laser, *IEEE J. Quantum Electron.,* QE-14, 3, 170, 1978.

19. **Dumanchin, R., Michon, M., Farcy, J. C., Boudinet, G., and Rocca-Serra, J.,** *IEEE J. Quantum Electron.,* QE-8, 1, 163, 1972.

20. **Baranov, V. Yu, Kazsakov, S. A., Malyuta, D. D., Mezhevov, V. S., Napartovich, A. P., Nisiev, V. G., Orlov, M. Yu, Starodubstev, A. I., and Starostin, A. N.,** Average power limitations in high repetition rate pulsed gas lasers, *Appl. Opt.,* 19, 930, 1980.

21. **Newman, L. A. and Hart, R. A.,** Recent R and D advances in sealed-off CO_2 lasers, *Laser Focus/Electro-optics,* 23(6), 80, 1987.

22. **Rhodes, C. K., Ed.,** Excimer lasers, in *Topics in Applied Physics,* Vol. 30, Springer-Verlag, Berlin, 1979.

23. **Menown, H. and Neale, C. V.,** Thyratrons for short pulse laser circuits, 13th Pulse Power Modulator Symposium, Buffalo, NY, June 1978.

24. **Ewing, J. J. and Brau, C. A.,** High efficiency UV lasers, in *Tunable Lasers and Applications,* (Springer Series in Optical Sciences), Springer-Verlag, Berlin, 1976, 21.

25. **Walter, W. T., Piltch, H., Solimene, N., and Gould, G.,** *IEEE J. Quantum Electron.,* 2, 474, 1966.

26. **Piper, J. A.,** Recent advances in metal vapour lasers, in *Lasers '78,* Proceedings of the International Conference on Lasers, Corcoran, V. J., Ed., STS Press, Orlando, FL, 1979.

27. **Isaev, A. A. and Kazaryan, M. A.,** Investigation of pulsed copper vapour laser, *Sov. J. Quantum Electron.,* 7, 253, 1977.

28. **Smilanski, I., Kerman, A., Levin, L. A., and Erez, G.,** Scaling of the discharge heated copper vapor laser, *Opt. Commun.,* 25, 79, 1978.

29. **Isaev, A. A., Kasakov, V. V., Lesnoi, M. A., Markova, S. V., and Petrash, G. G.,** *Sov. J. Quantum Electron.,* 16(11), 1517, 1986.

30. **Hammerling, P., Budgor, A. B., and Pinto, A., Eds.,** *Tunable Solid State Lasers,* Springer-Verlag, Berlin, 1985;**Moulton, P. F.,** Tunable solid-state lasers targeted for a variety of applications, *Laser Focus/Electro-optics,* 23, 56, 1987.

31. **Walling, J. C.,** Properties of Alexandrite lasers in physics of new laser sources, in *Physics,* Abraham, N. B., Arechhi, F. T., and Sona, A., Eds., (NATO ASI Series — Series B), Vol. 132, Plenum Press, New York, 1985.

32. **Hargrove, L. E., Fork, R. L., and Pollack, M. A.,** Locking of He-Ne laser modes induced by synchronous intracavity modulation, *Appl. Phys. Lett.,* 5, 4, 1964.

33. **De Maria, A. J., Stetser, D. A., and Heynau, H.,** Self mode-locking of lasers with saturable absorbers, *Appl. Phys. Lett.,* 8, 174, 1966.

34. **De Maria, A. J., Stetser, D. A., and Glenn, W. H.,** Ultrashort light pulses, *Science,* 156, 1557, 1967.

35. **De Maria, A. J., Glenn, W. H., Brienza, M. J., and Mack, M. E.,** Picosecond laser pulses, *Proc. IEEE,* 57, 2, 1969.

36. **Laubereau, A. and Kaiser, W.,** Generation and applications of passively mode-locked picosecond light pulses, *Opto-electronics,* 6, 1, 1974.

37. **Eckardt, R. C., Lee Chi, H., and Bradford, J. N.,** Effect of self-phase modulation on the evolution of picosecond pulses in a Nd-glass laser, *Opto-electronics,* 6, 67, 1974.

38. **Zindt, W., Laubereau, A., and Kaiser, W.,** Generation of chirp-free picosecond pulses, *Opt. Commun.,* 22, 161, 1977.

39. **Taylor, J. R., Sibbett, W., and Cormier, A. J.,** Bandwidth-limited picosecond pulses from a neodymium-phosphate glass oscillator, *Appl. Phys. Lett.,* 31, 184, 1977.

40. **Kaufmann, K. J.,** Picosecond spectroscopy applied to the study of chemical and biological reactions, *Crit. Rev. Solid State Mater. Sci.,* 8(3), 265, 1978.

41. **Kopainsky, B., Kaiser, W., and Drexhage, K. H.,** New ultrafast saturable absorbers for Nd:lasers, *Opt. Commun.,* 32, 451, 1980.

42. **Vasil'Eva, M. A., Vishchakas, J., Gulbinas, V., Malyshev, V. I., Masalov, A. V., Kabelka, V., and Syrus, V.,** Amplitude and phase nonlinear response of bleachable dyes using picosecond excitation, *IEEE J. Quantum Electron.,* 724, 1983.

43. **Lin, C.,** Polymethine IR laser dyes for passive mode-locking of ruby lasers, *Opt. Commun.,* 13, 106, 1975.

44. **Horowitz, L., Papanestor, P., and Heller, D. F.,** Mode-locked performance of tunable alexandrite lasers, Proc. Int. Conf. Lasers '83, December 12 to 16, 1983, 170.

45. **Siegman, A. E. and Kuizenga, D. J.,** Reviews: active mode-coupling phenomena in pulsed and continuous lasers, *Opto-electronics,* 6, 43, 1974.
46. **Yariv, A. and Yeh, P.,** Acousto-optic devices, in *Optical Waves in Crystals,* John Wiley & Sons, New York, 1984, 366.
47. **Goldberg, L. S. and Schoen, P. E.,** Stable active-passive mode-locking of a Nd-phosphate glass laser using Eastman $\neq$ 5 saturable dye, in *Ultrafast Phenomena Part IV,* (Springer Series in Chemical Physics 38), Auston, D. M. and Eisenthal, K. B., Eds., Springer-Verlag, Berlin, 1984, 87.
48. **Chan, C. K. and Sari, S. O.,** Tunable dye laser pulse converter for production of picosecond pulses, *Appl. Phys. Lett.,* 25, 403, 1974.
49. **Eckstein, J. N., Ferguson, A. I., Hansch, T. W., Minard, C. A., and Chan, C. K.,** Production of deep blue tunable picosecond light pulses by synchronous pumping of a dye laser, *Opt. Commun.,* 27, 466, 1978.
50. **Johnson, A. M., Stolen, R. H., and Simpson, W. M.,** 80 X single-stage compression of frequency doubled Nd:yttrium aluminum garnet laser pulses, *Appl. Phys. Lett.,* 44, 729, 1984.
51. **Kolner, B. H., Bloom, D. M., Kafka, J. D., and Baer, T. M.,** Compression of mode-locked Nd:YAG Pulses to 1.8 picoseconds, in *Ultrafast Phenomena Part IV,* (Springer Series in Chemical Physics 38), Auston, D. H. and Eisenthal, K. B., Eds., Springer-Verlag, Berlin, 1984, 19.
52. **Voss, D. F. and Goldberg, L. S.,** Simultaneous amplification and compression of continuous-wave mode-locked Nd:YAG laser pulses, *Opt. Lett.,* 11, 210, 1986.
53. **Strickland, D. and Mourou, G.,** Compression of amplified chirped optical pulses, *Opt. Commun.,* 56, 219, 1985.
54. **Maine, P., Strickland, D., Bouvier, M., and Mourou, G.,** Amplification of picosecond pulses to the terawatt level by chirped pulse amplification and compression, in Conference on Lasers and Electro-Optics, Digest of Technical Papers, April 26 to May 1, 1987, Optical Society of America, Baltimore, p. 366.
55. **Watanabe, S., Watanabe, M., and Endoh, A.,** Passive mode-locking of a long pulse XeCl laser, *Appl. Phys. Lett.,* 43, 533, 1983.
56. **Szatmari, S. and Schafer, F. P.,** Simple generation of high power, picosecond tunable excimer laser pulses, *Opt. Commun.,* 48, 279, 1983.
57. **Szatmari, S. and Schafer, F. P.,** Picosecond gain dynamics of KrF,* *Appl. Phys., B,* 33, 219, 1984.
58. **Szatmari, S., Schafer, F. P., Muller-Horsche, E., and Muckenheim, W.,** Hybrid dye-excimer laser system for the generation of 80 fs, 900 GW pulses at 248 nm, *Opt. Commun.,* 63, 305, 1987.
59. **Szatmari, S., Racz, B., and Schafer, F. P.,** Bandwidth-limited amplification of a 230 fs pulse in XeCl, in Conference on Lasers and Electro-Optics, Digest of Technical Papers, OSA-IEEE, Baltimore, MD, April 26 to May 1, 1987, 212.
60. **Franken, P. A. and Ward, J. F.,** Optical harmonics and nonlinear phenomena, *Rev. Mod. Phys.,* 35, 23, 1963.
61. **Zernike, F.,** Refractive indices of ammonium dihydrogen phosphate and potassium dihydrogen phosphate between 2000 Å and 1.5 μ, *J. Opt. Soc. Am.,* 54, 1215, 1964.
62. **Hitz, C. B.,** Choosing a doubling crystal for YAG, *Laser Focus,* 32, 1976.
63. **Nikogosyan, D. N.,** Nonlinear optic crystals (review and summary of data), *Sov. J. Quantum Electron,* 7, 1, 1977.
64. **Dewey, C. F., Cook, W. R., Hodgson, R. T., and Wynne, J. J.,** Frequency doubling in KB_5O_8, $_4H_2O$ and $NH_4B_5O_8$, $_4H_2O$ to 217.3 nm, *Appl. Phys. Lett.,* 26, 714, 1975.
65. **Kato, K.,** Second harmonic generation to 2048 Å in β-BaB_2O_4, *IEEE J. Quantum Electron.,* QE-22, 1013, 1986.
66. **Eimerl, D., Davis, L., Velsko, S., Graham, E. K., and Zalkin, A.,** Optical, mechanical and thermal properties of barium borate, *J. Appl. Phys.,* 62, 1968, 1987.
67. **Moody, S. E., Eggleston, J. M., and Seamans, J. F.,** Long-pulse second harmonic generation in KTP, *IEEE J. Quantum Electron.,* QE-23, 335, 1987.
68. **Pixton, R.,** Tripling YAG frequency, *Laser Focus,* 66, 1968.
69. **Johnson, F. M.,** Stimulated Raman scattering, in *Handbook of Lasers,* CRC Press, Boca Raton, FL, 1971, 526.
70. **Frey, R. and Pradere, F.,** Powerful tunable infrared generation by stimulated Raman scattering, *Opt. Commun.,* 12, 98, 1974.
71. **Young, M.,** *Optics and Lasers,* Springer-Verlag, Berlin, 1977, 185.
72. **Baumgartner, R. A. and Byer, R. L.,** Optical parametric amplification, *IEEE J. Quantum Electron,* QE-15, 432, 1979.
73. **Laubereau, A., Greiter, L., and Kaiser, W.,** Intense tunable picosecond pulses in the infrared, *Appl. Phys. Lett.,* 25, 87, 1974.
74. **Seilmeier, A., Spanner, K., Laubereau, A., and Kaiser, W.,** Narrow band tunable infrared pulses with sub-picosecond time resolution, *Opt. Commun.,* 24, 237, 1978.

75. **Laubereau, A., Seilmeier, A., and Kaiser, W.,** A new technique to measure ultrashort vibrational relaxation times in liquid systems, *Chem. Phys. Lett.,* 36, 232, 1975.

76. **Brearley, A. M., Strandjord, A. J. G., Flom, S. R., and Barbara, P. F.,** Picosecond time-resolved emission spectra: techniques and examples, *Chem. Phys. Lett.,* 113, 43, 1985.

77. **Schafer, F. P.,** Dye lasers, *Topics in Applied Physics,* Vol. 1, 2nd revised ed., Springer-Verlag, Berlin, 1977.

78. **Meier, J. V., Barnes, N. P., Remelius, D.K., and Kokta, M. R.,** Flashlamp-pumped Cr^{3+}: GSAG lasers, *IEEE J. Quantum Electron.,* QE-22, 2058, 1986.

79. **Lacovara, P., Esterowitz, L., and Allen, R.,** Flashlamp-pumped Ti: Al_2O_3 laser using fluorescent conversion, *Opt. Lett.,* 10, 273, 1985.

80. **Nair, L. G.,** Dye lasers, *Prog. Quantum Electron.,* 7, 153, 1982.

81. **Maeda, M.,** *Laser Dyes. Properties of Organic Compounds for Dye Lasers,* Academic Press, Tokyo, 1984.

82. **Polland, H. J., Elsaesser, T., Seilmeier, A., Kaiser, W., Kussler, M., Marx, N. J., Sens, B., and Drexhage, K. H.,** Picosecond dye laser emission in the infrared between 1.4 and 1.8 μm, *Appl. Phys.,* 32, 53, 1983.

83. **Brackmann, U.,** Lambdachrome Laser Dyes, Lambda Physik GmbH, Göttingen, 1986.

84. Kodak Laboratory and Research Products Cat. No. 53, Eastman Kodak Co., Rochester, NY, 1987.

85. Laser Dye Catalog, Exciton Chemical Co., Dayton, OH, 1986.

86. **Kuhn, H.,** A quantum-mechanical theory of light absorption of organic dyes and similar compounds, *J. Chem. Phys.,* 17, 1198, 1949.

87. **Strickler, S. J. and Berg, R. A.,** *J. Chem. Phys.,* 37, 814, 1962.

88. **Drexhage, K. H.,** Structure and properties of laser dyes, in *Dye Lasers Topics in Applied Physics,* Vol. 1, 2nd revised ed., Schäfer, F. P., Ed., Springer-Verlag, Berlin, 1977, chap. 4.

89. **Meyer, M. and Mialocq, J. C.,** Ground state and singlet excited state of laser dye DCM: dipole moments and solvent induced spectral shifts, *Opt. Commun.,* 64, 264, 1987.

90. **Winston, H. and Gudmundsen, R. A.,** Refractive gradient effects in proposed liquid lasers, *Appl. Opt.,* 3, 143, 1964.

91. **Snavely, B. B.,** Flashlamp-excited organic dye lasers, *Proceedings IEEE,* 57, 1374, 1969.

92. **Burlamacchi, P., Pratesi, R., and Vanni, U.,** Refractive index gradient effects in a superradiant slab dye laser, *Opt. Commun.,* 9, 31, 1973.

93. **Wellegehausen, B., Laepple, L., and Welling, H.,** High power CW dye lasers, *Appl. Phys.,* 6, 335, 1975.

94. **Teschke, O., Whinnery, J. R., and Dienes, A.,** Thermal effects in jet-stream dye lasers, *IEEE J. Quantum Electron.,* QE-12, 513, 1976.

95. **Leutwyler, S., Schumacher, E., and Woste, L.,** Extending the solvent palette for cw jet stream dye lasers, *Opt. Commun.,* 19, 197, 1976.

96. **Mc Intyre, I. A. and Dunn, M. H.,** Measurement of the temperature dependence of refractive index of dye laser solvents, *J. Phys. E.,* 18, 19, 1985.

97. **Shank, C. V., Bjorkholm, J. E., and Kogelnik, H.,** Tunable distributed-feedback dye laser, *Appl. Phys. Lett.,* 18, 395, 1971.

98. **Olson, J. D. and Horne, F. H.,** Direct determination of temperature dependence of refractive index of liquids, *J. Chem. Phys.,* 58, 2321, 1973.

99. **Erokhin, A. I., Morachevskii, N. V., and Faizullov, F. S.,** Temperature dependence of the refractive index in condensed media, *Sov. Phys. JETP,* 47, 699, 1978.

100. *Handbook of Chemistry and Physics,* 62nd ed., CRC Press, Cleveland, 1981-82, E 380.

101. **Reisler, E., Eisenberg, H., and Minton, A. P.,** Temperature and density dependence of the refractive index of pure liquids, *J. Chem. Soc. Faraday Trans. 2,* 68, 1001, 1972.

102. Physical and thermodynamic properties of aliphatic alcohols, *J. Phys. Chem. Ref. Data,* 2, 1, 1973.

103. **Beysens, D. and Calmettes, P.,** Temperature dependence of the refractive indices of liquids: deviation from the Lorentz-Lorenz formula, *J. Chem. Phys.,* 66, 766, 1977.

104. **Johnston, T. F., Brady, R. H., and Proffitt, W.,** Powerful single frequency ring dye laser spanning the visible spectrum, *Appl. Opt.,* 21, 2307, 1982.

105. **Marowsky, G. and Tittel, F. K.,** Single mode operation of a tunable cw dye laser, *Appl. Phys.,* 5, 181, 1974.

106. **Adrain, R. S., Arthurs, E. G., Bradley, D. J., Roddie, A. G., and Taylor, J. R.,** Amplification of picosecond dye laser pulses, *Opt. Commun.,* 12, 140, 1974.

107. **Peterson, O. G., Tuccio, S. A., and Snavely, B. B.,** cw operation of an organic dye solution laser, *Appl. Phys. Lett.,* 17, 245, 1970.

108. **Kuznetsov, R. T. and Fofonova, R. M.,** Photostability of rhodamine 6G with respect to spectral characteristics, *J. Appl. Spectrosc. U.S.S.R.,* 40, 380, 1983.

109. **Viktorova, A. A., Savikin, A. P., and Tsaregradskii, V. B.,** Influence of thiourea on the emission characteristics of a laser based on an aqueous solution of rhodamine 6G, *Sov. J. Quantum Electron,* 13, 1140, 1983.

110. **Snavely, B. B. and Schafer, F. P.**, Feasibility of cw operation of dye lasers, *Phys. Lett.*, 28A, 728, 1969.
111. **Keller, R. A.**, Effect of quenching of molecular triplet states in organic dye lasers, *IEEE J. Quantum Electron.*, QE-6, 411, 1970.
112. **Marling, J. B., Gregg, D. W., and Thomas, S. J.**, Effect of oxygen on flashlamp-pumped organic dye lasers, *IEEE J. Quantum Electron.*, QE-6, 570, 1970.
113. **Pappalardo, R., Samelson, H., and Lempicki, A.**, Long pulse laser emission from rhodamine 6G using cyclooctatetraene, *Appl. Phys. Lett.*, 16, 267, 1970.
114. **Pappalardo, R., Samelson, H., and Lempicki, A.**, Long pulse laser emission from rhodamine 6G, *IEEE J. Quantum Electron.*, QE-6, 716, 1970.
115. **Tuccio, S. A., Drexhage, K. H., and Reynolds, G. A.**, cw laser emission from coumarin dyes in the blue and green, *Opt. Commun.*, 7, 248, 1973.
116. **Levin, M. B., Reznikova, I. I., Cherkasov, A. S., and Shirokov, V. I.**, Effect of diphenylpolyenes on the lasing characteristics of some rhodamine solutions, *Opt. Spectrosc.*, 40, 413, 1976.
117. **Levin, M. B., Cherkasov, A. S., and Shirokov, V. I.**, Improvement of the lasing efficiency of dyes when 1,4-diphenylbutadiene is added to their solutions, *Opt. Spectrosc.*, 41, 82, 1976.
118. **Asimov, M. M., Gavrilenko, V. N., and Rubinov, A. N.**, Effect of chemical impurities on the characteristics of induced absorption of xanthene and oxazine dye solutions, *Opt. Spectrosc.*, 54, 263, 1983.
119. **Von Trebra, R. J. and Koch, T. H.**, Photochemistry of coumarin laser dyes: the role of singlet oxygen in the photo-oxidation of coumarin 311, *J. Photochem.*, 35, 33, 1986.
120. **Von Trebra, R. J. and Kock, T. H.**, DABCO stabilization of coumarin dye lasers, *Chem. Phys. Lett.*, 93, 315, 1982.
121. **Snavely, B. B.**, Continuous-wave dye lasers, in *Dye Lasers, Topics in Applied Physics*, Vol. 1, 2nd revised ed., Schäfer, F. P., Ed., Springer-Verlag, Berlin, 1977, chap. 2.
122. **Tuccio, S. A. and Strome, F. C.**, Design and operation of a tunable continuous dye laser, *Appl. Opt.*, 11, 64, 1972.
123. **Stepanov, B. I. and Rubinov, A. N.**, Lasers based on solutions of organic dyes, *Sov. Phys. Usp.*, 11, 304, 1968.
124. **McDonald, D. B. and Jonah, C. D.**, Amplification of picosecond pulses using a copper vapor laser, *Rev. Sci. Instrum.*, 55, 1166, 1984.
125. **Arai, Y., Niki, H., Adachi, S., Takeda, T., Yamanaka, T., and Yamanaka, C.**, Development of a single-mode dye laser pumped by a copper vapor laser, *Technol. Rep. Osaka Univ.*, 36, 361, 1986.
126. **Rosenthal, I.**, Photochemical stability of rhodamine 6G in solution, *Opt. Commun.*, 24, 164, 1978.
127. **Soffer, B. H. and Mc Farland, B. B.**, Continuously tunable, narrow band organic dye lasers, *Appl. Phys. Lett.*, 10, 266, 1967.
128. **Hansch, T. W.**, Repetitively pulsed tunable dye laser for high resolution spectroscopy, *Appl. Opt.*, 11, 895, 1972.
129. **Hanna, D. C., Karkkainen, P. A., and Wyatt, R.**, A simple beam expander for frequency narrowing of dye lasers, *Opt. Quantum Electron.*, 7, 115, 1975.
130. **Sorokin, P. P., Lankard, J. R., Moruzzi, V. L., and Hammond, E. C.**, Flashlamp-pumped organic dye lasers, *J. Chem. Phys.*, 48, 4726, 1968.
131. **Furumoto, H. W. and Ceccon, H. L.**, Optical pumps for organic dye lasers, *Appl. Opt.*, 8, 1613, 1969.
132. **Furumoto, H. W. and Ceccon, H. L.**, Ultraviolet organic liquid lasers, *IEEE J. Quantum Electron.*, QE-6, 262, 1970.
133. **Schafer, F. P.**, Principles of dye laser operation, in *Dye Lasers, Topics in Applied Physics*, Vol. 1, 2nd revised ed., Schäfer, F. P., Ed., Springer-Verlag, Berlin, 1977, 54.
134. **Basov, Yu, G.**, Flashlamp pumping systems for dye lasers (Review), *Instrum. Exp. Tech. U.S.S.R.*, 29, 6/1, 1239, 1986.
135. **Bradley, D. J., Caughey, W. G. I., and Vukusic, J. I.**, High efficiency interferometric tuning of flashlamp pumped dye lasers, *Opt. Commun.*, 4, 150, 1971.
136. **Strome, F. C. and Webb, J. P.**, Flashtube-pumped dye laser with multiple-prism tuning, *Appl. Opt.*, 10, 1348, 1971.
137. **Mack, M. E.**, 0.2 W repetitively pulsed flashlamp pumped dye laser, *Appl. Phys. Lett.*, 19, 108, 1971.
138. **Hirth, A., Lasser, Th., Meyer, R., and Schetter, K.**, Comparison of coaxial-and preionized linear flashlamps as pumping sources for high power repetitive pulsed dye lasers, *Opt. Commun.*, 34, 223, 1980.
139. **Schmidt, W. and Schafer, F. P.**, Self mode-locking of dye lasers with saturable absorbers, *Phys. Lett.*, 26A, 558, 1968.
140. **Bradley, D. J. and O'Neill, F.**, Passive mode-locking of flashlamp pumped rhodamine dye lasers, *Opto-electronics*, 1, 69, 1969.
141. **Arthurs, E. G., Bradley, D. J., and Roddie, A. G.**, Build up of picosecond pulse generation in passively mode-locked rhodamine dye lasers, *Appl. Phys. Lett.*, 23, 88, 1973.
142. **Bradley, D. J.**, Generation and measurement of frequency-tunable picosecond pulses from dye lasers, *Opto-electronics*, 6, 25, 1974.

143. **Bradley, D. J.**, Methods of generation, in *Ultrashort Light Pulses, Picosecond Techniques and Applications, Topics in Applied Physics,* Vol. 18, Shapiro, S. L., Ed., Springer-Verlag, Berlin, 1977, chap. 2.
144. **Schneider, S.**, Flashlamp-pumped mode-locked dye lasers, *Philos. Trans. R. Soc. London Ser. A,* 298, 233, 1980.
145. **Arthurs, E. G., Bradley, D. J., and Roddie, A. G.**, Frequency-tunable transform-limited picosecond dye laser pulses, *Appl. Phys. Lett.,* 19, 480, 1971.
146. **Mialocq, J. C. and Goujon, P.**, Tunable blue picosecond pulses from a flashlamp pumped dye laser, *Appl. Phys. Lett.,* 33, 819, 1978.
147. **Mialocq, J. C. and Goujon, P.**, Impulsions picosecondes à 520 nm dans un laser à colorant coumarine excité par éclair, *Opt. Commun.,* 24, 255, 1978.
148. **Lill, E., Schneider, S., and Dorr, F.**, Passive mode-locking of a flashlamp pumped fluorol 7GA dye laser in the green spectral region, *Opt. Commun.,* 20, 223, 1977.
149. **Arthurs, E. G., Bradley, D. J., and Roddie, A. G.**, Passive mode-locking of flashlamp-pumped dye lasers between 580 and 700 nm, *Appl. Phys. Lett.,* 20, 125, 1972.
150. **Mialocq, J. C. and Goujon, P.**, Pinacyanol as a mode-locking saturable absorber for rhodamine dye lasers, *Opt. Commun.,* 20, 342, 1977.
151. **Taylor, J. R.**, Passive mode-locking of DCM and rhodamine 101 flashlamp-pumped dye laser systems, *Opt. Commun.,* 57, 117, 1986.
152. **Arthurs, E. G., Bradley, D. J., Puntambekar, P. N., Ruddock, I. S., and Glynn, T. J.**, The effect of saturable absorber lifetime in picosecond pulse generation. II. The cresyl-violet laser, *Opt. Commun.,* 12, 360, 1974.
153. **Hirth, A., Vollrath, K., Faure, J., and Lougnot, D.**, Etude et réalisation d'un laser à colorant á modes couplés excité par tube à éclair et émettant dans le proche infrarouge, *C. R. Acad. Sci.,* Sér. B, 276, 153, 1973.
154. **Glenn, W. H., Brienza, M. J., and Maria, A. J.**, Mode-locking of an organic dye laser, *Appl. Phys. Lett.,* 12, 54, 1968.
155. **Fan, B. and Gustafson, T. K.**, Narrow band picosecond pulses from an ultrashort-cavity dye laser, *Appl. Phys. Lett.,* 28, 202, 1976.
156. **Kranitzky, W., Kopainsky, B., Kaiser, W., Drexhage, K. H., and Reynolds, G. A.**, A new infrared laser dye of superior photostability tunable to 1.24 μm with picosecond excitation, *Opt. Commun.,* 36, 149, 1981.
157. **Lobentanzer, H. and Polland, H. J.**, Generation of tunable picosecond pulses between 1.18 μm and 1.53 μm in a ring laser configuration using dye No. 5, *Opt. Commun.,* 62, 35, 1987.
158. **Bor, Zs.**, Tunable picosecond pulse generation by an N₂ laser pumped self Q-switched distributed feedback dye laser, *IEEE J. Quantum Electron.,* QE-16, 517, 1980.
159. **Bor, Zs., Muller, A., Racz, B., and Schafer, F. P.**, Ultrashort pulse generation by distributed feedback dye lasers. I. Temporal characteristics, *Appl. Phys., B* 27, 9, 1982.
160. **Bor, Zs., Muller, A., Racz, B., and Schafer, F. P.**, Ultrashort pulse generation by distributed feedback dye lasers. II. Energy characteristics, *Appl. Phys., B* 27, 77, 1982.
161. **Bor, Zs., Muller, A., and Racz, B.**, UV and blue picosecond pulse generation by a nitrogen-laser-pumped distributed feedback dye laser, *Opt. Commun.,* 40, 294, 1982.
162. **Bor, Zs. and Racz, B.**, Picosecond dye laser pumped by an excimer laser, *Appl. Opt.,* 24, 1910, 1985.
163. **Bor, Zs. and Muller, A.**, Picosecond distributed feedback dye lasers, *IEEE J. Quantum Electron.,* QE-22, 1524, 1986.
164. **Watanabe, M., Watanabe, S., and Endoh, A.**, Mode-locking of excimer laser pumped dye lasers, *Appl. Phys. Lett.,* 45, 929, 1984.
165. **Kortz, H. P., Pax, P., and Aubert, R.**, Generation of high energy tunable picosecond pulses, Proceedings of SPIE, 533, 32, 1985.
166. **Dienes, A.**, Mode-locked cw dye lasers, *Opto-electronics,* 6, 99, 1974.
167. **Adams, M. C., Bradley, D. J., Sibbett, W., and Taylor, J. R.**, Synchronously pumped continuous wave dye lasers, *Philos. Trans. R. Soc. London Ser. A,* 298, 217, 1980.
168. **O'Connor, D. V. and Phillips, D.**, Time-Correlated Single-Photon Counting, Academic Press, London, 1984, 55.
169. **Seilmeier, A., Kaiser, W., Sens, B., and Drexhage, K. H.**, Tunable picosecond pulses around 1.3 μm generated by a synchronously pumped infrared dye laser, *Opt. Lett.,* 8, 205, 1983.
170. **Roskos, H., Optiz, S., Seilmeier, A., and Kaiser, W.**, Operation of an infrared dye laser synchronously pumped by a mode-locked cw Nd:YAG laser, *IEEE J. Quantum Electron.,* QE-22, 697, 1986.
171. **Wisoff, P. J. K., Caro, R. G., and Mitchell, G.**, A high power picosecond dye oscillator synchronously pumped by a Q-switched, mode-locked Nd:YAG laser, *Opt. Commun.,* 54, 353, 1985.
172. **Kafka, J. D. and Baer, T.**, A synchronously pumped dye laser using ultrashort pump pulses, Proceedings of SPIE 533, 38, 1985.

173. **Johnson, A. M. and Simpson, W. M.,** Tunable femtosecond synchronously mode-locked dye-laser pumped by the compressed second harmonic of Nd:YAG Proceedings of SPIE 533, 52, 1985.

174. **Johnson, A. M. and Simpson, W. M.,** Optically biased tunable femtosecond dye laser and spectral windowing of the compressed second harmonic of Nd:YAG, *IEEE J. Quantum Electron.,* QE-22, 133, 1986.

175. **Ippen, E. P., Shank, C. V., and Dienes, A.,** Passive mode-locking of the cw dye laser, *Appl. Phys. Lett.,* 21, 348, 1972.

176. **Shank, C. V. and Ippen, E. P.,** Subpicosecond kilowatt pulses from a mode-locked cw dye laser, *Appl. Phys. Lett.,* 24, 373, 1974.

177. **Ippen, E. P. and Shank, C. V.,** Dynamic spectroscopy and subpicosecond pulse compression, *Appl. Phys. Lett.,* 27, 488, 1975.

178. **Diels, J. C., Van Stryland, E., and Benedict, G.,** Generation and measurement of 200 femtosecond optical pulses, *Opt. Commun.,* 25, 93, 1978.

179. **Dietel, W., Fontaine, J. J., and Diels, J. C.,** Intracavity pulse compression with glass: a new method of generating pulses shorter than 60 fs, *Opt. Lett.,* 8, 4, 1983.

180. **Fork, R. L., Greene, B. I., and Shank, C. V.,** Generation of optical pulses shorter than 0.1 ps by colliding pulse mode locking, *Appl. Phys. Lett.,* 38, 671, 1981.

181. **Valdmanis, J. A., Fork, R. J., and Gordon, J. P.,** Generation of optical pulses as short as 27 femtoseconds directly from a balancing self-phase modulation, group velocity dispersion, saturable absorption, and saturable gain, *Opt. Lett.,* 10, 131, 1985.

182. **Tomlinson, W. K., Stolen, R. H., and Shank, C. V.,** Compression of optical pulses chirped by self-phase modulation in fibers, *J. Opt. Soc. Am., B* 1, 139, 1984.

183. **Fujimoto, J. G., Weiner, A. M., and Ippen, E. P.,** Generation and measurement of optical pulses as short as 16 fs, *Appl. Phys. Lett.,* 44, 832, 1984.

184. **Knox, W. H., Fork, R. L., Downer, M. C., Stolen, R. H., and Shank, C. V.,** Optical pulse compression to 8 fs at a 5 KHz repetition rate, *Appl. Phys. Lett.,* 46, 1120, 1985.

185. **Brito-Cruz, C. H., Fork, R. L., and Shank, C. V.,** Compression of optical pulses to 6 fs using cubic phase distorsion compensation in Conference on Lasers and Electro-optics, Digest of Technical Papers, OSA-IEEE, Baltimore, MD, April 26 to May 1, 1987, 12.

186. **Palfrey, S. L. and Grischkowsky, D.,** Generation of 16 fs frequency tunable pulses by optical pulse compression, *Opt. Lett.,* 10, 562, 1985.

187. **French, P. M. W. and Taylor, J. R.,** Passive mode-locking of a continuous wave dye laser operating in the blue-green spectral region, *Appl. Phys. Lett.,* 50, 1708, 1987.

188. **French, P. M. W. and Taylor, J. R.,** The passively mode-locked and dispersion compensated rhodamine 110 dye laser, *Opt. Commun.,* 61, 224, 1987.

189. **French, P. M. W. and Taylor, J. R.,** Passive mode-locked continuous-wave rhodamine 110 dye laser, *Opt. Lett.,* 11, 297, 1986.

190. **French, P. M. W., Dawson, M. D., and Taylor, J. R.,** Passive mode-locking of a cw dye laser in the yellow spectral region, *Opt. Commun.,* 56, 430, 1986.

191. **French, P. M. W. and Taylor, J. R.,** The passive mode-locking of the continuous wave rhodamine B dye laser, *Opt. Commun.,* 58, 53, 1986.

192. **French, P. M. W. and Taylor, J. R.,** Passive mode-locking of the continuous wave DCM dye laser, *Appl. Phys.,* B41, 53, 1986.

193. **French, P. M. W. and Taylor, J. R.,** Passive mode-locking of an energy transfer continuous-wave dye laser, *IEEE J. Quantum Electron.,* QE-22, 1162, 1986.

194. **Smith, K., Langford, N., Sibbett, W., and Taylor, J. R.,** Passive mode-locking of a continuous wave dye laser in the red-near-infrared spectral region, *Opt. Lett.,* 10, 559, 1985.

195. **French, P. M. W., Williams, J. A. R., and Taylor, J. R.,** Passive mode-locking of a cw energy transfer dye laser operating in the near infrared around 750 nm, private communication.

196. **Ryan, J. P., Goldberg, L. S., and Bradley, D. J.,** Comparison of synchronous pumping and passive mode-locking of cw dye lasers for the generation of picosecond and subpicosecond light pulses, *Opt. Commun.,* 27, 127, 1978.

197. **Fehrenback, G. W., Gruntz, K. J., and Ulbrich, R. G.,** Subpicosecond light pulses from a synchronously mode-locked dye laser with composite gain and absorber medium, *Appl. Phys. Lett.,* 33, 159, 1978.

198. **Ishida, Y., Nakamura, A., Naganuma, K., and Yajima, T.,** Internal amplification of subpicosecond pulses from a cw mode-locked dye laser, *Opt. Commun.,* 39, 411, 1981.

199. **Ishida, Y., Naganuma, K., and Yajima, T.,** Tunable subpicosecond pulse generation from 535 to 590 nm by a hybridly mode-locked dye laser, *Jpn. J. Appl. Phys.,* 21, L312, 1982.

200. **Diels, J. C., Jamasbi, N., and Sarger, L.,** Passive and hybrid femtosecond operation of a linear astigmatism compensated dye laser, in *Ultrafast Phenomena Part V,* (Springer Series in Chemical Physics), Vol. 46, Fleming, G. R. and Siegman, A. E., Eds., Springer-Verlag, Berlin, 1986, 2.

201. **Chesnoy, J. and Fini, L.,** Stabilization of a femtosecond dye laser synchronously pumped by a frequency-doubled mode-locked YAG laser, *Opt. Lett.,* 11, 635, 1986.

202. **Dawson, M. D., Boggess, T. F., Garvey, D. W., and Smirl, A. L.,** Generation of 55 fs pulses and variable spectral windowing in a linear cavity synchronously pumped cw dye laser, in *Ultrafast Phenomena Part V,* (Springer Series in Chemical Physics), Vol. 46, Fleming, G. R. and Siegman, E. A., Eds., Springer-Verlag, Berlin, 1986, 5.

203. **Dobler, J., Schulz, H. H., and Zinth, W.,** Generation of femtosecond light pulses in the near infrared around λ = 850 nm, *Opt. Commun.,* 57, 407, 1986.

204. **French, P. M. W., Gomes, A. S. L., Gouveia-Neto, A. S., and Taylor, J. R.,** A femtosecond hybrid mode-locked rhodamine B dye laser, *Opt. Commun.,* 59, 366, 1986.

205. **Dawson, M. D., Boggess, T. F., Garvey, D. W., and Smirl, A. L.,** Femtosecond pulse generation in the red/deep red spectral region, *IEEE J. Quantum Electron.,* QE-23, 290, 1987.

206. **Langford, N., Smith, K., Sibbett, W., and Taylor, J. R.,** The hybrid mode-locking of a cw rhodamine 700 dye laser, *Opt. Commun.,* 58, 56, 1986.

207. **Aoyagi, Y., Segawa, Y., and Namba, S.,** Tunable subpicosecond pulse generation in near IR region from OX-725 dye laser by passive and synchronous hybrid modelocking method, *Jpn. J. Appl. Phys.,* 20, 1595, 1981.

208. **Smith, K., Sibbett, W., and Taylor, J. R.,** Subpicosecond generation via hybrid mode-locking of styryl 9 in the near infrared, *Opt. Commun.,* 49, 359, 1984.

209. **Flamant, P. and Meyer, Y. H.,** Absolute gain measurements in a multistage dye amplifier, *Appl. Phys. Lett.,* 19, 491, 1971.

210. **Schmidt, A. J.,** Generation of a single tunable ultrashort light pulse, *Opt. Commun.,* 14, 287, 1975.

211. **Ganiel, U., Hardy, A., Neumann, G., and Treves, D.,** Amplified spontaneous emission and signal amplification in dye-laser systems, *IEEE J. Quantum Electron.,* QE-11, 881, 1975.

212. **Hnilo, A. A. and Martinez, O. E.,** An expression for the small signal gain of pulsed dye laser amplifiers, *Opt. Commun.,* 60, 87, 1986.

213. **Hnilo, A. A. and Martinez, O. E.,** On the design of pulsed dye laser amplifiers, *IEEE J. Quantum Electron.,* QE-23, 593, 1987.

214. **Urisu, T. and Kajiyama, K.,** Single pulse extraction from a mode-locked pulse train using a dye-laser amplifier, *Opt. Commun.,* 20, 34, 1977.

215. **Salour, M. M.,** Powerful dye laser oscillator-amplifier system for high resolution and coherent pulse spectroscopy, *Opt. Commun.,* 22, 202, 1977.

216. **Rolland, C. and Corkum, P. B.,** Amplification of 70 fs pulses in a high repetition rate XeCl pumped dye laser amplifier, *Opt. Commun.,* 59, 64, 1986.

217. **Lavi, S., Bialslanker, G., Amit, M., Belker, D., Erez, G., and Miron, E.,** Characterization of a pulse amplifier for cw dye lasers, *Opt. Commun.,* 60, 309, 1986.

218. **Hargrove, R. S. and Tehman, K.,** High power efficient dye amplifier pumped by copper vapor lasers, *IEEE J. Quantum Electron.,* QE-16, 1108, 1980.

219. **Hirlimann, C., Seddiki, O., Morhange, J. F., Mounet, R., and Goddi, A.,** Femtosecond jet laser preamplifier, *Opt. Commun.,* 59, 52, 1986.

220. **Erickson, L. E. and Szabo, A.,** Spectral narrowing of dye laser output by injection of monochromatic radiation into the laser cavity, *Appl. Phys. Lett.,* 18, 433, 1971.

221. **Bhattacharyya, S. K., Eggett, P., and Thomas, L.,** Locking of a flashlamp-pumped dye laser at small injection powers, *Opt. Commun.,* 39, 387, 1981.

222. **Boquillon, J. P., Quazzany, Y., and Chaux, R.,** Injection-locked flashlamp-pumped dye lasers of very narrow linewidth in the 570-720 nm range, *J. Appl. Phys.,* 62, 23, 1987.

223. **Deacon, D. A. G., Elias, L. R., Madey, J. M., Ramian, G. J., Schwettman, H. A., and Smith, T. I.,** *Phys. Rev. Lett.,* 38, 892, 1977.

224. **Madey, J. M. et al.,** The Stanford Mark III Infrared Free Electron Laser, 7th Int. Conf. on Free Electron Lasers, Lake Tahoe, CA, September, 1985.

225. **Renieri, A.,** The free electron laser, *Physics of New Laser Sources,* NATO ASI Series B: Physics, Plenum Press, New York, 132, 255, 1985.

226. **Bazin, C., Billardon, M., Deacon, D. A. G., Elleaume, P., Farge, Y., Madey, J. M. J., Ortega, J. M., Petroff, Y., Robinson, K. E., and Velghe, M. F.,** Results of the first phase of the ACO storage ring laser experiments, in *Physics of Quantum Electronics,* Jacobs, S. F., Moore, G. T., Pilloff, H. S., Sargent, M., III, Scully, M. O., and Spitzer, R., Eds., Addison-Wesley, Reading, MA, 1982, 8.

Chapter 2

PHOTOPHYSICAL AND PHOTOCHEMICAL PRIMARY PROCESSES IN POLYMERS

Shigeo Tazuke and Tomiki Ikeda

TABLE OF CONTENTS

I.   Introduction 54

II.   Specificity of Polymers in Photophysics/Chemistry 55

III.   Brief Description of Initial Photophysical Processes in General 56
     A.   Photoabsorption 56
     B.   Nonradiative Unimolecular Decay Paths 57
     C.   Unimolecular Radiative Process 57
     D.   Bimolecular Dissipation Processes 58
     E.   Energy Transfer and Migration 58
     F.   Excimer and Exciplex 59
     G.   Electron Transfer 60

IV.   Energy Transfer and Energy Migration Processes in Polymers 60
     A.   Theoretical Overview 61
           1.   Strong Interaction 61
           2.   Medium Interaction 62
           3.   Weak Interaction 62
     B.   Theories of Excitation Transfer in Molecular Aggregates 64
     C.   Experimental Observation of Energy Transfer and Energy
          Migration in Polymers 67
           1.   Polystyrene 68
           2.   Polymers Containing Naphthalene Moieties and
                Phenanthrene Moieties 71
           3.   Polymers Containing Anthracene Moieties 73

V.   Excimer and Exciplex Formation in Polymers 74
     A.   Excimer 74
     B.   Exciplex 78

VI.   Photoinduced Electron Transfer Processes in Polymers 80
     A.   Homogeneous Systems 80
     B.   Solid-State and Solid-Liquid Interfacial Systems 81

I. INTRODUCTION

Interaction between electromagnetic wave of variable energies and molecules is the fundament of photophysical/chemical effects on materials. A basic understanding of photoenergy absorption by molecules and subsequent energy dissipation processes, either radiative or nonradiative, have been best investigated in the gas phase, particularly under low pressure. In a low-pressure gas phase, molecules are practically isolated and the complicated problem of molecular interactions is minimized. Consequently, the photophysial/chemical process can be treated more fundamentally and cleanly in the gas phase than in the condensed phase. Polymer photophysics/chemistry locates in the extreme opposite of the gas-phase study, that is, practically most important whereas least unequivocal.

While detailed analysis of photophysical/chemical processes in the gas phase provides a sound basis for photochemistry in general, practically important gas phase studies are rather limited. Understanding the cosmic photochemical processes, approach to air pollution problem (photochemical smog) and to photochemical vapor deposition (PCVD) technology may be the main contribution. With respect to the recent booming interest in anything relevant to optical technology with an expectation to be of the mainstream of future information media, photophysical/chemical events are being investigated with special reference to material science. It is easy to comment but difficult to practice an unequivocal analysis in molecular aggregate systems, particularly in solid polymers. Since most materials are used as solid, it is urgent to establish experimental and analytical methods for photon-molecule interactions when the molecules are strongly interacting with each other and their motions are greatly reduced. These systems are far more complicated than the gas phase and dilute fluid solution. Some of the advanced instrumentations useful for gas phase or solution are hardly applicable to solid-state study. For example, high-power laser photolysis with a short pulse duration is difficult to apply to the analysis of solid-state photochemistry because of the unsolved problem of heat accumulation. The authors should be reminded in the beginning that the basic research on solid-state photochemistry/physics most imminent to practical application is most underdeveloped.

Among all materials covering metals, ceramics, and organic materials, polymers are essentially transparent, most easily processable, and of relatively low cost. Use of polymers is increasing year by year, either substituting the existing materials or exploring novel applications. At this moment, the majority of applications are as *passive* materials which are defined as the materials useful because of minimized photoenergy absorption such as optical fibers, transparent glass, lens, and so forth. Demand for *active* materials is now emerging. Such a group of materials may be called *photoresponsive materials*. Apart from photopolymers already in the market, many new uses are actively being surveyed including photomemory, photoswitching, second harmonic and third harmonic generator (THG), and so on[1]. Although all of these fancy organic materials are not necessarily photosensitive polymers, polymers are used very often as matrix alone and the understanding of photophysical/chemical events in (not necessarily, *of*) polymer systems is essential.

When the mechanical or physical properties of the polymer itself are to be modified by a photochemical process, the photochemistry of the polymer is the target of research. Special materials falling in this class are positive and negative photoresists, photocurable paint, ink, adhesive, and photopolymer in general. Photopolymerization of monofunctional monomers in polymer matrix will lead to curing of the mixture and therefore usable as photopolymer. While this process does not involve photochemistry of polymers, polymer matrix effects on photochemical processes are the central subject of research.

More sophisticated photosensitive materials such as photoconductors, photochromic materials, materials for nonlinear optics, photochemical hole burning (PHB), etc. are in general the dispersion of small key molecules in the polymer matrix. For example, although photoconductive polymers have been extensively studied, all of the contemporary commercial organic photoconductors (OPC) consist of the combination of small molecular organic donors in polymer matrix[2]. The purpose of basic research is consequently to investigate how photo-induced carrier generation and carrier transport occur under the influence of the polymer matrix. Photochromic materials and organic materials for nonlinear optics will probably be developed along the same line; small molecular photochromic compounds will be coated or cast with the aid of polymer matrix. A recently announced organic second harmonic gen-eration (SHG) is an oriented poly(oxyethylene) film containing nitroaniline molecules[3]. The role of polymer is to provide the favorable mechanical properties as well as the favorable alignment of the functional molecules by means of stretching. A common basic unsolved problem is the comparison of the photophysical/chemical behaviors of polymer-bonded functional molecules with those simply dispersed in the polymer matrix.

The largest practical subject will be the photodegradation problem. While accelerated photodegradation is a subject of growing concern from an environmental viewpoint, the prevention of photodegradation is presently a more imminent subject. This is genuine polymer photochemistry in solids suitable for exhaustive study. It is rather unfortunate that the subject is commercially so important that the majority of data are with practical samples containing many additives. In addition, the recipe of the samples is behind the veil of secrecy. Fur-thermore, as a trace amount of contaminants such as catalyst residues is known to trigger photodegradation, it is difficult to translate the data into the intrinsic properties of the pure polymers. On the other hand, purely academic research is often conducted with polymers in solution. Although the degradation study in solution is more reproducible and quantitative, it is too remote from photodegradation of practical materials.

Analysis of photophysical/chemical events of biological interest is another subject of basic research.[5] Understanding, rather than application, is the target of the research. Since the amount of the available biological sample is small, spectroscopic methods are the main stream. Among these, fluorescence technique is most commonly used. The fluorescence probe method was originally developed in biology and biochemistry to obtain information on the microenvironment of a particular site in biological polymers. Molecular mobility, polarity, and hydrophobicity can be discussed with a trace amount of samples.[6] Without fluorescence spectroscopy, such information can never be expected. The underlying principle is the detection of molecular interactions and molecular motions in the excited state of a fluorophore as the static and dynamic changes in fluorescence reflecting the molecular environment.

The fluorescence probe method is now widely applied for the study of segment mobility, miscibility, phase separation, free volume, polymerization, and curing processes of synthetic polymers.[6] This is a rapidly growing field because of its handiness to obtain the local molecular information. The photophysics of fluorescence probe in polymer either bonded or mixed is the fundament of the method.

We have briefly reviewed the area relevant to polymer photophysics/chemistry. A ques-tion arises concerning what is new in the initial photophysical/chemical processes in polymers when a number of excellent books are available for small molecular systems.[7-9]

II. SPECIFICITY OF POLYMERS IN PHOTOPHYSICS/ CHEMISTRY

The difference between small molecules and polymers in general may be summarized as follows:[10]

1. Nonhomogeneous distribution of chromophores in polymer systems. Even in extremely dilute solutions, the microscopic chromophore distribution in a scale of polymer chain

size is nonhomogeneous if the polymer bears many chromophores per chain. The local chromophore concentration can be very high even in dilute polymer solution. Many so-called polymer effects can be interpreted in terms of the high local concentration. The similar situation can be realized in micelles and other molecular aggregates.

2. Because of the interaction with the polymer main chain and the side chain, the motion of a chromophore bonded to polymer is retarded. Even in a dilute solution, solvation of chromophores both in the ground and excited states is reduced. The polymer chain may participate in solvation instead of solvent molecules.

3. Interchromophore interactions increase in polymer either by intrapolymer association or by interpolymer association. As exemplified by DNA, interpolymer association could occur in very dilute solution if the condition of structural matching for association is satisfied.[11] Intrapolymer interaction of chromophores is a common trend as shown by the strong excimer emission from aromatic vinyl polymers.[12]

4. When a polymer solution is concentrated eventually to give a neat polymer, the polymer chains start to interact at a certain critical concentration. The correlation length representing the size of the network decreases with concentration. Concentrated solutions of small molecules do not exhibit such behaviors. The critical point typically observed in neat polymer as T_g or other transition temperature corresponds to a certain free volume at the temperature with a fixed concentration or at the concentration at a fixed temperature. Below and above the critical point, the molecular motion is critically different and therefore the photophysical behaviors as well as the chemical reactions requiring diffusion or activation free volume are subjected to sudden changes.[13,14]

5. Orientation effect of chromophores may be expected in a polymer. When a solid polymer is stretched and oriented, the chromophores will be arranged in a regular order. This could be expected even in solution if the polymer chain takes a regular conformation such as helix in polypeptides.[15] A specific chromophore arrangement is not limited to polymers. A better chromophore alignment may be expected in bilayer membranes or liquid crystals.

This is a critical view of *polymer effects*. So far we are handling the synthetic polymers having structural and conformational distributions, very few effects are specific to the polymer. Micelles, vesicles, bilayer membranes, and liquid crystals could do the same except for the excellent mechanical properties of polymer materials.

In the following, we will discuss the primary photophysical processes and some subsequent chemistry in polymers in comparison with those of dilute solutions of small molecules. Since the description of authentic photophysical/chemical processes of small molecules have been better established than in polymers, we will put emphasis on the difference between small molecules and polymers. Although multiphoton processes such as interaction between excited species and multiphoton ionization are important in laser photochemistry of polymers, these topics which will be discussed by other authors are omitted.

III. BRIEF DESCRIPTION OF INITIAL PHOTOPHYSICAL PROCESSES IN GENERAL[7-9]

A. PHOTOABSORPTION

When a light beam passes through material, the photon energy may be absorbed and converted to electronic excitation energy on the condition that the energy gap between the ground state and one of excited levels matches to the photon energy. The transition probability of interstate (n → m) transition is decided by the integral transition moment shown below.[7] In the wave function, the nuclear part and the electronic part can be separated (the Bohn-Oppenheimer approximation).

$$M_{nm} = \int \Phi_n^*(r) \sum er\Phi_m(r)dv = e <\Psi_n^{el}(r, R)\phi_{n\nu''}(R)| \sum er|$$
$$\times \Psi_m^{el}(r, R)\phi_{m\nu}, (R)> <\Theta_m(s)|\Theta_n(s)> \tag{1}$$

where r and R are electron and vibration coordinates. Ψ^{el}, ϕ, and Θ are electronic, vibrational, and spin wavefunctions. The magnitude of the integral is a function of the degree of spatial overlap, symmetry, and spin states of the interacting ground- and excited-state orbitals including the vibrational states. The electronic transition is forbidden to a different degree if any of the conditions is violated. When the spatial overlap is very small, as in the case of $n - \pi^*$ transition, the transition is symmetry forbidden as in the case of d-d transition, or spin forbidden represented by S-T transition, and the photoabsorption intensity is greatly reduced from the fully allowed case of $\pi - \pi^*$ or $\sigma - \sigma^*$ transition.

Besides photoabsorption by an isolated molecule, charge transfer interaction between two molecules having different electron-donating and electron-accepting tendencies brings about a new mechanism of photoabsorption so-called charge transfer band.[16] Any molecular interactions strong enough to cause perturbation on electronic energy levels influence the electronic transition energy and the probability. Thus, association of dye molecules or aromatic hydrocarbons exhibits new absorption bands. Even if the new absorption bands do not appear, minor changes of molecular environment such as solvation comparable in energy to the vibrational energy affect the vibrational structure of the spectrum. Changes in the transition moment as a result of molecular interactions are known as hyperchromism or hypochromism.[17]

B. NONRADIATIVE UNIMOLECULAR DECAY PATHS

In organic compounds, the formation of S_1 or S_n state is a common primary act. The excited energy is nonradiatively lost to produce the zero vibrational level of S_1. Emission from higher singlet states or higher vibrational levels of S_1 is hardly observed in condensed systems. This process called internal conversion is purely intramolecular and too fast to be observed within the time scale of 10^{-12} s. The nonradiative path to the state of different spin multiplicity (intersystem crossing) is essentially a spin-forbidden process. The molecular environmental effects may induce singlet/triplet mixing to facilitate the process as shown by external heavy atom and magnetic field effects as results of enhanced spin-orbit interaction, electron spin-external magnetic field interaction, and/or electron spin-nuclear spin interaction.

C. UNIMOLECULAR RADIATIVE PROCESS

Fluorescence ($S_1 \rightarrow S_0$) and phosphorescence ($T_1 \rightarrow S_0$) are very informative processes. When the molecular structure of fluorescence state is not much different from the ground state, the vibrational levels of S_1 reflecting the shape of absorption band is comparable with those of S_0 deciding the vibrational structures of fluorescence; thus, the fluorescence spectrum is a mirror image of the absorption spectrum. After vertical excitation to the Franck-Condon excited state (i.e., the molecular structure is the same as the ground state and not yet adjusted to meet the new electronic state with an equilibrium geometry), relaxation to the fluorescence state where the molecular structure and the solvation sphere have been reorganized to minimize the excited state energy is generally a fast process. However, the relaxation path could be directly observed if molecular motions are suppressed in viscous media or by the use of an ultrafast time-resolved measurement.

Phosphorescence is a much slower process than fluorescence. Since the probability of emission is small and competes with nonradiative energy dissipation via coupling with molecular vibration, the measurement is better performed in a rigid matrix in the absence of air.

D. BIMOLECULAR DISSIPATION PROCESSES

In addition to the intrinsic dissipation processes of an isolated molecule, both singlet and triplet excited states are thermalized by the interactions with surrounding solvent or matrix molecules, transfer energy or electron to a particular molecule, or proceed to chemical reactions. Various vibrational modes of solvent or matrix molecules facilitate the conversion of electronic energy to vibrational energy. These bimolecular processes depend on the fluidity of the medium as well as on the concentration of surrounding molecules. As a consequence, the excited state lifetime is shorter in fluid solution than in dilute gas phase or in frozen matrix.

When an excited state is quenched by the interaction with well-defined solute molecule (quencher), the emission intensity (I) relative to that in the absence of quencher (I_0) is expressed as functions of the quencher concentration ([Q]) in two ways. The first is for the case of dynamic quenching in which the rate of diffusion of light emitting species and Q is fast enough in comparison with the quenching rate. The second is for the case of static quenching in which the diffusion of quencher molecules is too slow to keep the homogeneous distribution of quencher during the lifetime of excited state and the quenching occurs only when at least one quencher molecule is present within a critical distance from the excited species. An important difference between dynamic and static quenching is the fact that the determination of the degree of quenching either by depression of emission intensity or by decrease in emission lifetime is identical for the former whereas the emission intensity alone decreases in the latter case, the emission lifetime being kept constant.

E. ENERGY TRANSFER AND MIGRATION

Energy transfer is a potential pathway of luminescence quenching when the excited state of the energy acceptor (quencher) is nonluminescent. The mechanism of energy transfer is classified as follows:

1.   Long-range energy transfer via dipole-dipole interaction. The mechanism is operative at a distance of 10 to 100 Å for singlet state energy transfer; $S_1 + S_0' \rightarrow S_0 + S_1'$. For the energy transfer between M_1^* and M_2 at a distance R, the orientation factor (κ) for relative alignment of energy donor and energy acceptor is usually averaged over all orientation angles (i.e., a random orientation of chromophores is assumed). This may be a doubtful assumption for polymers and molecular aggregates. We should be careful that this is not an exclusive mechanism of singlet energy transfer. If an energy donor and an acceptor are much closer than the distance of dipole-dipole interaction at the time of excitation, other mechanisms mentioned below may be also available.

$$W = \frac{9000\kappa^2(\ln 10)k_e}{128\pi^5 n^4 NR^6} \int f_{M_1}(\nu)\epsilon_{M_2}(\nu)\,\frac{d\nu}{\nu^4} \tag{2}$$

where k_e is the radiative transition probability of $M_1^* \rightarrow M_1$, N is the Avogadro number, $f_{M_i}(\nu)$ is the emission spectrum of M_1 normalized as $\int f_{M_1}(\nu)d\nu = 1$, ϵ_{M_2} is the molar absorption coefficient of M_2, and n is the refractive index of the medium.

2.   Short-range energy transfer via orbital overlap. For the triplet energy transfer, $T_1 + S_0' \rightarrow S_0 + T_1'$, the dipole-dipole interaction term disappears and a much closer interaction between orbitals at a distance of $10^0 - 10^1$ Å is required. This process is more collisional than (1). Dependence of energy-transfer efficiency on the direction of collision is a problem to be solved for the systems of restricted molecular motion.

When the excited state of energy acceptor is luminescent as well, energy transfer is observed as sensitized luminescence. Direct observation of energy migration where energy donor and energy acceptor are identical is more difficult. The unequivocal direct evidence can be obtained by the measurement of luminescence depolarization in a rigid matrix in which chromophores are randomly distributed, which will be described in the next section.

3.   Energy migration via complex formation (Voltz mechanism).[18]

$$M_1 + M_0' \rightleftharpoons M_1...M_0' \rightleftharpoons M_0 + M_1' \tag{3}$$

When the interacting pair consists of different chemical species, dissociation to the local excited state to complete energy transfer is either an energy uphill process unlikely to occur or an energy downhill process which should be completed prior to exciplex formation. Distinction between (2) and (3) is practically difficult unless the intermediate complex is emissive. In this case, however, the intermediate complex will act as an energy trap rather than promote energy migration.

F. EXCIMER AND EXCIPLEX

Intermolecular forces are enhanced in the excited state. This is in part due to reduced ionization potential and enhanced electron affinity so that the formation of electron donor acceptor (EDA) complex is facilitated. The EDA complex formed in the excited state is named exciplex or heteroexcimer. We should remind that the excited state of ground state EDA complex is excluded from the category of exciplex. By definition, exciplex exists only in the excited state and dissociates immediately when the excitation energy is dissipated. The ground state energy surface of an exciplex is dissociative and, consequently, the emission spectrum is broad and structureless. The main binding force in exciplex is charge resonance so that the energy state depends on solvent polarity.

$$D^* + A \text{ (or } D + A^*) \rightleftharpoons (D...A)^* \leftrightarrow (D^+A^-)^* \tag{4}$$

Excimer is formed between like molecules. The name comes from *excited dimer*. The binding force is mainly of exciton resonance. Similar to exciplex, excimer exists in the excited state alone and emits broad and structureless emission appearing at longer wavelength region of the emission from the excited monomer (M^*). Solvent polarity does not affect the emission characteristics. The excited state of ground state dimer is again excluded.

$$M^* + M \rightleftharpoons (M^*M) \leftrightarrow (MM^*) \tag{5}$$

Although excimer and exciplex look similar, distinct differences are as follows:

1.   More intensive orbital overlap is required for excimer than for exciplex. The distance between D and A in exciplex is larger than that between two identical molecules in excimer ($3 - 4$ Å).
2.   Because of less interaction in exciplex than in excimer, steric effect is more prominent in excimer formation than in exciplex formation. Also the mutual orientation of chromophores is more strictly required for excimer. For excimers of aromatic hydrocarbons, the face-to-face parallel alignment of the π-planes is a necessary condition for excimer formation.
3.   More than one excimer may be formed for a chromophore whereas exciplex is so far considered to be a single species characteristic of a pair of a donor and an acceptor.
4.   Electron transfer reaction proceeds via exciplex in polar media whereas excimer proceeds often to photodimerization and possibly to energy migration.

G. ELECTRON TRANSFER

This is another widely prevailing mechanism of luminescence quenching. An excited molecule has a lower ionization potential owing to the excitation of an electron to a vacant orbital from an occupied orbital. Simultaneously, the electron affinity of an excited molecule is higher than the ground state because it accepts an electron to the formerly occupied level. As a consequence, the excited molecules are more susceptible to both oxidation and reduction than the ground state molecules.

$$D + A \xrightarrow{h\nu} \begin{Bmatrix} D^* + A \\ D + A^* \end{Bmatrix} \underset{k_{21}}{\overset{k_{12}}{\rightleftharpoons}} \begin{Bmatrix} (D^*...A) \\ (D...A^*) \end{Bmatrix} \underset{k_{32}}{\overset{k_{23}}{\rightleftharpoons}} (D^+A^-) \xrightarrow{k_{34}} Ds^+ + As^-$$

$$\quad\quad\quad\quad\quad\quad 1 \quad\quad\quad\quad\quad 2 \quad\quad\quad\quad\quad 3 \quad\quad\quad\quad 4 \quad\quad (6)$$

The electron transfer process is essentially an equilibrium process as in the case of energy transfer. However, when the reaction is sufficiently downhill in free energy, the back reaction can be neglected. On the condition of taking the forward processes (k_{23} and k_{34}) alone into account, except for the encounter complex formation (k_{12} and k_{21}), the overall rate of electron transfer is analyzed by the Marcus theory.[19] Although there are more sophisticated theories,[20] the classic Marcus theory alone can be compared with experimental rate constant based on available physical constants of the reacting pairs and the solvent.

While the relation between ΔG (the overall free energy change of the reaction) and the overall rate constant (k_q) agrees with the theory when ΔG is more negative than $^-$5 kcal/mol, the reaction rate at highly exoergic region is controversial. The theory indicates the drop of the rate constant with increasing exoergicity (inverted region). The trend has not been proved experimentally in homogeneous fluid solutions. For intramolecular electron transfer between two terminal groups attached to a rigid hydrocarbon spacer, the presence of the inverted region has been clearly demonstrated.[21] When two chromophores are connected, electron transfer could proceed through the spacer.[22] Recently, the presence of inverted region was also explained as a result of changing the shape of energy surfaces while the Marcus theory assumes constancy of the shape of energy surfaces.[23,24]

From a more practical viewpoint of redox photochemistry, the quantum yield of reactive ion radicals is of concern. In general, the product quantum yield is much smaller than that of luminescence quenching. There are a number of deactivation paths from 2, 3, and 4. Coulombic interaction before and after electron transfer is a decisive factor deciding the charge separation yield.[25,26] For example, when the back reaction (k_{32}) to the excited state prevails, we cannot expect a high yield of charge separation. The kinetic features of photoinduced electron transfer including back electron transfer have been discussed in detail.[27,28]

In application to polymer systems, the role of polyelectrolyte providing Coulombic interaction to reacting species is a point of interest. As will be discussed later, a fast electron transfer quenching (static quenching) can be realized by attracting quencher molecule to luminescent species by Coulombic interaction while in such cases the charge separation yield is low because of enhanced back electron transfer reaction.[29]

IV. ENERGY TRANSFER AND ENERGY MIGRATION PROCESSES IN POLYMERS

Nonradiative energy migration is a primary process occurring in molecular aggregates where identical molecules are packed in close proximity of each other. Organic crystals may be classified into this category, but "molecular aggregate" is usually used for such systems as micelles, liquid crystals (lyotropic and thermotropic), monolayers, Langmuir-Blodgett's films, and polymers. They range from highly ordered systems (LC, LB films, and monolayers) to disordered systems (micelles and polymers).

Nonradiative energy migration (EM) was first recognized by Gaviola and Pringsheim,[30] Weigert and Kapper,[31] and Levshin[32] in concentrated solutions of dye molecules. They noticed that emission from the concentrated dye solution was depolarized whereas emission from the viscous dilute dye solution retained polarization. Formulation of the nonradiative energy transfer by Förster (Equation 2) allowed the "concentration depolarization phenomena" to be understood on some quantitative basis.[33] Energy transfer by dipole-dipole interaction (Förster's mechanism) was later developed to involve multipole-multipole interaction and electron exchange mechanism.[34] Some detailed description is given here on energy transfer and migration processes in polymers.

A. THEORETICAL OVERVIEW[7,35]

Electronic excitation transfer is a phenomenon occurring essentially between a pair of molecules, M_1 and M_2, and depends on the strength of interaction between the two molecules. If we assume that M_1 is excited initially and excitation is transferred to M_2, wavefunctions for the initial (ψ_i) and the final (ψ_f) states of the system can be written as

$$\psi_i = \psi_{M1}{}^*(\mathbf{r}_1)\psi_{M2}(\mathbf{r}_2) \qquad \psi_f = \psi_{M1}(\mathbf{r}_1)\psi_{M2}{}^*(\mathbf{r}_2) \tag{7}$$

where $\mathbf{r}_1$ and $\mathbf{r}_2$ represent electron coordinate in M_1 and M_2, respectively. The matrix element mixing the initial and final vibronic states, β, may be written as

$$\beta = \int \psi_i H(\mathbf{r}_1, \mathbf{r}_2)\psi_f d\tau \tag{8}$$

where H is the perturbation Hamiltonian. Three limiting cases are considered: (1) strong interaction where $\beta \gg \Delta E \gg \alpha$, (2) medium interaction where $\Delta E \gg \beta \gg \alpha$, and (3) weak interaction where $\Delta E \gg \alpha \gg \beta$. Here, ΔE is the electronic energy gap between the corresponding zero-order states and α is the average coupling energy between the lattice modes and the molecular vibrational modes.

1. Strong Interaction

The time variation of finding the excitation is described by

$$P(t) = (\cos^2\beta t/\hbar)\psi_i^2 + (\sin^2\beta t/\hbar)\psi_f^2 \tag{9}$$

if we assume an initial condition that the system is in the initial state ψ_i at t = 0. Thus, for the strong interaction, a decay of the initial state occurs in a time interval t = h/4β, and a subsequent buildup of the initial state again occurs after another interval of t = h/4β, providing that M_1 = M_2 and there is no quenching of M_1^* or M_2^*. However, the time scale of this type of transfer is extraordinarily short as expected by the expression t = h/4β. Assuming β = 1 eV (~8100 cm^{-1}), then we get a result of t ~ 10^{-15} s. Thus, when M_1 is excited, excitation transfer takes place before the nuclei in the molecules adopt their equilibrium geometry. $\psi(M_1^*M_2)$ and $\psi(M_1M_2^*)$ are no longer distinguishable, and their resonance state should be regarded as one excited state.

2. Medium Interaction

In this limiting case, the nuclear motion competes sufficiently with the excitation transfer process, thus vibronic levels must be taken into account. As already described, the total wavefunction can be separated into electronic (θ) and vibrational (ϕ) parts,

$$\psi = \theta \cdot \phi_{v'}, \qquad \psi^* = \theta^* \cdot \phi_{v'} \tag{10}$$

Thus,

$$\beta_v = \beta \cdot <\phi_v|\phi_{v'}>^2 \tag{11}$$

The time of excitation transfer is

$$t = \frac{h}{4\beta \cdot <\phi_v|\phi_{v'}>^2} \tag{12}$$

Equation 12 indicates that the time scale of excitation transfer in the medium interaction is longer than that in the strong interaction. In fact, the time of transfer strongly depends on the overlap of the vibrational wavefunctions. This type of excitation transfer is observed in benzene crystals and anthracene crystals.

3. Weak Interaction

In this case, the interaction is so small that ψ_i and ψ_f can almost be regarded as independent eigenstates. The time of the excitation transfer is expressed by

$$t = \frac{\hbar\alpha}{2\pi\beta^2 \cdot <\phi_v|\phi_{v'}>^4} \tag{13}$$

The transition probability is dependent on the term, $\beta^2 \cdot <\phi_v|\phi_{v'}'>^4$, thus with increasing distance (R) between M_1 and M_2 it decreases rapidly. Vibrational energy is completely redistributed before the transfer takes place. If we take the dipole-dipole interaction term alone (higher order interaction terms are truncated),

$$\beta \sim \frac{\kappa|\mu|^2}{n^2R^3} \tag{14}$$

where μ is the transition moment, n is the refractive index of the medium, and κ is the orientational factor describing the mutual arrangement of M_1 and M_2. Equation 14 leads finally to the well-known expression (Equation 2) for the dipole-dipole electronic energy transfer proposed by Förster.

Most of the singlet-singlet energy transfer phenomena have been analyzed by the dipole-dipole mechanism.[36] First experimental confirmation of the inverse sixth-power dependence of the transfer rate on the distance between M_1 and M_2 was conducted by Stryer and Hangland[37] using oligomers of poly-L-proline, labeled at one end with a naphthyl (donor) and at the other end with a dansyl (acceptor) group (Figure 1).

Poly-L-proline forms a rigid helix, providing a fixed distance between the donor and acceptor groups ranging from 12 to 46 Å. They used the sensitized fluorescence from the acceptor and obtained $R^{-5.9 \pm 0.3}$ dependence, which is in excellent agreement with the theory. Excitation transfer between two nonidentical chromophores was, in this way, successfully observed, however, excitation transfer between two identical chromophores is much more difficult to confirm experimentally.

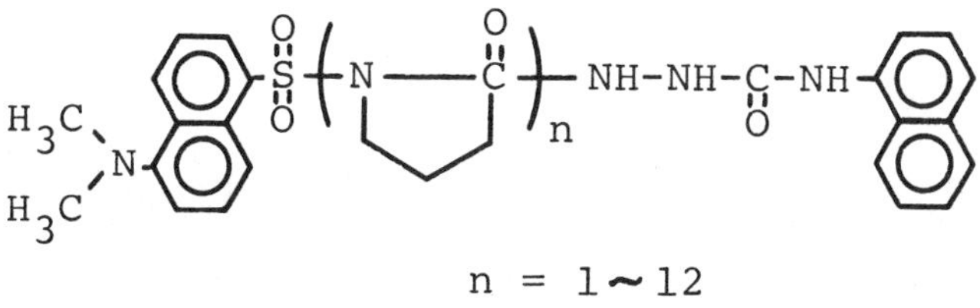

n = 1~12

FIGURE 1. Structure of poly-L-proline labeled with a naphthyl and a dansyl groups. (From Stryer, L. and Haugland, R. P., *Proc. Natl. Acad. Sci.*, 58, 719, 1967. With permission.)

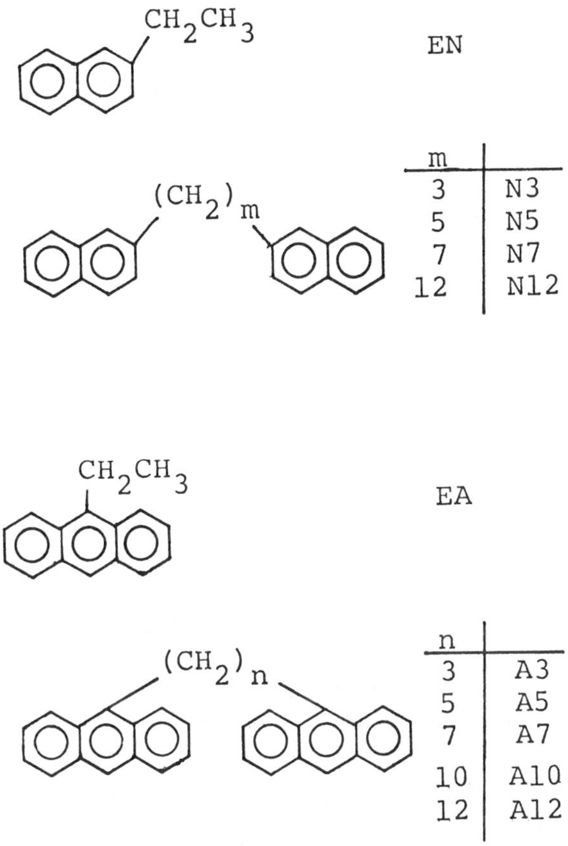

FIGURE 2. Structure of alkanes end-capped with two identical aromatic moieties. (From Ikeda, T., Lee, B., Kurihara, S., Tazuke, S., Ito, S., and Yamamoto, M., *J. Am. Chem. Soc.*, 110, 8299, 1988. With permission.)

Direct observation of excitation transfer between two identical chromophores has been recently done by picosecond time-resolved fluorescence anisotropy measurements.[38] In order to study the excitation hopping behavior in *purely isolated* two identical chromophoric systems, various alkanes end-capped with two identical aromatic groups were used (Figure 2).

In Figure 3, the anisotropy decays for Nn compounds are shown. Here, all measurements were performed at 77 K in rigid matrix in order to eliminate any contribution of rotational diffusion of the chromophores to fluorescence depolarization. It is clear that anisotropy, r(t),

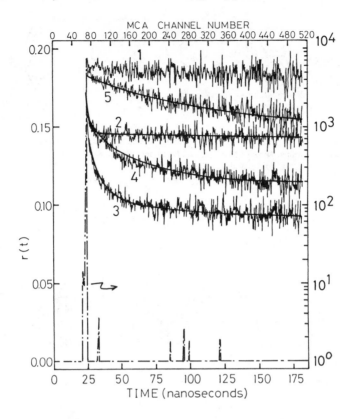

FIGURE 3.   Anisotropy decays of Nn and EN. 1, EN; 2, N3; 3, N5; 4, N7; and 5, N12. λ_{ex} = 318 nm and λ_{em} = 335 nm. (From Ikeda, T., Lee, B., Kurihara, S., Tazuke, S., Ito, S., and Yamamoto, M., *J. Am. Chem. Soc.*, 110, 8299, 1988. With permission.)

for EN does not change with time, which is an explicit piece of evidence that excitation remains at the initially excited sites during its lifetime. In contrast, in Nn, decay of r(t) is clearly observed, indicating that excitation hopping takes place between the two naphthyl groups attached to both ends of the alkyl chain. These results were analyzed by assuming forward and backward excitation hopping between the two naphthyl moieties. This sort of experiment is expected to provide a basis for excitation migration in molecular aggregate systems.

B. THEORIES OF EXCITATION TRANSFER IN MOLECULAR AGGREGATES

Exact treatment of excitation migration in disordered systems (e.g., concentrated dye solution and polymers) has been a challenge for many theorists for many years. Difficulties arise from the fact that in molecular aggregates the location of the initially excited sites is not clearly defined. For example, in polymers having many chromophores along the polymer chain, the probability of excitation at the chain end is equal to that of the center of the polymer chain. However, if excitation migration operates, efficiency is definitely different between the two sites. Thus, for the analysis of the experimental results, account must be taken of distribution of the initially excited sites. Another difficulty, much more serious for the exact analysis, is that excitation hops around in the system like a random walk; thus, to follow the locus of the excitation hopping is quite difficult even if we assume the dipole-dipole transfer *a priori* for the excitation hopping mechanism. Assume that excitation hops in a polymer coil. We can draw many pathways for the excitation movements: between nearest neighbors and between a pair which is far apart from each other along a polymer chain. For such complicated systems, exact solution of the problem may be impossible in principle, but approximation may be done.

The most promising theoretical approach to this problem was developed by Fayer et al.[39-43] and others.[44-53] An excellent review article on the theories of excitation hopping is available;[54] hence only an outline of the theories is described here.

We assume a simplified model in which chromophores are distributed in space randomly and Brownian diffusion of each chromophore is slow enough to permit the assumption that chromophores remain stationary on the time scale of excitation transfer. Furthermore, we assume incoherent excitation transfer in which Förster's equation (Equation 2) is applicable and polymer chains consisting of N chromophores whose locations can be represented by

$$\{\mathbf{R}\} = \{\mathbf{r}_1, \mathbf{r}_2, ..., \mathbf{r}_N\} \tag{15}$$

In the absence of traps (e.g., excimer-forming sites or any acceptor), the probability that an excitation is found on the jth chromophore at time t, $p_j'(R,t)$, satisfies the master equation

$$\frac{dp_j'(R, t)}{dt} = -p_j'(R, t)/\tau + \sum_k w_{jk}[p_k'(R, t) - p_j'(R, t)] \tag{16}$$

where τ is the measured lifetime of the excited species and w_{jk} is the transfer rate between chromophores j and k ($w_{jj} = 0$), which may be written as a simplified form:

$$w_{jk} = \frac{1}{\tau}\left(\frac{R_0}{r_{jk}}\right)^6 \tag{17}$$

It is very convenient to make the following substitution in order to eliminate the decay term in the master equation

$$p_j(R, t) = p_j'(R, t)\exp(t/\tau) \tag{18}$$

Then we have

$$\frac{dp_j(R, t)}{dt} = \sum_k w_{jk}[p_k(R, t) - p_j(R, t)] \tag{19}$$

as the master equation. If we define the matrix $\mathbf{W}$ by

$$\mathbf{W}_{jk} = w_{jk} - \delta_{jk}\sum_l w_{kl} \tag{20}$$

then, we have the Pauli master equation

$$\frac{d}{dt}\mathbf{p}(R, t) = \mathbf{W} \cdot \mathbf{p}(R, t) \tag{21}$$

where $\mathbf{p}(R,t)$ is a vector with components $p_1(R,t)$, $p_2(R,t)$, . . . , $p_N(R,t)$. Equation 21 has the formal solution

$$\mathbf{p}(R, t) = \exp(t\mathbf{W}) \cdot \mathbf{p}(R, 0) \tag{22}$$

The time and distance dependent ensemble averaged density of excitation is given by

$$P(\mathbf{r}, t) = <\sum_j \delta(\mathbf{r}_j - \mathbf{r})p_j(R, t)> \tag{23}$$

where the ensemble average of a function A(R) is given by

$$<A(R)> = \int d\mathbf{r}_1 \dots \int d\mathbf{r}_N A(R) \tag{24}$$

The Green function G($\mathbf{r},\mathbf{r}'$, t) which is related to the experimental observables is defined by

$$P(\mathbf{r}, t) = \int d\mathbf{r}'G(\mathbf{r}, \mathbf{r}', t)P(\mathbf{r}', 0) \tag{25}$$

The Green function gives the conditional probability of finding an excitation at position **r** at time t, given that it originated at position **r**' at t = 0, and is represented by

$$G(\mathbf{r}, \mathbf{r}', t) = <\sum_j \sum_k \delta(\mathbf{r}_j - \mathbf{r})\delta(\mathbf{r}_k - \mathbf{r}')[\exp(t\mathbf{W})]_{jk}> \tag{26}$$

The Green function can be separated into two terms: G^s($\mathbf{r},\mathbf{r}'$,t) and G^m($\mathbf{r},\mathbf{r}'$, t), thus,

$$G(\mathbf{r}, \mathbf{r}', t) = G^s(\mathbf{r}, \mathbf{r}', t) + G^m(\mathbf{r}, \mathbf{r}', t) \tag{27}$$

where

$$G^s(\mathbf{r}, \mathbf{r}', t) = \delta(\mathbf{r} - \mathbf{r}')<[\exp(t\mathbf{W})]_{11}> \tag{28}$$

$$G^m(\mathbf{r}, \mathbf{r}', t) = (N - 1)<\delta(\mathbf{r}_{12} - \mathbf{r} + \mathbf{r}')[\exp(t\mathbf{W})]_{12}> \tag{29}$$

The integral of G^s($\mathbf{r},\mathbf{r}'$, t) over a small volume about **r**' is the probability of finding the excitation on the site of initial excitation at time t and that of G^m($\mathbf{r},\mathbf{r}'$, t) about **r** is the probability of finding the excitation at **r** at time t, provided that the excitation is initially at **r**'.

Introducing the Laplace transform, we obtain

$$\hat{G}^s(\epsilon) = <[(\epsilon\mathbf{I} - \mathbf{W})^{-1}]_{11}> \tag{30}$$

where $\hat{G}^s(\epsilon)$ is the transform of G^s($\mathbf{r} - \mathbf{r}'$, t), ϵ is the Laplace variable, and **I** is the unit matrix. Expanding the matrix $(\epsilon\mathbf{I} - \mathbf{W})^{-1}$ in powers of ϵ and **W**, we get

$$\hat{G}^s(\epsilon) = \sum_{n=0}^{\infty} \epsilon^{-(n+1)}<(\mathbf{W}^n)_{11}> \tag{31}$$

The simplified form of the Green function was then treated by various mathematical ways to get G^s(**r**, t) which is directly related to fluorescence anisotropy or other observables.[39-43,48]

C. EXPERIMENTAL OBSERVATION OF ENERGY TRANSFER AND ENERGY MIGRATION IN POLYMERS

Experimental observations of EM in polymer systems have been done by many workers. There are two types of EM occurring in polymers, depending on the type of the excited states; singlet and triplet. In the case of high-density excitation, more than one S* can be formed along a polymer chain, thus EM of S* or T* can lead to S-S or T-T annihilation. These processes will be described in detail in the chapters by Masuhara and Burkhart, and we will not refer to these topics here in more detail. Furthermore, with respect to triplet EM (triplet exciton migration), Burkhart will give a full description in Chapter 4; thus we will focus our attention on singlet EM process here in which contribution of multiphotonic processes is negligible.

Singlet EM is evaluated mainly by three methods: (1) fluorescence quenching by external quenchers, (2) fluorescence polarization, and (3) energy transfer to traps (excimer forming sites (EFS) and acceptor chromophores incorporated into a polymer chain). In method 1, such effective fluorescence quenchers as oxygen and CCl_4 are employed, which are in fact known to quench fluorescence of some aromatic chromophores by diffusion-controlled reaction. The quenching rate constant, k_q, can be determined by the Stern-Volmer equation,

$$\frac{I_0}{I} = 1 + k_q\tau[Q] \tag{32}$$

where I_0 and I are fluorescence intensities in the absence and presence of the quencher, $[Q]$, and τ is the lifetime of S* in the absence of Q. In early works, the decay of S* was assumed to obey a single-exponential function. However, recent advancement in instrument on lifetime measurements has enabled us to measure the lifetimes in nanosecond and even in picosecond time region quite precisely (e.g., single-photon counting apparatus equipped with a ps pulse source). Results of such precise lifetime measurements have revealed that the single exponential decay is rather an exception in polymer systems where many chromophores are chemically bonded to a polymer chain. Rationale to use a weighted average lifetime, $<\tau>$, for the Stern-Volmer equation in multiexponential decay systems has been recently given by Bai et al.[55] If we assume $I(t) = \Sigma A_i \exp(-t/\tau_i)$, $<\tau>$ is given by

$$<\tau> = \frac{\int_0^\infty t I(t)dt}{\int_0^\infty I(t)dt} = \frac{\sum_i A_i\tau_i^2}{\sum_i A_i\tau_i} \tag{33}$$

The relative contribution of the ith component to the overall decay is expressed by

$$\phi_i = \frac{\int_0^\infty A_i e^{-t}\tau_i dt}{\int_0^\infty I(t)dt} = \frac{A_i\tau_i}{\sum_i A_i\,\tau_i} \tag{34}$$

Provided that each component is independently quenched by Q with the same k_q value, Equation 32 can be modified as follows:

$$\frac{I_0}{I} = \sum \phi_i(1 + k_q\tau_i[Q]) = 1 + k_q[Q]\frac{\sum_i A_i\tau_i^2}{\sum_i A_i\tau_i}$$

$$= 1 + k_q<\tau>[Q] \tag{35}$$

Evaluation of k_q values is followed by the application of Smoluchowski-Einstein equation which is usually modified to involve EM diffusion coefficient Λ,

$$k_q = 4\pi N(D_S + D_Q + \Lambda)pR \times 10^{-3} \tag{36}$$

where D_S and D_Q are diffusion coefficients of the chromophore bound to a polymer chain and the quencher, respectively, p is the quenching probability per collision, R is the sum of collisional radii for the chromophore and the quencher, and N is Avogadro's number. Taking some appropriate values for D_S, D_Q, p and R, Λ can be estimated. In the case of down-chain EM, the average EM length, l, can be evaluated on the basis of a one-dimensional random walk as

$$l = (2\Lambda\tau)^{1/2} \tag{37}$$

The advantage of method 1 over other methods is its simplicity both in experiments and in analysis. Without complicated mathematics like those described in the previous section, we can obtain the Λ and l values. However, its shortcoming is clear: reliability of the Λ and l values obtained. Many assumptions and approximations are made in the calculation. For example, D_S is usually taken as 0 in spite of the fact that local motion of a polymer chain is known to occur on the time scale of less than 1 ns.[56] Furthermore, in addition to the lifetime heterogeneity, local concentration of Q in the vicinity of the polymer coil might be different from that of bulk phase, although [Q] is usually estimated from the bulk concentration. Therefore, the Λ and l values obtained from method 1 should be regarded as a rough estimate.

As described in the previous section, fluorescence polarization (method 2) provides unequivocal evidence for EM, provided that other depolarization factors, such as rotational diffusion of the chromophores, is eliminated. Qualitative verification of EM can be easily achieved by steady-state measurements of fluorescence polarization. However, evaluation of EM rate like Λ is quite difficult from the steady-state measurements. Quantitative evaluation of EM rate can only be achieved with the aid of time-resolved measurements of fluorescence polarization (or anisotropy) and the complicated mathematics outlined in the previous section. Because of the time-resolved measurements of polarization and strict theoretical treatment, the EM rates evaluated by this method are most accurate.

From a conceptional point of view, method 3 is simplest. One acceptor chromophore can be incorporated into a polymer chain, say, at the end. By measuring the rise of the acceptor fluorescence, one can estimate the EM rate in principle. However, analysis of the time-resolved measurements is quite complicated due to concomitant occurrence of single-step energy transfer to the acceptor chromophores. Similar mathematical treatment to that described in the previous section is required.

Excimer formation is often observed in many aromatic polymers in solution and even in solid state (film), and EFS usually acts as a trap of migrating excitation. In other words, excimer formation efficiency is always enhanced when EM is operative, since efficient photon harvest is achieved.

1. Polystyrene

In polystyrene (PS), EM has been always discussed in relation to the excimer formation kinetics. A generally accepted view of the excimer formation in PS is that EM occurs effectively along a polymer chain (one-dimensional random walk) or in three-dimensional manner as suggested by Klöpffer[57] and the migrating excitation is trapped by EFS, leading to the formation of excimer.

Ishii et al.[58-60] studied the excimer formation in PS with various molecular weights

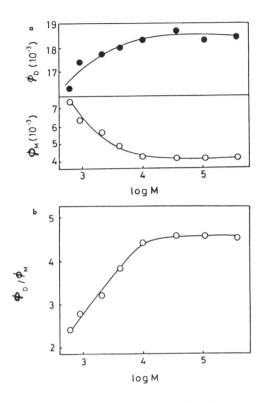

FIGURE 4. (a) Molecular weight dependence of the fluorescence quantum yields of excimer (ϕ_D) and monomer (ϕ_M) emissions in PS: (●), ϕ_D; (○), ϕ_M; (b) Molecular weight dependence of ϕ_D/ϕ_M. (From Ishii, T., Handa, T., and Matsunaga, S., *Macromolecules*, 11, 40, 1978. With permission.)

ranging from 600 to 4×10^5 in dilute solution of dichloroethane. They found a strong dependence of ϕ_D/ϕ_M on molecular weight as shown in Figure 4. Here, ϕ_D and ϕ_M represent fluorescence quantum yield of excimer and monomer emission, respectively. Below MW $\sim 10^4$, ϕ_D/ϕ_M increased almost linearly with MW and above MW $\sim 10^4$ the ratio remained constant. They also estimated the EM rate by method 1 by using oxygen as a quencher. The l value for a sample with MW = 600 was found to be ~ 50 Å and decreased with increasing MW and reached to a constant value of ~ 26 Å above MW $\sim 10^4$. Their experiments were based on the steady-state measurements of fluorescence, but their reported values of l agreed fairly well with those obtained by the time-resolved measurements.[56]

Itagaki et al.[56] conducted a pulse radiolysis study of PS oligomers ranging from dimer to tridecamer in dilute solution of cyclohexane. Their apparatus for the lifetime measurements comprised a linear accelerator and a streak camera, allowing a picosecond time resolution (~ 10 ps). They observed a similar molecular weight dependence of ϕ_D/ϕ_M on molecular weight (Figure 5) to that reported by Ishii et al., but saturation of ϕ_D/ϕ_M took place in a much lower molecular weight region; DP (degree of polymerization) $\leqq 8$. They further observed that the decay of the monomer emission obeyed a single-exponential function regardless of molecular weight and the excimer emission was composed of two components: (1) a rise corresponding to the monomer emission decay and (2) a slow decay of the excimer. They concluded that the excimer formation in PS could be analyzed by the Birks' kinetics where dissociation process was negligible (see Section V). From the molecular weight dependence of k_a, the rate constant for excimer formation, they also concluded that at DP $\leqq 8$, internal rotation leading to a face-to-face conformation of the two phenyl rings was

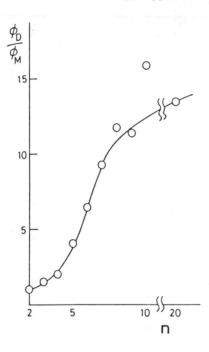

FIGURE 5.   Dependence of ϕ_D/ϕ_M on the degree of polymerization (n). (From Itagaki, H., Horie, K., Mita, I., Washio, M., Tagawa, S., and Tabata, Y., *J. Chem. Phys.*, 79, 3996, 1983. With permission.)

the rate-determining step in excimer formation, whereas at DP > 8, EM was rate-determining, for which they estimated as $1 \sim 50$ Å and $\Lambda \sim 1 \times 10^{-4}$ cm²/s.

It is quite important to know whether excimer formation in PS is due to the nearest-neighbor interaction or nonnearest-neighbor interaction. Kyle and Kilp[61] and other groups[62,63] synthesized head-to-head PS (HH-PS) (Figure 6) and examined the excimer formation in HH-PS.

Kilp et al. found that no excimer formation was observed in HH-PS even if solvent was changed from good solvent to poor solvent. They estimated the EM rate by method 1 using CCl₄ as a quencher as $\Lambda \sim 7.5 \times 10^{-5}$ cm²/s for HT-PS (head-to-tail) whereas $\Lambda \sim 0$ for HH-PS. They concluded that excimer formation in PS was exclusively due to the nearest-neighbor interaction (n = 3 rule proposed by Hirayama[64]) and EM occurred by excitation hopping through phenyl rings separated by three carbon atoms.

Frank et al.[65-68] studied EM and excimer formation behavior of isolated PS in PS/poly(vinyl methyl ether) (PVME) miscible blend in the solid state. The PS/PVME blends are characteristic in view of the fact that they form miscible blends at any ratio of PS to PVME when prepared by casting from toluene solution, whereas they form immiscible blends (phase separation) when prepared from tetrahydrofuran solution. They examined various PSs with MW ranging from 2×10^3 to 4×10^5. They observed a similar MW dependence of ϕ_D/ϕ_M to that observed by Ishii et al., that is, below MW $\sim 2 \times 10^4$ ϕ_D/ϕ_M increased with MW and above that MW ϕ_E/ϕ_M remained nearly constant. They analyzed their results on the basis of the one-dimensional EM with low concentration of traps (EFS) which were assumed as *meso* tt dyads. According to their analysis, in the small MW region (MW < 2×10^4) the probability for the migrating excitation to find the traps decreases with decreasing MW, leading to low ϕ_D/ϕ_M. In the large MW region (MW > 2×10^4), on the other hand, because of finite lifetime of the excitation and finite number of EFS, length of the polymer chain which can be sampled by the migrating excitation is limited and further the long polymer chains are practically divided into shorter segments, thus

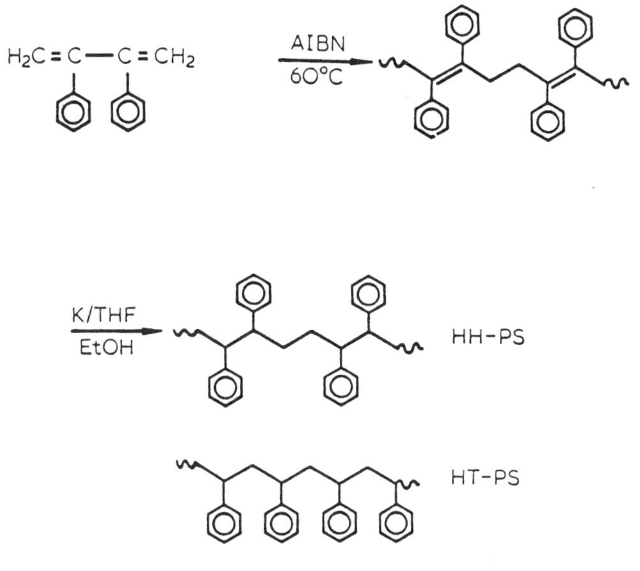

FIGURE 6. Synthesis of head-to-head PS. (From Kyle, B. R. M. and Kilp, T., *Polymer*, 25, 989, 1984. With permission.)

resulting in the excimer formation efficiency independent of MW. They extended their study to include three-dimensional EM and further to monitor phase separation behavior of PS/PVME blends from excimer formation efficiency.

Most of the studies reported so far on photophysical processes of PS have demonstrated occurrence of EM along a polymer chain. However, the experiments conducted by MacCallum et al.[69-72] showed no sign of EM in PS. They measured the degree of fluorescence polarization (P) of the monomer emission and the excimer emission on various PS derivatives in films and in solutions (method 2). Their experiments were based on the expectation that if EM was effectively operative and the excitation was trapped by EFS, both the monomer and the excimer emission should be completely depolarized since spacial orientation of the phenyl ring as well as EFS should be random. The results showed that P of the excimer emission (330 nm) was around 0.2 both in film and in dilute solution of cyclohexane at room temperature (RT). Furthermore, P of the monomer emission (280 nm) was ~0.1 in the rigid glass matrix at 77 K where no excimer emission was detected. The results on poly(α-methylstyrene) were so similar they concluded that EM did not take place in PS derivatives.

2. Polymers Containing Naphthalene Moieties and Phenanthrene Moieties

Soutar et al.[73-78] studied the EM and excimer formation behavior of copolymers of 1-vinylnaphthalene (1VN) with methyl acrylate (MA) and methyl methacrylate (MMA) and of acenaphthylene (ACENA) with MA and MMA with various compositions. Using method 2, they showed that EM was operative in the copolymers containing 1VN or ACENA, and excitation was effectively transported to EFS. According to their analysis, P and ϕ_D/ϕ_M are related to such parameters as the mole fraction of naphthalene unit in the copolymer, f_n, and that of naphthalene-naphthalene sequence, f_{nn}, and average length of successive naphthalene units, l_n, by the equations,

$$\frac{1}{P} = k'l_1 \tag{38}$$

$$\frac{\phi_D}{\phi_M} = k''l_n f_{nn} \tag{39}$$

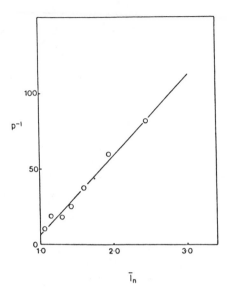

FIGURE 7.   Depolarization as a function of mean path length of naphthyl chromophores in the copolymer of 1VN and MA. Measured in 2-methyltetrahydrofuran at 77 K. (From Anderson, R. A., Reid, R. F., and Soutar, I., *Eur. Polym. J.*, 15, 925, 1979. With permission.)

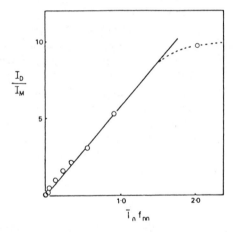

FIGURE 8. ϕ_D/ϕ_M as a function of $1_n f_{nn}$ in the copolymer of 1VN and MA. Measured in dichloromethane at 298 K. (From Anderson, R. A., Reid, R. F., and Soutar, I., *Eur. Polym. J.*, 15, 925, 1979. With permission.)

In the copolymers of 1VN with MA or MMA, Equations 38 and 39 were valid as shown in Figures 7 and 8, indicating that EM occurs via nearest-neighbor interaction. On the other hand, in the copolymers of ACENA with MA or MMA, plots of 1/P vs. 1_n or ϕ_D/ϕ_M vs. $1_n f_{nn}$ did not give a straight line. Instead, plots of 1/P vs. f_n and of ϕ_D/ϕ_M vs. Σf_n gave straight lines, where Σ represents the fraction of pentads which contain next-to-nearest neighbor naphthalene units. These results indicate that in the copolymer containing ACENA, EM occurs via three-dimensional excitation hopping within the polymer coil and excimer formation results from the next-to-nearest neighbor interaction.

In the attempt to achieve effective down-chain EM by reducing the number of EFS, Nakahira et al.[79-81] investigated the photophysical behavior of polymers of 2-*tert*-butyl-6-

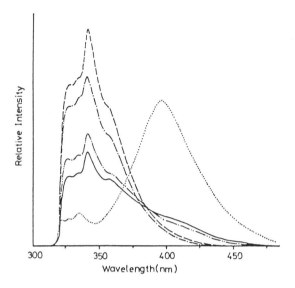

FIGURE 9. Fluorescence spectra of polymers of 2-*tert*-butyl-6-vinyl-naphthalene (BVN) and its model compound measured in 2-methyltetra-hydrofuran at 20°C. (——), homopolymer of BVN; (-··-), copolymer with St [BVN mole fraction, 0.80]; (----), copolymer with St [BVN mole fraction, 0.04]; (-··-), 2,6-di-*tert*-butylnaphthalene (model compound); (·····), poly(2-vinylnaphthalene). (From Nakahira, T., Sakuma, T., Iwabuchi, S., and Kojima, K., *J. Polym. Sci. Polym. Phys. Ed.*, 20, 1863, 1982. With permission.)

vinylnaphthalene (BVN) and St. As shown in Figure 9, they observed no or little excimer formation even in the homopolymer of BVN due to the bulky substituent introduced into naphthalene ring. However, Λ estimated by method 1 was $\sim 10^{-6}$ cm^2/s, one order of magnitude smaller than those of other naphthalene polymers. They concluded that although the introduction of the bulky group was quite effective in reducing the number of EFS, it concomitantly resulted in unfavorable orientational factor (κ in Equation 2) in energy transfer.

Relevant to "antenna effect" in polymeric systems, Guillet et al.[82-91] reported a series of studies on EM in naphthalene and phenanthrene polymers after the first report by one of the authors[92] on anthracene polymers used as electron transfer sensitizers. Using methods 1 to 3, they showed that EM was operative both in naphthalene- and phenanthrene-containing polymers and Λ was estimated as $\sim 10^5$ cm^2/s for both polymers. Emphasis was placed on advantage of the phenanthrene-containing polymers (polymers containing (9-phenan-thryl)methyl methacrylate units) in the study of EM in polymers. Since the phenanthrene-containing polymers do not form excimers nor undergo self quenching, the fluorescence decay was simply single exponential even for the homopolymer and its lifetime was nearly identical with that of the small molecular weight model compound. This situation eliminates the complexity arising from lifetime heterogenity in the analysis of the time-resolved behavior of EM.

3. Polymers Containing Anthracene Moieties

EM in anthracene-containing polymers was extensively studied by Webber et al.[93-97] as "photon-harvesting polymers". Using methods 1 and 3, they explored the EM behavior in connection with excimer formation in such polymers as shown in Figure 10.

In PAMMA and PAE, only broad structureless emission was observed due to partially or fully eclipsed excimer, regardless of the condition of the measurements, namely, in dilute solution at RT, in glass matrix at 77K, and in film at RT and at 77 K. On the other hand,

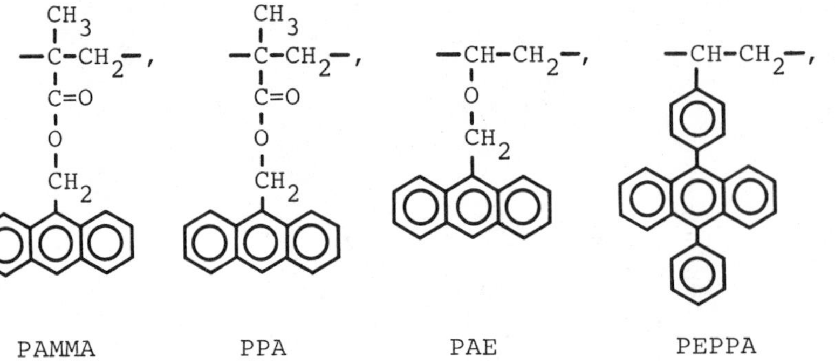

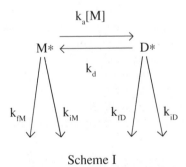

FIGURE 10.   Structure of anthracene-containing polymers.[93-95]

in PPA and PEPPA, no or little excimer formation was observed. Furthermore, photostability was found to be much better in PPA and PEPPA than in PAMMA and PAE. Λ was estimated by method 1 using CCl_4. A value of 48 Å was obtained for PPA, while $\Lambda = 0$ for PAMMA and PAE. They ascribed this difference to the excimer formation efficiency in those polymers.

V. EXCIMER AND EXCIPLEX FORMATION IN POLYMERS

A. EXCIMER

Many studies on excimer formation in polymers have been performed on vinyl polymers with pendant aromatic chromophores such as PS, poly(vinylnaphthalene), and so forth. In these polymers, because of the high local concentration of the chromophores along the polymer chain, another bichromophoric process, EM, occurs with high efficiency as described in the previous section. As we have seen, in many cases excimer formation follows EM and excitation is effectively transported to the EFS. The EFS is regarded as specific conformation in which face-to-face arrangement of the two chromophores is achieved. In fluid solution, location of EFS is a function of time since local segmental motion of the polymer chain occurs on a competitive time scale. Thus, in fluid solution, both excitation and EFS are mobile and their encounter results in the formation of excimer. Examples of excimer formation, followed by EM, are described in some detail in the previous section, thus we will not discuss here.

Scheme I

In the early stage of studies on excimer formation in polymers, the Birks' kinetic scheme (Scheme 1) was exclusively applied and thermodynamic parameters were evaluated by steady-state fluorescence measurements.[8] Particular attention was paid to temperature variation of the fluorescence quantum yield of excimer (ϕ_D) and monomer (ϕ_M). Under photostationary condition, assuming that the Birks' scheme is applicable, the ratio of ϕ_D to ϕ_M can be expressed by[8]

$$\frac{\phi_D}{\phi_M} = \frac{k_{fD}k_a[M]}{k_{fM}(k_d + k_D)}$$ (40)

where [M] is the concentration of chromophore in the ground state and $k_D = k_{fD} + k_{iD}$. The temperature variation of ϕ_D/ϕ_M was analyzed by Arrhenius-type equation (ln ϕ_D/ϕ_M vs. l/T plots) and these plots gave bell-shaped profiles in many cases when a wide range of temperature was examined. Two straight line portions in these plots were analyzed independently on the basis of the following treatment and called "high-temperature region" and "low-temperature region", respectively. In the high-temperature region where $k_a[M]$, $k_d \gg k_M$, k_D, Equation 40 becomes

$$\frac{\phi_D}{\phi_M} = k \frac{k_a}{k_d}$$ (41)

if we assume that k_{fM}, k_{fD}, and [M] are independent of temperature. These assumptions lead to an equilibrium between M* and D* in the high-temperature region, thus a slope in the lnϕ_D/ϕ_M vs. 1/T plot gives the binding energy of excimer. On the other hand, in the low-temperature region we can assume that $k_D \gg k_d$, and k_{fM}, k_{fD}, and k_D are invariant with temperature, Equation 40 becomes

$$\frac{\phi_D}{\phi_M} = kk_a[M]$$ (42)

If we further assume that [M] is constant over the temperature range examined, the plot of lnϕ_D/ϕ_M vs. 1/T should give a straight line and from the slope of the line we can estimate the activation energy of excimer formation.

These treatments are, in fact, oversimplified. However, they provide a basis of the thermodynamical analyses of the excimer formation in polymer systems. One example is given here: activation enthalpies of excimer formation ($\Delta H_a^\ddagger$) and of excimer dissociation ($\Delta H_d^\ddagger$) were compared between polyesters tagged with pyrenylmethyl groups and their dimer model compounds (Figure 11).[98] Excimer formation in the polyesters was five to ten times more efficient than that in the corresponding dimer models. The favored excimer formation in the polymers was ascribed to both enthalpy and entropy terms. The restricted segmental mobility in the polymers was an origin of the higher $\Delta H_a^\ddagger$ and $\Delta H_d^\ddagger$ values for intramolecular excimer formation and dissociation than those of dimer model compounds. Eventually, in polymer, excimer formation and dissociation are both energetically less favorable than those in dimer models. However, the dissociation process is much more unfavorable in polymer, thus resulting in a higher $-\Delta H$ value. Contribution of the entropy term to the excimer formation is characteristic in polymer. Chromophores bound to a polymer chain are essentially in a low-entropy state, since the mobility of the polymer side groups is limited. This situation reduces the entropy loss accompanied by the face-to-face arrangement of chromophores in excimer formation, leading to entropically more favored excimer formation in polymer systems.

The most prominent feature seen in the polymers is the effect of tacticity. As is well known, there are two types of the excimer states, each corresponding to different configuration of the relevant chromophores. One possesses a face-to-face arrangement (sandwich excimer) and its emission maximum lies at longer wavelength. The other arises from partial overlap of the two chromophores and its emission maximum is at shorter wavelength. The latter is usually called "second" or "high energy" excimer. Many examples of the two types of excimer states have been so far reported for naphthalene,[99-101] anthracene,[102,103] and carbazole[104-108] derivatives. Poly(N-vinylcarbazole) (PNCz) shows both sandwich-type and

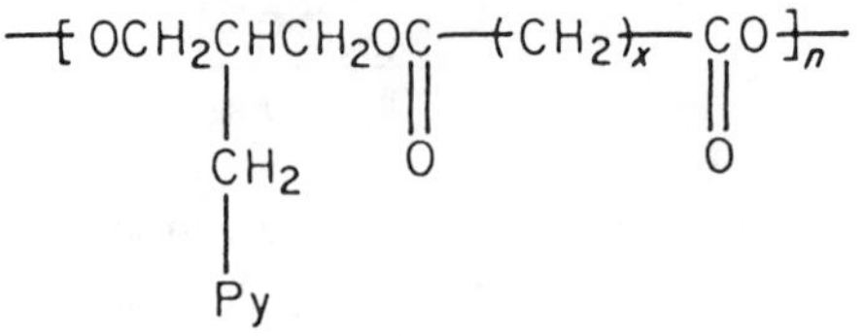

$$\text{---}\!\!\{\,OCH_2CHCH_2OC\text{---}\!(CH_2)_{\overline{x}}\text{---}CO\}_{\overline{n}}$$

$$\underset{\underset{Py}{|}}{\underset{CH_2}{|}}\qquad\underset{O}{\|}\qquad\underset{O}{\|}$$

PE(Py-X)

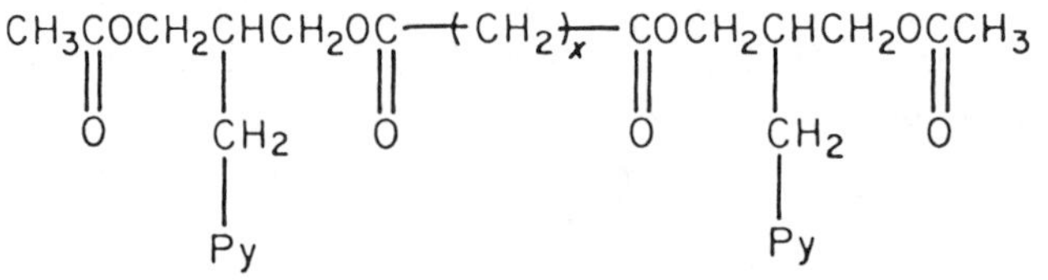

$$CH_3COCH_2CHCH_2OC\text{---}\!(CH_2)_{\overline{x}}\text{---}COCH_2CHCH_2OCCH_3$$

DE(Py-X)

$$CH_3COCH_2CHCH_2OCCH_3$$

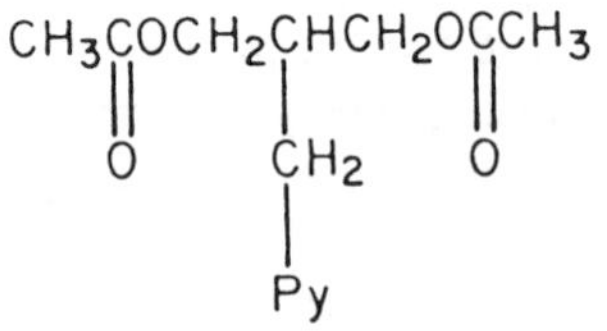

MEPy

FIGURE 11. Structures of polyesters tagged with pyrenylmethyl groups and their dimer model compounds. Py = 1-pyrenyl. (From Tazuke, S., Ooki, H., and Sato, K., *Macromolecules*, 15, 400, 1982. With permission.)

second excimer emissions. However, their intensity ratio strongly depends on the preparation method of the polymer.[104,106,108] PVCz prepared by radical polymerization is rich in syndiotactic sequence, thus favors partical overlap of the two neighboring carbazolyl chromophores at tt dyad in the syndiotactic sequence. The radically prepared PVCz exhibits higher emission intensity of the second excimer than that of PVCz prepared by cationic polymerization.

Effect of tacticity on excimer configuration was extensively explored using diastereomeric bichromophoric model compounds.[109,110] DeSchryver et al. revealed that racemic isomer of 2,4-bis(9-carbazolyl)pentane (tt, gg) showed excimer emission with maximum at 370 nm in isooctane while meso isomer (tg, gt) exhibited emission with a peak at 420 nm. Furthermore, lifetime of the former was shorter than that of the latter. These two excimer species exactly correspond to the second and the sandwich-type excimers observed in PVCz. Such model experiments, with the aid of compounds with well-defined configuration, are inevitable for the understanding of excimer in the polymer systems. In fact, the generally observed phenomena that polymers with isotactic rich consequence favor the sandwich-type excimer and those with syndiotactic rich consequence favor the second excimer are rationalized on the basis of such model experiments.

Excimer formation in the polymer is affected by flexibility of the polymer chain. This is reasonable requirement in view of the fact that overlap of the two chromophores must be achieved during the lifetime of the excited chromophore. The excimer formation in polymer is further affected by solvent in fluid solution. Rationale for this lies in the same line as the effect of the polymer chain flexibility. In good solvents the polymer chain tends to expand, while in poor solvents they take a rather shrunken form with entanglement. In view of local concentration of chromophores, polymer/poor solvent system is favored for excimer formation, and in fact this effect has been observed.[62,111,112] Excimer formation has been shown to be drastically enhanced in aqueous solutions. This was interpreted in terms of favored hydrophobic interaction between the excimer-forming pairs in aqueous solutions.[113,114]

Finally, it must be mentioned here that applicability of the simple Birks' kinetics (Scheme I) to polymer systems has been questioned on the basis of precise time-resolved measurements on excimer and monomer emission. A combination of reliable light source with extremely short pulse duration (e.g., a ps laser), photomultipliers with ultrafast response (e.g., a microchannel plate PM), and advanced data processing techniques has enabled us to evaluate critically the adequacy of a mathematical function to describe the decay behavior over the intensity range of 10^4.[115,116] In particular, data acquisition by means of a single photon counting method has been shown to be most suitable for the strict analyses of the decay behavior.[116] The Birks' scheme predicts that the time variation in monomer (i_M) and excimer (i_D) emission intensity can be described by the functions,[8]

$$i_M = A_1 \exp(-t/\tau_1) + A_2 \exp(-t/\tau_2) \tag{43}$$

$$i_D = A_3 [-\exp(-t/\tau_1) + \exp(-t/\tau_2)] \tag{44}$$

Based on these equations, the time profile of the excimer emission can be expressed by two exponential terms with the same but opposite sign of the preexponential factors. Furthermore, the fast decay component in the monomer emission (τ_1) should be the same as the rise component in the excimer emission, and both species should have a common slow decay component (τ_2). However, the more precise the analyses of the decay behavior, the more deviation from such an ideal scheme became apparent.[117]

Strict evaluation of the decay behavior of excimer and monomer emission in polymers by Phillips et al. revealed that in most polymers the simple Birks' scheme cannot be applied to excimer formation dynamics and further a generalized scheme which is applicable to excimer formation in a variety of polymers should be ruled out.[78,117-120] Soutar and Phillips emphasized that a scheme which describes the decay behavior most appropriately was varied from system to system.[121] One example of schemes is given here which was proposed by Phillips et al. to describe the excimer formation dynamics in polymers containing naphthalene moieties (vinylnaphthalene,[78,117,119,120] naphthyl methacrylate,[117] and acenaphthylene[118] in fluid solutions). They found that the time variation in monomer emission intensity was expressed more adequately by Equation 45 which consists of three exponential terms.

$$i_M = A_1 \exp(-t/\tau_1) + A_2 \exp(-t/\tau_2) + A_3 \exp(-t/\tau_3) \tag{45}$$

Furthermore, the time-resolved emission spectra showed only two spectrally distinct species. These observations led them to a proposal of the following scheme,

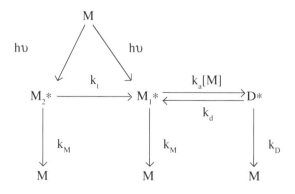

Scheme II

in which there are two monomeric excited states, M_1^* and M_2^*, but excimer formation is only possible through M_1^*, and M_1^* and M_2^* cannot be distinguished by spectral method, and their unimolecular decay processes are assumed identical. However, one-to-one correspondence of each lifetime component with individual species is the most difficult task in these analyses.

B. EXCIPLEX

Exciplex in polymer provides an additional information on micropolarity around the chromophores. As in the case of excimer, two chromophores (donor and acceptor) must encounter during the lifetime of the excited species (D^* or A^*). However, due to Coulombic nature of exciplex, exciplex formation is strongly affected by solvent polarity. Although there have been many studies reported so far on exciplex formation in low molecular weight systems, exciplex in polymer has been limited partially because sample preparation is somewhat difficult.

The most significant feature of exciplex in polymer may be summarized as follows: (1) exciplex formation is enhanced in polymer, (2) exciplex emission is concentration dependent even at such a low concentration as 10^{-5} M, while in relevant low molecular weight systems such concentration dependence is never observed, (3) the concentration dependence is further a function of molecular weight of the polymer and optimal molecular weight region is often observed, (4) weak EDA interaction in the ground state is visualized sometimes between D and A bound to a polymer chain by absorption spectra, and (5) solvation by the polymer chain is predominant; thus, exciplex emission can be observed even in highly polar solvents.

In the case of excimer in polymer, interpolymer association is negligibly small; thus, intrapolymer interaction including nonneighboring interaction is enough to be taken into accounts in dilute solutions. The striking difference between excimers and exciplexes in polymer system is exemplified by the samples shown in Figure 12.[122] Each pair (I and II, III and IV, and V and VI) possesses the identical main chain structure and a similar molecular weight. The polymers having *N,N*-dimethylanilino groups (II, IV, and VI) show exciplex emission, while those with phenyl groups (I, III, and V) form excimer. The intensity ratio of the exciplex (or excimer) emission (F_e) to the monomer emission (F_m) is shown in Figure 13 as a function of the concentration of the functional groups. It is clearly shown that in the exciplex-forming polymers the ratio F_e/F_m strongly depends on the concentration of the functional groups in the range below $\sim 5 \times 10^{-5}$ M. On the other hand, in the excimer-forming polymers F_e/F_m value is kept constant in the same concentration region. Because of the same structure in the main chain between the two types of the polymers, such effects as hydrogen bonding can be ruled out as the origin of the concentration dependence on F_e/F_m. The only possible explanation for this difference is the interpolymer EDA interaction in the ground state exerted by the exciplex-forming polymers. This interpretation is rationalized by the fact that, in the exciplex-forming polymers, remarkable broadening of the absorption spectra occurs in comparison with those of the excimer-forming polymers. The ground state association of the exciplex-forming polymers was found to be affected by molecular weight and solvents.[123] Since the ground state association of the polymer chains is enhanced by the "zipping" effect of the polymer chains, it is quite reasonable that the degree of association is sensitively influenced by the length of the polymer chains and regularity in the sequence of the chromophores. Thus, in a random copolymer whose structure is similar to IV but the spacer length in the main chain is randomly altered, the ground state association was found to be very small.

The most common feature observed in exciplexes is their strong dependence on solvent polarity. Usually, with increasing solvent polarity, the emission lifetime becomes shorter with simultaneous weakening of the emission intensity. Furthermore, a shift of the emission maximum to a longer wavelength is always accompanied by the increase in the solvent

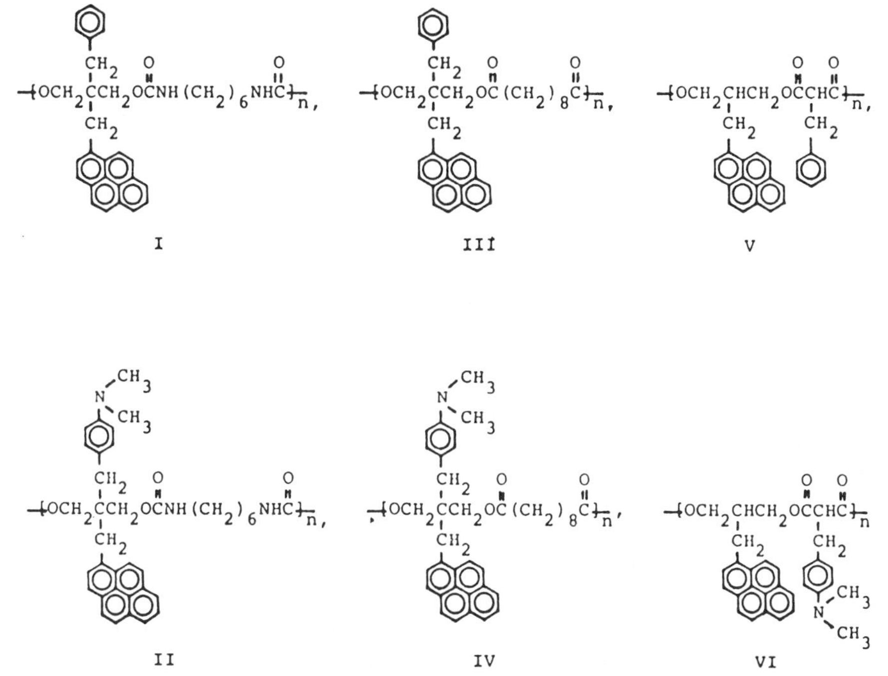

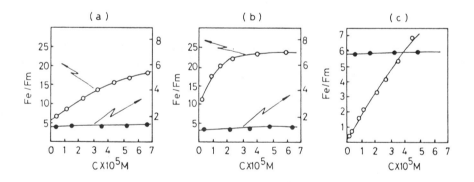

FIGURE 12. Structures of excimer-forming polymers (I, III, V) and exciplex-forming polymers (II, IV, VI). (From Tazuke, S. and Yuan, H. L., *J. Phys. Chem.*, 86, 1250, 1982. With permission.)

FIGURE 13. Concentration dependence of F_e/F_m in 1,2-dichloroethane. (From Tazuke, S. and Yuan, H. L., *J. Phys. Chem.*, 86, 1250, 1982. With permission.) (a) I(●) and II(○); (b) III(●) and IV(○); and (c) V(●) and VI(○).

polarity. However, when the exciplex-forming pairs are bound to polymer chains, the resulting exciplexes were found to be drastically shielded from the influence of the neat solvents by the polymer chains. In isolated systems, exciplex emission cannot be observed in such polar solvents as dimethylsulfoxide and water; however, in polymer systems the emission can be observed with sufficient intensity in those solvents.[124] An interesting result obtained by the Lippert-Mataga analysis of the exciplex in polymer was that apparent dipole moment was smaller than that of the isolated systems.[123] This was ascribed to hindered solvation by the polymer chain from neat solvent molecules.[123]

Effect of conformation of the polymer chain on exciplex formation was examined for PDPM in dimethylformamide (DMF) (Figure 14).[124] PDPM is very soluble in DMF. On adding water to PDPM/DMF, the solvent polarity increased, but at the same time the polymer

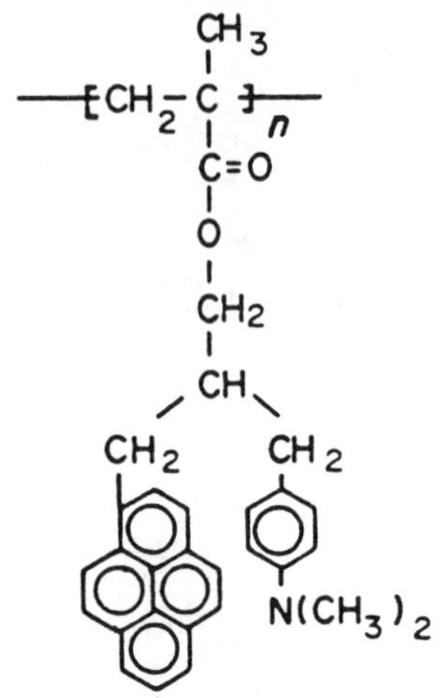

FIGURE 14.   Structure of poly[2-(4-*N,N*-dimethylaminobenzyl-2(1-pyr-
enylmethyl)ethyl methacrylate)] (PDPM). (From Tazuke, S., Iwaya, Y.,
and Hayashi, R., *Photochem. Photobiol.*, 35, 621, 1982. With permission.)

chain was shrunken since water is nonsolvent to PDPM. Contrary to the isolated system,
exciplex emission was enhanced with increasing amount of water. This is exactly the effect
of polymer conformation on exciplex in polymer system.

To close this section, an elegant application of the excimer and exciplex formation in
polymeric systems is described: cyclization dynamics of polymer chains.[125] Cyclization
properties of flexible polymer chains in fluid solutions have been a subject of extensive
research in relation to dynamic properties of the polymer chains. Only spectroscopic methods
on polymers whose both ends are covalently labeled with *appropriate* dye molecules allow
precise measurements on cyclization equilibrium constants and cyclization rate constants,
since for the precise evaluation labeled molecules must react on every encounter, that is,
the reaction must be diffusion-controlled. Two types of bimolecular photophysical processes
have been applied to the study of the cyclization kinetics of the polymers: (1) excimer
formation in polymers having pyrene (Py) moieties at both ends[126-128] and (2) exciplex
formation in polymers containing a Py moiety at one end and a dimethylanilino (DMA)
group at the other end.[129] They can probe a different time scale, depending on the nature
of the chromophores and the nature of the phenomena observed; exciplex formation between
Py and DMA at the ends of the polymer chain provides a time scale for the encounter of
the both ends ranging from 10^8 to 10^5 s^{-1}. Excimer formation between Py moieties attached
at the both ends allows determination of the cyclization rate down to 10^4 s^{-1}. For details,
the readers are referred to the review articles.[125,128]

VI. PHOTOINDUCED ELECTRON TRANSFER PROCESSES IN POLYMERS

A. HOMOGENEOUS SYSTEMS

In polymer solutions, the photoinduced electron transfer process itself is nothing different
from small molecular systems. Smaller diffusion constants and reduced dielectric constants
around polymer chain are disadvantageous for the electron transfer process. However, modification of the electron transfer process by means of introducing Coulombic repulsion is
very effective in polymers. Thus, a number of attempts have been reported for the effect of
polyelectrolytes on photoinduced electron transfer reactions.[130-135] Electron transfer sensitizers surrounded by ionically charged groups behave differently from the sensitizers in a

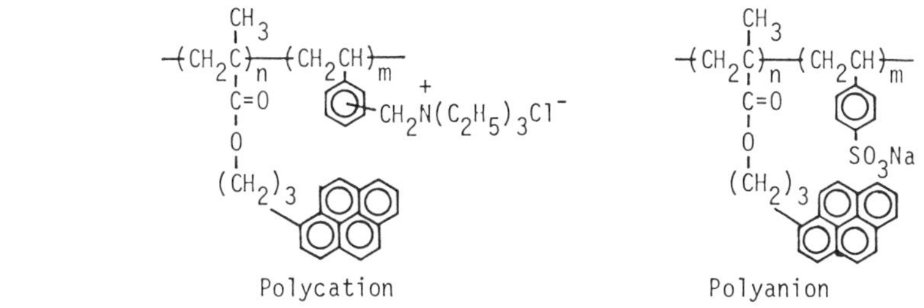

FIGURE 15. Polycation and polyanion bearing pyrenyl group. Polycation: n/m = 7/93 and polyanion: n/m = 10/90.

neutral environment. An earlier example is shown in Figure 15[132] in the photoreduction of methyl viologen (MV^{2+}) by the excited pyrenyl group in the presence of triethanolamine as a sacrificial reagent. The pyrenyl group bonded to polycation is a better sensitizer than relevant neutral or anionically charged sensitizers while the quenching constant (K_q) of MV^{2+} towards pyrene fluorescence is smaller than the case of polyanion bearing pyrenyl groups. This sequence is explainable as a result of repulsive Coulombic effect before electron transfer so that the approach of MV^{2+} to the cationically charged environment is retarded. After forward electron transfer, the electron transferring pair gives a high yield of MV^+. In the case of anionic polymers, the situation is reversed. While fluorescence quenching of pyrene by MV^{2+} is static and very efficient owing to ion pair formation of MV^{2+} with the anionic polymer, the quantum yield of MV^+ formation is negligibly small. The role of back electron transfer is the governing factor to decide the overall yield. The overall efficiency can be expressed by the product of the quenched fraction of the excited sensitizer and the efficiency of charge separation, the former being improved by increasing quencher concentration whereas the latter being decided by the nature of electron transferring pair.

The role of Coulombic interaction is essentially the same for small molecular systems as discussed in Section III for photoinduced electron transfer processes of benzophenone, phenothiazine, and ruthenium complexes.[25-28]

Precise analysis of electron transfer processes in polymer by means of fluorescence quenching is more difficult than in small molecular systems because (1) excited state decay profile is often expressed by a multi- or nonexponential function[135] so that unequivocal quenching constant cannot be determined and (2) in multichromophoric polymers, energy migration along polymer chain may occur prior to electron transfer even in extremely dilute solution and therefore the definition of diffusion constant is obscured. Therefore, the diffusion of excited state and excited species are not identical.

All the publications dealing with electron transfer in polyelectrolyte systems either did not treat properly the problem mentioned above or simply neglected the complexities.

B. SOLID-STATE AND SOLID-LIQUID INTERFACIAL SYSTEMS

Photoinduced electron transfer in solid polymer is relevant to photoconductors. The recent trend of organic photoconductive composition is the two-layered system, that is, carrier generation layer (CGL) and carrier transport layer (CTL) are stacked. CGL is composed of an electron accepting dye whereas CTL is made of electron donors either bonded to polymer or mixed in polymer matrix. While the rate of photoinduced electron transfer from CTL to CGL is not known, the produced hole in CTL is transferred by a hopping mechanism. This rate (hole drift mobility) has been measured in a number of systems. In comparison with organic single crystals, carrier mobility in relevant polymers is smaller by a factor of 10^{-6} or more. Furthermore, our recent measurement revealed that carrier mobility of photoconductive polymers bearing *trans*-1,2-bis(9-carbazolyl)cyclobutane is smaller than the chromophore molecules dispersed in polycarbonate matrices as shown in Figure 16.[136]

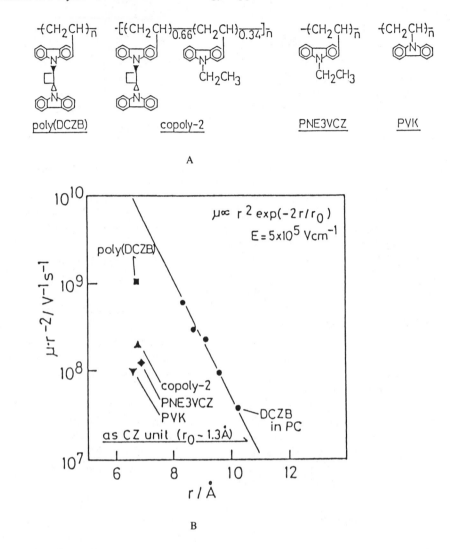

FIGURE 16. (A) Hole drift mobility polymers and DCZB dispersed in polycarbonate. (B) r: mean distance between hopping sites, E: applied potential (volt), and μ: hole drift mobility.

The polymer effect on carrier mobility is therefore negative in comparison with not only single crystals but also molecularly dispersed systems. The bulk carrier mobility of polymer seems to be somewhat better than the corresponding dispersed systems. However, this is attributed to the concentration effect. Provided that the chromophore concentration in a dispersed system could be increased as high as that in the polymer bonded system, carrier mobility is apparently better for the former.

Photoinduced interfacial electron transfer is another area of interest. Relevant to photocatalytic action of semiconductor particles, polymers bearing viologen groups have been investigated by several groups.[137-139] A series of polymers shown in Figure 17 illustrates excellent stabilizers of colloidal TiO_2 particles acting as protective colloid.

Owing to close contact of viologen groups with TiO_2, electron transfer from TiO_2 to polymer-bonded viologen is instantaneous and the quantum yield is much higher than the case of small molecular viologens in solution. The high rate and yield of electron transfer are explained by tight binding of the polymer to the particle surface brought about by electrostatic and hydrophobic interactions. However, the high local concentration of viologen

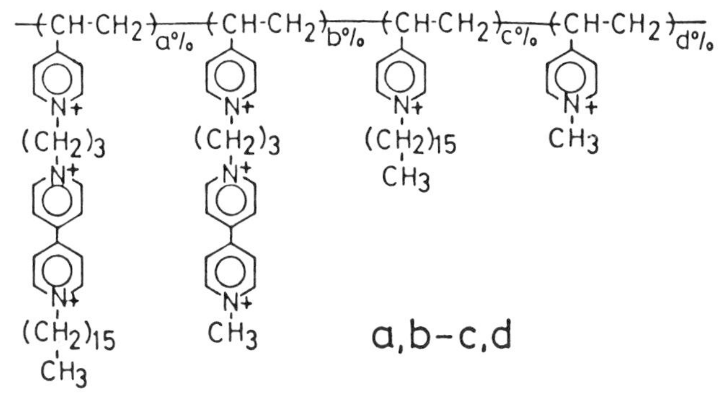

FIGURE 17.   Examples of viologen polymers. a, b-c, d: 6, 0-0,94; 6, 18-0, 76; 6, 60-0, 34; 0, 9-6, 85; and 0, 0-8, 92. (From Nakahira, T. and Graetzel, M., *J. Phys. Chem.*, 88, 4006, 1984. With permission.)

cation radical, in particular when viologen is hydrophobic, induces rapid dimerization so that the turnover number as electron relay is reduced.[137]

Earlier, Matsuo et al.[140] showed that photoinjected electrons on the viologen polymer are delocalized over the polymer chain owing to electron exchange between neighboring viologen units. This characteristic will be favorable to reduce the probability of back electron transfer immediately after primary electron transfer and, consequently, the viologen polymers may be suitable as electron relays. Thus, the combination of zwitterionic viologen with viologen polymer acts as a more efficient electron relay than the single use of the components for photoinduced electron transfer between tris(2,2'-bypyridine)ruthenium(II) or tris(1,10-phenanthroline)ruthenium(II) and platinum colloid.[138] In the absence of Pt colloid, the cation radical on cationically charged viologen polymer is long living, indicating a slow back electron transfer to the Ru(III) species.

These polymer effects are attributable to local high concentration and electrostatic charge. Photoinduced electron transfer from leuko Crystal Violet to pyrene at polymer-solution interface depends very much on surface charge of polymer, surface density, and depth of the sensitizer bonded on polymer surface.[29,141,142] Poly(ethylene-*g*-acrylic acid) was esterified by 1-hydroxymethylpyrene either in tetrahydrofuran (film 1) or in acetonitrile (film 2). Because of lower affinity of the polymer to acetonitrile, film 2 has a more condensed but thinner pyrene-containing surface layer than film 1. The depth of possible diffusion of cationically charged LCV^+ into polymer film decreases with increasing the degree of esterification and at the same time the penetrating depth of the incident light decreases with the local concentration of pyrene group as shown in Figure 18. The balance between the number of available LCV^+ and that of excited pyrenyl groups determines the efficiency of electron transfer sensitization. For film 2, the location of photoreaction is limited to thin surface so that the efficiency is independent of the degree of esterification. For film 2, there seems to be a mismatch between the depth of photoabsorption and diffusion of LCV^+ showing a complex dependence of the efficiency on the degree of esterification.

Photoreduction of LCV^+ by charged polymers containing pyrenyl groups were investigated both in homogeneous solution and at solution-polymer film interface. The effects of charge on polymer are much stronger in heterogeneous systems. The reactivity is in the order of polycation > neutral polymer > polyanion. High charge density on the polymer surface and enhanced polymer-solvent affinity account for the high reactivity of the polycation.

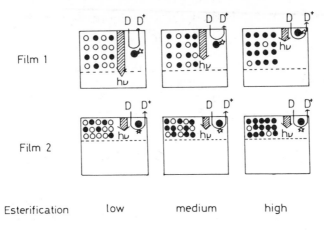

FIGURE 18.   Sketches of films 1 and 2. ○: acrylic acid site, ●: pyrenylated site, ☆: excited state, and D: LCV.

REFERENCES

1. **Ichimura, K., Ed.,** *Synthesis and Application of Photofunctional Polymers,* CMC, Tokyo, 1984.
2. **Loutfy, R. O.,** Organic Photoconductive Materials, invited lecture at the 11th IUPAC Symp. Photochemistry, Lisbon, 1986, 17.
3. **Watanabe, T., Miyata, S., Miyazaki, T., and Yoshinaga, K.,** Second harmonic generation by means of controlling higher order structure of *p*-nitroaniline/polyethylene oxide system, *Polym. Prepr., Jpn.,* 36, 2522, 1987.
4. **Guillet, J. E.,** *Polymer Photophysics and Photochemistry,* Cambridge University Press, London, 1985.
5. **Cundall, R. B. and Dale, R. E., Eds.,** *Time-Resolved Fluorescence Spectroscopy in Biochemistry and Biology,* (NATO-ASI series 69), D. Reidel, Dordrecht, Netherlands, 1981.
6. **Winnik, M. A., Ed.,** *Photophysical and Photochemical Tools in Polymer Science,* (NATO-ASI Series 182), D. Reidel, Dordrecht, Netherlands, 1986.
7. **Mataga, N. and Kubota, T.,** *Molecular Interactions and Electronic Spectra,* Marcel Dekker, New York, 1970.
8. **Birks, J. B.,** *Photophysics of Aromatic Molecules,* Wiley-Interscience, New York, 1970.
9. **Turro, N. J.,** *Modern Molecular Photochemistry,* Benjamin Cummings, Menlo Park, CA, 1978.
10. **Tazuke, S.,** Polymer bonded excited complexes — a powerful approach to study polymer effects, *Makromol. Chem. Suppl.,* 14, 145, 1985.
11. **Tazuke, S. and Yuan, H. L.,** Importance of structure regularity in polymer association, *Macromolecules,* 17, 1878, 1984.
12. **Phillips, D. and Roberts, A. J., Eds.,** *Photophysics of Synthetic Polymers,* Science Review, Northwood, 1982.
13. **Hayashi, R., Tazuke, R., and Frank, C. W.,** Twisted intramolecular charge transfer phenomenon as a fluorescence probe of microenvironment. Effect of polymer concentration on local viscosity and microscopic polarity around a polymer chain of Poly(methyl methacrylate), *Macromolecules,* 20, 983, 1987.
14. **Mita, I. and Horie, K.,** Diffusion controlled reactions in polymer systems, *J. Macromol. Sci., Rev. Macromol. Chem. Phys.,* C27, 91, 1987.
15. **Sisido, M.,** Electronic interactions in one-dimensional chromophoric assemblies, *Macromol. Chem. Suppl.,* 14, 131, 1985.
16. **Mulliken, R. S. and Person, W. B.,** *Molecular Complexes,* Wiley-Interscience, New York, 1969.
17. **Barnes, D. G. and Rhodes, W.,** Generalized susceptibility theory. II. Optical absorption properties of helical polypeptides, *J. Chem. Phys.,* 48, 817, 1968.
18. **Birks, J. B.,** *Photophysics of Aromatic Molecules,* Wiley-Interscience, New York, 1970, chap. 4 and 11.
19. **Marcus, R. A.,** On the theory of oxidation-reduction involving electron transfer. I, *J. Chem. Phys.,* 24, 966, 1956; On the theory of oxidation-reduction reactions involving electron transfer. II, *J. Chem. Phys.,* 26, 867, 1957; Application to data on the rates of isotropic exchange reactions, *J. Chem. Phys.,* 43, 679, 1965.

20. **Cannon, R. D.,** *Electron Transfer Reactions,* Butterworths, London, 1980, 219.
21. **Miller, J. R., Calcaterra, L. T., and Closs, G. L.,** Intramolecular long-distance electron transfer in radical anions. The effects of free energy and solvent on the reaction rates, *J. Am. Chem. Soc.,* 106, 3047, 1984.
22. **Verhoeven, J. W.,** Through-bond charge transfer interaction and photoinduced charge separation, *Pure Appl. Chem.,* 58, 1285, 1986.
23. **Kakitani, T. and Mataga, N.,** Photoinduced electron transfer in polar solutions. I. New aspects of the role of the solvent mode in electron-transfer processes in charge-separation reactions, *Chem. Phys.,* 93, 381, 1985.
24. **Kakitani, T. and Mataga, N.,** Photoinduced electron transfer in polar solvents. II. New aspects of the role of the solvent mode in electron-transfer processes in charge-recombination reactions and comparison with charge-separation reactions, *J. Phys. Chem.,* 89, 4752, 1985.
25. **Tazuke, S., Kawasaki, Y., Kitamura, N., and Inoue, T.,** Unexpected high photooxidizing efficiency of benzophenone derivatives having tetraalkylammonium substituents, *Chem. Lett.,* 251, 1980.
26. **Tazuke, S., Kitamura, N., and Kim, H. B.,** Photoinduced looping electron transfer. What occurs between electron transfer and charge separation? *Supramolecular Photochemistry,* (NATO ASI series C: 214), Balzani, V., Ed., D. Reidel, Dordrecht, Netherlands, 1988, 87.
27. **Kim, H. B., Kitamura, N., Kawanishi, Y., and Tazuke, S.,** Bell-shaped temperature dependence in quenching of excited Ru(bpy)$_3^{2+}$ by organic acceptor, *J. Am. Chem. Soc.,* 109, 2509, 1987.
28. **Kitamura, N., Obata, R., Kim, H. B., and Tazuke, S.,** Back electron transfer to the excited state in photoinduced electron transfer reactions of Ruthenium(II) complexes, *J. Phys. Chem.,* 91, 2033, 1987.
29. **Tazuke, S., Takasaki, R., Iwaya, Y., and Suzuki, Y.,** Polymer bonded electron transfer sensitizers, *Materials for Microlithography,* (ACS Symp. Ser.), American Chemical Society, Washington, D.C., 266, 187, 1984.
30. **Gaviola, E. and Pringsheim, P.,** The influence of concentration on the polarization of the fluorescence of dye solutions, *Z. Phys.,* 24, 24, 1924.
31. **Weigert, F. and Kapper, G.,** Polarized fluorescence in dye solutions, *Z. Phys.,* 25, 99, 1924.
32. **Levshin, W. L.,** The polarized fluorescence light of dye solutions, *Z. Phys.,* 26, 274, 1924.
33. **Forster, Th.,** Zwischenmolekulare Energiewanderung und Fluoreszenz, *Ann. Phys. (Leipzig),* 2, 55, 1948.
34. **Dexter, D. L.,** A theory of sensitized luminescence in solids, *J. Chem. Phys.,* 21, 836, 1953.
35. **Robinson, G. W. and Frosch, R. P.,** Electron excitation transfer and relaxation, *J. Chem. Phys.,* 38, 1187, 1963.
36. **Berlman, I. B.,** *Energy Transfer Parameters of Aromatic Compounds,* Academic Press, New York, 1973.
37. **Stryer, L. and Haugland, R. P.,** Energy transfer: a spectroscopic ruler, *Proc. Natl. Acad. Sci. U.S.A.,* 58, 719, 1967.
38. **Ikeda, T., Lee, B., Kurihara, S., Tazuke, S., Ho, S., and Yamamoto, M.,** Time-resolved observation of excitation hopping between two identical chromophores attached to both ends of alkanes, *J. Am. Chem. Soc.,* 110, 8299, 1988.
39. **Gochanour, C. R., Andersen, H. C., and Fayer, M. D.,** Electronic excited state transport in solution, *J. Chem. Phys.,* 70, 4254, 1970.
40. **Loring, R. F., Andersen, H. C., and Fayer, M. D.,** Electronic excited state transport and trapping in solution, *J. Chem. Phys.,* 76, 2015, 1982.
41. **Loring, R. F., Andersen, H. C., and Fayer, M. D.,** Hopping transport on a randomly substituted lattice for long range and nearest neighbor interactions, *J. Chem. Phys.,* 80, 5731, 1984.
42. **Baumann, J. and Fayer, M. D.,** Excitation transfer in disordered two-dimensional and anisotropic three-dimensional systems: effects of spatial geometry on time-resolved observables, *J. Chem. Phys.,* 85, 4087, 1986.
43. **Peterson, K. A. and Fayer, M. D.,** Electronic excitation transport on isolated, flexible polymer chains in the amorphous solid state randomly tagged or end tagged with chromophores, *J. Chem. Phys.,* 85, 4702, 1986.
44. **Pearlstein, R. M.,** Impurity quenching of molecular excitons. I. Kinetic comparison of Forster-Dexter and slowly quenched Frenkel excitons in linear chains, *J. Chem. Phys.,* 56, 2431, 1972.
45. **Lakatos-Lindenberg, K., Hemenger, R. P., and Pearlstein, R. M.,** Solutions of master equations and related random walks on quenched linear chains, *J. Chem. Phys.,* 56, 4852, 1972.
46. **Hemenger, R. P. and Pearlstein, R. M.,** Time-dependent concentration depolarization of fluorescence, *J. Chem. Phys.,* 59, 4064, 1973.
47. **Hemenger, R. P., Lakatos-Lindenberg, K., and Pearlstein, R. M.,** Impurity quenching of molecular excitons. III. Partially coherent excitons in linear chains, *J. Chem. Phys.,* 60, 3271, 1974.
48. **Haan, S. W. and Zwanzig, R.,** Förster migration of electronic excitation between randomly distributed molecules, *J. Chem. Phys.,* 68, 1879, 1978.
49. **Godzik, K. and Jortner, J.,** Electronic energy transport in substitutionally disordered molecular crystals, *J. Chem. Phys.,* 72, 4471, 1980.

50. **Blumen, A., Klafter, J., and Silbey, R.,** Theoretical studies of energy transfer in disordered condensed media, *J. Chem. Phys.,* 72, 5320, 1980.
51. **Fredrickson, G. H., Andersen, H. C., and Frank, C. W.,** Electronic excited-state transport on isolated polymer chains, *Macromolecules,* 16, 1456, 1983.
52. **Fredrickson, G. H., Andersen, H. C., and Frank, C. W.,** Electronic excited-state transport and trapping on polymer chains, *Macromolecules,* 17, 54, 1984.
53. **Fredrickson, G. H., Andersen, H. C., and Frank, C. W.,** Electronic excitation transport as a probe of chain flexibility, *Macromolecules,* 17, 1496, 1984.
54. **Frank, C. W., Fredrickson, G. H., and Andersen, H. C.,** Electronic excitation transport as a tool for the study of polymer chain statics, in *Photophysical and Photochemical Tools in Polymer Science,* Winnik, M. A., Ed., (NATO-ASI Series 182), D. Reidel, Dordrecht, Netherlands, 1986.
55. **Bai, F., Chang, C. H., and Webber, S. E.,** Fluorescence and energy migration in 2-vinylnaphthalene alternating and random copolymers with methyl methacrylate and methacrylic acid, *Macromolecules,* 19, 588, 1986.
56. **Itagaki, H., Horie, K., Mita, I., Washio, M., Tagawa, S., and Tabata, Y.,** Intramolecular excimer formation of oligostyrenes from dimer to tridecamer: the measurements of rate constants for excimer formation, singlet energy migration, and relaxation of internal rotation, *J. Chem. Phys.,* 79, 3996, 1983.
57. **Klopffer, W.,** Energy transfer in films of polymers with aromatic side-groups, *Ann. N.Y. Acad. Sci.,* 366, 373, 1981.
58. **Ishii, T., Handa, T., and Matsunaga, S.,** A study of mechanism of excimer formation through fluorescence quenching for isotactic and atactic polystyrene and poly(*p*-methylstyrene) in solution, *Makromol. Chem.,* 178, 2351, 1977.
59. **Ishii, T., Handa, T., and Matsunaga, S.,** Effect of molecular weight on excimer formation of polystyrene in solution, *Macromolecules,* 11, 40, 1978.
60. **Ishii, T., Handa, T., and Matsunaga, S.,** Steric effects in excimer formation in isotactic and atactic polystyrene and poly(α-methylstyrene) in solution, *J. Polym. Sci. Polym. Phys. Ed.,* 17, 811, 1979.
61. **Kyle, B. R. M. and Kilp, T.,** Intramolecular excimer formation and energy migration in head-to-head polystyrene, *Polymer,* 25, 989, 1984.
62. **Torkelson, J. M., Lipsky, S., and Tirrell, M.,** Polystyrene fluorescence: effect of molecular weight in various solvents, *Macromolecules,* 14, 1601, 1981.
63. **Lindsell, W. E., Robertson, F. C., and Soutar, I.,** Intramolecular excimer formation in macromolecules. V. Head-to-head polystyrene, block copolymers of styrene and butadiene, and regular condensation polymers of α-methylstyrene, *Eur. Polym. J.,* 17, 203, 1981.
64. **Hirayama, F.,** Intramolecular excimer formation. I. Diphenyl and triphenyl alkanes, *J. Chem. Phys.,* 42, 3162, 1965.
65. **Fitzgibbon, P. D. and Frank, C. W.,** Energy migration in the aromatic vinyl polymers. I. A one-dimensional random walk model, *Macromolecules,* 15, 733, 1982.
66. **Gelles, R. and Frank, C. W.,** Energy migration in the aromatic vinyl polymers. II. Miscible blends of polystyrene with poly(vinyl methyl ether), *Macromolecules,* 15, 741, 1982.
67. **Gelles, R. and Frank, C. W.,** Energy migration in the aromatic vinyl polymers. III. Three-dimensional migration in polystyrene/poly(vinyl methyl ether), *Macromolecules,* 15, 747, 1982.
68. **Gelles, R. and Frank, C. W.,** Effect of molecular weight on polymer blend phase separation kinetics, *Macromolecules,* 16, 1448, 1983.
69. **MacCallum, J. R. and Rudkin, L.,** Non-radiative energy transfer in polymers, *Nature,* 266, 338, 1977.
70. **MacCallum, J. R.,** Energy migration in poly(vinyl arenes), *Eur. Polym. J.,* 17, 209, 1981.
71. **MacCallum, J. R. and Rudkin, A. L.,** Quenching of fluorescence of solutions of polystyrene and copolymers with methyl methacrylate, *Eur. Polym. J.,* 17, 953, 1981.
72. **MacCallum, J. R.,** Polarization of excimer emission in solution, *Polymer,* 23, 175, 1982.
73. **Reid, R. F. and Soutar, I.,** Energy migration and excimer formation in copolymers of 1-vinylnaphthalene and methyl methacrylate, *J. Polym. Sci. Polym. Lett.,* 15, 153, 1977.
74. **Kettle, G. J. and Soutar, I.,** Fluorescence depolarization studies of micro-Brownian relaxation of poly(methyl methacrylate) in solution, *Eur. Polym. J.,* 14, 895, 1978.
75. **Anderson, R. A., Reid, R. F., and Soutar, I.,** Intramolecular excimer formation in macromolecules. III. Energy migration and excimer formation in copolymers of vinylnaphthalene and methyl acrylate, *Eur. Polym. J.,* 15, 925, 1979.
76. **Anderson, R. A., Reid, R. F., and Soutar, I.,** Intramolecular excimer formation in macromolecules. IV. Energy migration and excimer formation in copolymers of acenaphthylene and methyl acrylate, *Eur. Polym. J.,* 16, 945, 1980.
77. **Reid, R. F. and Soutar, I.,** Intramolecular excimer formation in macromolecules. II. Energy migration and excimer formation in acenaphthylene-methyl methacrylate copolymers, *J. Polym. Sci. Polym. Phys. Ed.,* 18, 457, 1980.

78. **Philips, D., Roberts, A. J., and Soutar, I.,** Transient decay studies of photophysical processes in aromatic polymers. V. Temperature dependence of excimer formation and dissociation in copolymers of 1-vinyl-naphthalene and methyl methacrylate, *J. Polym. Sci. Polym. Phys. Ed.,* 20, 411, 1982.

79. **Nakahira, T., Ishizuka, S., Iwabuchi, S., and Kojima, K.,** Photoprocesses in poly(2-naphthylalkyl methacrylate). II. Steric effect on singlet energy migration and excimer formation, *Makromol. Chem. Rapid Commun.,* 1, 437, 1980.

80. **Nakahira, T., Sakuma, T., Iwabuchi, S., and Kojima, K.,** Singlet energy migration and excimer formation in polymers of 2-*tert*-butyl-6-vinylnaphthalene, *J. Polym. Sci. Polym. Phys. Ed.,* 20, 1863, 1982.

81. **Nakahira, T., Sasaoka, T., Iwabuchi, S., and Kojima, K.,** 2-*tert*-Butyl-6-vinylnaphthalene polymers. II. Steric effects of *tert*-butyl groups on intramolecular energy transfer from naphthalene to anthracene chromophore, *Makromol. Chem.,* 183, 1239, 1982.

82. **Holden, D. A. and Guillet, J. E.,** Singlet electronic energy transfer in polymers containing naphthalene and anthracene chromophores, *Macromolecules,* 13, 289, 1980.

83. **Ng, D. and Guillet, J. E.,** Studies of the antenna effect in polymer molecules. I. Singlet electronic energy transfer in poly[(9-phenanthryl)methyl methacrylate] and its copolymers, *Macromolecules,* 15, 724, 1982.

84. **Ng, D. and Guillet, J. E.,** Studies of the antenna effect in polymer molecules. II. Singlet electronic energy transfer in phenanthrene-containing polymers by transient fluorescence methods, *Macromolecules,* 15, 728, 1982.

85. **Holden, D. A. and Guillet, J. E.,** Studies of the antenna effect in polymer molecules. III. Role of singlet electronic energy migration in naphthalene polymer photophysics, *Macromolecules,* 15, 1475, 1982.

86. **Holden, D. A., Shephard, S. E., and Guillet, J. E.,** Studies of the antenna effect in polymer molecules. IV. Energy migration and transfer in 2-naphthyl methacrylate-aryl vinyl ketone copolymers, *Macromolecules,* 15, 1481, 1982.

87. **Holden, D. A., Ng, D., and Guillet, J. E.,** Studies of the antenna effect in polymer molecules. V. Determination of the extent of singlet energy migration in a phenanthrene-containing polymer by fluorescence quenching, *Br. Polym. J.,* 159, 1982.

88. **Ng, D., Yoshiki, K., and Guillet, J. E.,** Studies of the antenna effect in polymer molecules. VI. Singlet electronic energy transfer and migration in poly(2-vinylnaphthalene-co-9-anthrylmethyl methacrylate), *Macromolecules,* 16, 568, 1983.

89. **Holden, D. A., Ren, X.-X., and Guillet, J. E.,** Studies of the antenna effect in polymer molecules. VII. Singlet and triplet energy migration and transfer in 2-vinylnaphthalene-phenyl vinyl ketone copolymers, *Macromolecules,* 17, 1500, 1984.

90. **Ren, X.-X. and Guillet, J. E.,** Studies of the antenna effect in polymer molecules. IX. Energy transfer, migration, and photoreactivity of copolymers of 1-naphthylmethyl methacrylate and [2-(9,10-anthraqui-nonyl)]methyl methacrylate, *Macromolecules,* 18, 2012, 1985.

91. **Guillet, J. E. and Rendall, W. A.,** Studies of the antenna effect in polymer molecules. VIII. Photophysics of water-soluble copolymers of 1-naphthylmethyl methacrylate and acrylic acid, *Macromolecules,* 19, 224, 1986.

92. **Tazuke, S., Tomono, H., Kitamura, N., and Inoue, T.,** Unexpected high photooxidizing efficiency of benzophenone derivatives having tetraalkylammonium substituents, *Chem. Lett.,* 251, 1980.

93. **Hargreaves, J. S. and Webber, S. E.,** Singlet energy migration of anthracene polymers in polystyrene, *Macromolecules,* 16, 1017, 1983.

94. **Hargreaves, J. S. and Webber, S. E.,** Photophysics of anthracene polymers: fluorescence, singlet energy migration, and photodegradation, *Macromolecules,* 17, 235, 1984.

95. **Hargreaves, J. S. and Webber, S. E.,** Photophysics of diphenylanthracene polymers: fluorescence and singlet energy migration, *Macromolecules,* 17, 1741, 1984.

96. **Hargreaves, J. S. and Webber, S. E.,** Water-soluble photon-harvesting polymers: intracoil energy transfer in anthryl- and fluorescein-tagged poly(vinylpyrrolidinone), *Macromolecules,* 18, 734, 1985.

97. **Bai, F., Cheng, C.-H., and Webber, S. E.,** Photon-harvesting polymers: singlet energy transfer in anthracene-loaded alternating and random copolymers of 2-vinylnaphthalene and methacrylic acid, *Macromolecules,* 19, 2484, 1986.

98. **Tazuke, S., Ooki, H., and Sato, K.,** Inter- and intramolecular interactions of polymers as studied by fluorescence spectroscopy. X. Excimer formation by polyesters bearing pyrenylmethyl groups and their dimer model compounds, *Macromolecules,* 15, 400, 1982.

99. **Itagaki, H., Obutaka, N., Okamoto, A., Horie, K., and Mita, I.,** Kinetic studies on the formation of two intramolecular excimers in substituted dinaphthylpropanes, *J. Am. Chem. Soc.,* 104, 4469, 1982.

100. **Todesco, R., Gelan, J., Martens, H., Put, J., and DeSchryver, F. C.,** A kinetic scheme for intramolecular excimer formation in bis(α-naphtylmethyl)ether, involving different starting conformations, *J. Am. Chem. Soc.,* 103, 7304, 1981.

101. **Nakahira, T., Ishizuka, S., Iwabuchi, S., and Kojima, K.,** "Second excimer" luminescence from polymers with sterically hindered naphthalene chromophores, *Macromolecules,* 15, 1217, 1982.

102. **Hayashi, T., Suzuki, T., Mataga, N., Sakata, Y., and Misumi, S.,** Solvent-induced polarization phenomena in the excited state of composite systems with identical halves, *J. Phys. Chem.,* 81, 420, 1977.

103. **Itoh, M., Fuke, K., and Kobayashi, S.,** Direct observation of intramolecular anthracene excimer in 1,3-dianthrylpropane, *J. Chem. Phys.,* 72, 1417, 1980.

104. **Johnson, G. E.,** Emission properties of vinylcarbazole polymers, *J. Chem. Phys.,* 62, 4697, 1975.

105. **Klopffer, W.,** Transfer of electronic excitation energy in polyvinyl carbazole, *J. Chem. Phys.,* 50, 2337, 1969.

106. **Evers, F., Kobs, K., Memming, R., and Tirrell, D. R.,** Intramolecular excimer emission of PVK and rac- and meso-2,4-di-N-carbazolylpentane. Model substances for its syndiotactic and isotactic dyads, *J. Am. Chem. Soc.,* 105, 5988, 1983.

107. **Peter, G., Bassler, H., Schrof, W., and Port, H.,** Picosecond study of singlet exciton dynamics in polyvinylcarbazole (PVK) in the temperature range 5-300 K, *Chem. Phys.,* 94, 445, 1985.

108. **Itaya, A., Sakai, H., and Masuhara, H.,** Excimer dynamics of poly(N-vinylcarbazole) films revealed by time-correlated single photon counting measurements, *Chem. Phys. Lett.,* 138, 231, 1987.

109. **DeSchryver, F. C., Vandendriessche, Toppet, S., Demeyer, K., and Boens, N.,** Fluorescence of the diastereoisomers of 2,4-di(N-carbazolyl)pentane and the two excimers observed in poly(N-vinylcarbazole), *Macromolecules,* 15, 406, 1982.

110. **DeSchryver, F. C., Demeyer, K., and Toppet, S.,** Intramolecular photocyclomerization and excimer emission of 1,1'-di(1-naphthyl)diethyl ethers: model systems of poly(1-vinylnaphthalene), *Macromolecules,* 16, 89, 1983.

111. **Bokobza, L., Jasse, B., and Monnerie, L.,** Molecular dynamics of polystyrene model molecules. Solvent effects on intramolecular excimer formation in 2,4-diphenylpentanes, *Eur. Polym. J.,* 16, 715, 1980.

112. **Cuniberti, C. and Peico, A.,** Intramolecular excimer formation in polymers. Pyrene labeled poly(vinyl acetate), *Eur. Polym. J.,* 16, 887, 1980.

113. **Cheng, S.-T., Winnik, M. A., and Redpath, A. E. C.,** Cyclization dynamics of polymers. V. The effects of solvent on end-to-end cyclization of poly(ethylene oxide) probed by intramolecular pyrene excimer formation, *Makromol. Chem.,* 183, 1815, 1982.

114. **Suzuki, Y. and Tazuke, S.,** Functional polyionenes. II. Absorption and fluorescence properties of a polyionene and relevant model compounds bearing (9-anthryl)methyl groups, *Macromolecules,* 14, 1742, 1981.

115. **Demas, J. N.,** *Excited State Lifetime Measurements,* Academic Press, New York, 1983.

116. **O'Connor, D. V. and Phillips, D.,** *Time-Correlated Single Photon Counting,* Academic Press, New York, 1984.

117. **Phillips, D., Roberts, A. J., and Soutar, I.,** Transient decay studies of photophysical processes in aromatic polymers: intramolecular excimer formation in homopolymers of 1-vinylnaphthalene, 2-vinylnaphthalene and 1-naphtyl methacrylate, *Polymer,* 22, 427, 1981.

118. **Phillips, D., Roberts, A. J., and Soutar, I.,** A time-resolved fluorescence spectroscopic study of excimer formation in polyacenaphthylene and an alternating acenaphthylene/maleic anhydride copolymer, *J. Polym. Sci. Polym. Lett. Ed.,* 18, 123, 1980.

119. **Phillips, D., Roberts, A. J., and Soutar, I.,** Transient decay studies of photophysical processes in aromatic polymers. I. Multiexponential fluorescence decays in copolymers of 1-vinylnaphthalene and methyl methacrylate, *J. Polym. Sci. Polym. Phys. Ed.,* 18, 2401, 1980.

120. **Phillips, D., Roberts, A. J., and Soutar, I.,** Transient decay studies of photophysical processes in aromatic polymers. IV. Intramolecular excimer formation in homopolymers of 1-vinylnaphthalene, 2-vinylnaphthalene, and 1-naphtyl methacrylate, *Polymer,* 22, 427, 1981.

121. **Soutar, I. and Phillips, D.,** Singlet energy migration, trapping, and excimer formation in polymers, in *Photophysical and Photochemical Tools in Polymer Science,* (NATO-ASI Series 182), Winnik, M. A., Ed., D. Reidel, Dordrecht, Netherlands, 1986, 97.

122. **Tazuke, S. and Yuan, H. L.,** Comparison between excimer- and exciplex-forming polymers having an identical main chain structure, *J. Phys. Chem.,* 86, 1250, 1982.

123. **Tazuke, S., Sato, K., and Banba, F.,** Inter- and intramolecular interactions of polymers as studied by fluorescence spectroscopy. VI. Exciplex of poly(oxy-2-(9-anthryl)methylpropylene-oxy-(4-N,N-dimethylaminobenzyl)malony 1), *Macromolecules,* 10, 1224, 1977.

124. **Tazuke, S., Iwaya, Y., and Hayashi, R.,** Inter- and intramolecular interactions of polymers as studied by fluorescence spectroscopy. XII. Exciplex by pyrene-N,N-dimethylaniline pair bonded to polymers in aqueous media, *Photochem. Photobiol.,* 35, 621, 1982.

125. **Winnik, M. A.,** The cyclization of polymer chains in solution, in *Photophysical and Photochemical Tools in Polymer Science,* Winnik, M. A., Ed., (NATO-ASI Series 182), D. Reidel, Dordrecht, Netherlands, 1986, 293.

126. **Redpath, A. E. C. and Winnik, M. A.,** The dynamics of polymer cyclization. III. Excluded volume effects in the end-to-end cyclization of polystyrene probed by intramolecular pyrene excimer formation, *Ann. N.Y. Acad. Sci.,* 366, 75, 1981.

127. **Winnik, M. A., Redpath, A. E. C., Paton, K., and Dankelka, J.,** Cyclization dynamics of polymers. X. Synthesis, fractionation, and fluorescent spectroscopy of pyrene end-capped polystyrenes, *Polymer,* 25, 91, 1984.

128. **Winnik, M. A.,** End-to-end cyclization of polymer chains, *Acc. Chem. Res.,* 18, 73, 1985.

129. **Winnik, M. A., Sinclair, A. M., and Beinert, G.,** Cyclization dynamics of polymers. XVII. Probe effects on detection of polymer end-to-end cyclization, *Macromolecules,* 18, 1517, 1985.

130. **Sassoon, R. E. and Rabani, J.,** A novel effect of a polyelectrolyte on a photochemically induced electron transfer process involving a zwitterion, *Isr. J. Chem.,* 22, 138, 1982.

131. **Nosaka, Y., Kuwabara, A., and Miyama, H.,** Effect of ionic polymer environment on the photoinduced electron transfer from zinc porphyrin to viologen, *J. Phys. Chem.,* 90, 1465, 1986.

132. **Tazuke, S.,** Energy and electron transfer in micellar and polymeric systems. Kinetic mimicry of chlorophyll, *Contemp. Top. Polym. Sci.,* 4, 871, 1984.

133. **Morishima, Y. and Nozakura, S.,** Photoredox reaction of functionalized polyelectrolytes, *J. Polym. Sci., Polym. Symp.,* 74, 1, 1986.

134. **Guillet, J. E., Takahashi, T., McIntosh, A. R., and Bolton, J. E.,** A new polymeric model for the active site in artificial photosynthesis, *Macromolecules,* 18, 1788, 1985.

135. **Morishima, Y., Kobayashi, T., Furui, S., and Nozakura, S.,** Intramolecular compartmentalization of photoredox centers in functionalized amphiphilic polyelectrolytes: a model for collisionless electron transfer systems, *Macromolecules,* 20, 1707, 1987.

136. **Sasakawa, T., Ikeda, T., and Tazuke, S.,** Effect of polymer matrices on hole transport of 1.2-*trans*-bis(9H-carbazole-9-yl)-cyclobutane dissolved in a polymer binder, *J. Appl. Phys.,* 65, 2750, 1989.

137. **Nakahira, T. and Graetzel, M.,** Fast electron storage with colloidal semiconductors functionalized with polymeric viologen, *J. Phys. Chem.,* 88, 4006, 1984.

138. **Ohsako, T., Sakamoto, T., and Matsuo, T.,** Photoinduced charge separation, electron storage and electron transport to platinum colloid by the use of viologen polymer, *Chem. Lett.,* 1675, 1983.

139. **Sakamoto, T., Ohsako, T., Matsuo, T., Mulac, W. A., and Meisel, D.,** Pulse radiolysis studies of electron transfer between polymer and zwitterionic viologen radicals, *Chem. Lett.,* 1893, 1984.

140. **Nishijima, T., Nagamura, T., and Matsuo, T.,** Hydrogen generation by visible light irradiation by ruthenium complexes and colloidal platinum stabilized by viologen polymers in aqueous solutions, *J. Polym. Sci., Polym. Lett. Ed.,* 19, 65, 1981.

141. **Tazuke, S. and Takasaki, R.,** Photoinduced electron transfer at the polymer-solution interface. I. Surface property-reactivity relation, *J. Polym. Sci., Polym. Chem. Ed.,* 21, 1517, 1983.

142. **Tazuke, S. and Takasaki, R.,** Photoinduced electron transfer at the polymer-solution interface. II. Effects of ionic environment, *J. Polym. Sci., Polym. Chem. Ed.,* 21, 1529, 1983.

Chapter 3

TIME-RESOLVED LASER SPECTROSCOPY NANOSECOND AND PICOSECOND TECHNIQUES: PRINCIPLES, DEVICES, AND APPLICATIONS

David Phillips and Garry Rumbles

TABLE OF CONTENTS

I.   Introduction..92
        A.   Fates of Excited States 92
        B.   Classification of Chromophores in Polymers 95
        C.   Aims of Studies on Excited States of Synthetic Polymers 97
                1.   Fundamental Interests 97
                2.   Use of Probe Molecules 97

II.   Time-Resolved Fluorescence Techniques 97
        A.   Time-Correlated Single-Photon Counting (TCSPC)......................97
        B.   Convolution 99
                1.   Scattered Excitation Radiation 99
                2.   Wavelength Dependence 100
                3.   rf Interference..100
                4.   Impurity Fluorescence 100
                5.   Pulse Pile-Up Errors 100
        C.   Data Analysis 101
                1.   Straight Line Fitting 101
                2.   Phase-Plane Plot 101
                3.   The Method of Moments 101
                4.   Laplace Transforms 101
                5.   Modulating Functions 101
                6.   Fourier Transforms 101
                7.   Least Square Fitting 101
        D.   Lasers and TCSPC 103
        E.   Fitting Functions 106
                1.   Heterogeneity 106
                2.   Motion 112
                3.   Energy Migration 112
                4.   Relaxation 113
        F.   Time-Resolved Emission Spectra.....................................113
        G.   Fluorescence Anisotropy 115

III.   Alternative Methods of Measuring Fluorescence Decays.......................118
        A.   Optical Gating...119
                1.   Kerr Cell..119
                2.   Up Conversion..119
        B.   Streak Camera..119
        C.   Frequency Domain Phase-Modulation Fluorimetry 119

IV.   Alternative Methods of Studying Transient Species..........................129
        A.   Transient Absorption Methods.......................................129

B.   Transient Holographic Grating Techniques 136
C.   Time-Resolved Raman Methods 139

V.   Conclusion 140

Acknowledgments 140

References 142

I. INTRODUCTION

As in many fields, the introduction of pulsed lasers has caused quantum leaps forward in the methodology through which synthetic polymers can be investigated. Questions may now be asked with some prospect of sensible answers which were not feasible before the advent of reliable laser systems. It is the purpose of this Chapter to describe the experimental techniques using pulsed laser excitation which either have been used in the study of the excited state of synthetic polymers or have great potential uses. We include only time-resolved methods, ranging from the nanosecond to picosecond time scale, and are thus concerned with the fate of electronically excited states in polymer systems studied by a variety of techniques.

A. FATES OF EXCITED STATES

The options can be discussed with reference to Figure 1, which is the familiar Jablonskii diagram which illustrates the fates of a complex polyatomic molecule following excitation to the singlet manifold. Considering excitation to the first excited electronic state, S_1, the unimolecular electronic relaxation processes competing with fluorescence are intersystem crossing (ISC) to the triplet manifold, internal conversion (IC) to the ground state, and photochemical reaction. The time domain in which these and some other physical phenomena occur are summarized in Figure 2, together with the limitations of currently available pulsed laser technology. In condensed media, vibrational relaxation occurs on a picosecond time scale, and thus only chemical processes with rate constants in excess of $10^{12} s^{-1}$ will compete. Therefore, subsequent to excitation, vibrational relaxation is usually complete before electronic relaxation. IC is usually faster at higher excess energies, but is not of great importance for lower-lying vibrational levels of the first excited singlet state. The principal process competing with fluorescence is therefore ISC to the triplet manifold of levels. Writing a simple kinetic scheme permits definition of the quantum yield of fluorescence, ϕ_F, and decay time τ_F in terms of first-order rate constants:

$$M + h\nu \rightarrow {}^1M^* \qquad I \qquad \text{(absorption)} \qquad (1)$$

$$ {}^1M^* \rightarrow M + h\nu \qquad k_R \qquad \text{(fluorescence)} \qquad (2)$$

$$ {}^1M^* \rightarrow {}^3M^* \qquad k_{ISC} \quad \text{(intersystem crossing)} \qquad (3)$$

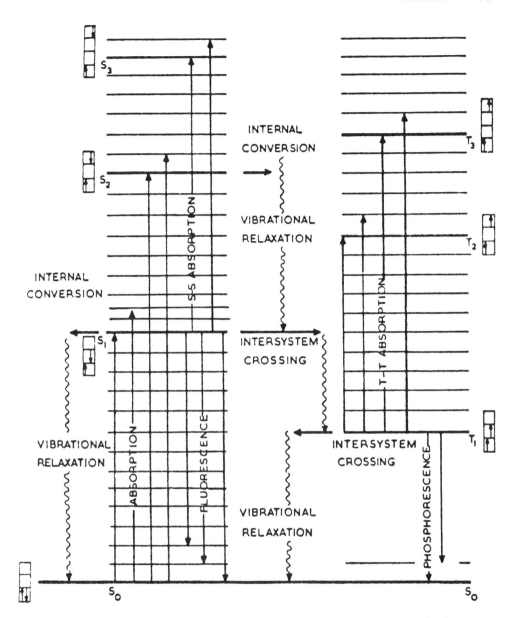

FIGURE 1.   Jablonskii state diagram depicting the fates of photoexcited polyatomic molecules.

$$^1M^* \rightarrow M \qquad\qquad k_{IC} \quad \text{(internal conversion)} \qquad\qquad (4)$$

$$^1M^* \rightarrow \text{products} \qquad k_D \quad \text{(dissociation)} \qquad\qquad (5)$$

From a steady-state analysis of the scheme, the quantum yield of fluorescence ϕ_F is given by

$$\phi_F = \frac{k_R}{(k_R + k_{ISC} + k_{IC} + k_D)} \qquad\qquad (6)$$

The fluorescence decay time, τ_F, is given by Equation 7 below as

$$\tau_F = (k_R + k_{ISC} + k_{IC} + k_D)^{-1} \qquad\qquad (7)$$

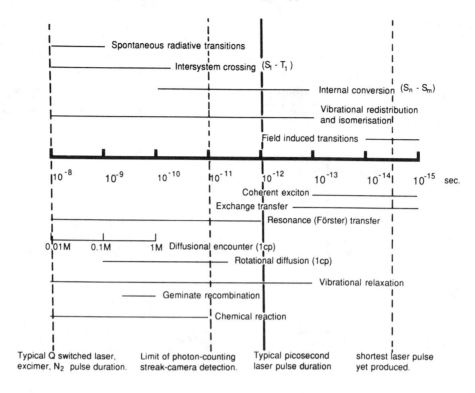

FIGURE 2.   Some physical and chemical processes which occur on the 10^{-6} to 10^{-15} s-time scale.

The intensity of fluorescence seen from any molecule, even before consideration of bimolecular interactions, depends upon the magnitude of the rate constant k_R relative to the sum (called Σk) of k_{ISC}, k_{IC} and k_D. It is clear from Equations 6 and 7 that rationalization of the photophysics and photochemistry of any singlet state molecular species in terms of the absolute magnitudes of rate constants for the various competing decay processes cannot be obtained with the sole knowledge of quantum yields, which are merely rate constant ratios. However, a knowledge of excited state lifetime through, for example, the measurement of fluorescence decay time together with quantum yields does in principle provide the absolute rate information required, since

$$k_R = \phi_F/\tau_F \tag{8}$$

$$k_{ISC} = \frac{\phi_{ISC}}{\tau_F} \tag{9}$$

and so on.

Measurement of fluorescence decay time is thus of immense importance. Alternatively, singlet state decay characteristics could be monitored by singlet-singlet absorption measurements. Both of these techniques will be discussed at length below. Additional information, particularly about molecular motion, can be obtained by anisotropy measurements and these will also be discussed. Referring again to Figure 1, the population of the triplet state of a molecule could be recorded as a function of time either by triplet-triplet absorption measurements or by monitoring phosophorescence, although the latter process is often of too low a yield to be useful.

The concentration of the products of photochemical reactions is also amenable to sampling by measurement of transient absorption which, for nonemitting species, is the conventional method. Thus, whereas conventional flash lamps are used to probe the 10 μs and longer time domains, excimer and Q-switched lasers are used to investigate the nanosecond domain and mode-locked lasers have allowed measurements in the picosecond and subpicosecond time region. In general, however, transient absorption spectra obtained by such methods are broad and featureless and offer little clue to the chemical structure of the species observed. When the technique is used to obtain kinetic information about the decay of excited electronic states and chemical species derived from these, the situation is frequently met where several transients are observed simultaneously because of overlapping absorption bands. It is in principle possible to improve on this situation and to obtain spectra which are more clearly identifiable with individual chemical entities by observing resonance Raman spectra using a second pulsed laser, and such spectra offer in principle the hope of obtaining structural information on the transient. We will describe in some detail this technique also.

B. CLASSIFICATION OF CHROMOPHORES IN POLYMERS

It is pertinent to consider a classification of species within pure and commercial polymers which act as absorbing and emitting chromophores.

Many simple synthetic polymers such as poly(ethylene) and poly(propylene) in a pure state will exhibit only σ-σ* absorptions in the high-energy UV region, where most organic molecules will absorb. Such excitations in general lead to photochemical reaction rather than luminescence, and excited states will thus be very short. Here we focus attention on species that absorb in the spectral region from, rather arbitrarily, say 250 nm to longer wavelengths, where luminescence may be an additional fate of photoexcited species, which will in general be longer lived.

A convenient classification of chromophores, has been suggested by Somersall and Guillet[1] in which distinction is made between polymers in which the repeat unit contains an absorbing (and emitting) unit, termed type B polymers, and those termed type A in which isolated chromophores are attached to a polymer chain as an end group or minor component of a copolymer. As a typical example of a type B polymer, poly(styrene) (I) may be cited, one member of a general class in which pendant aromatic groups are arranged along the backbone in a random (atactic) I_a, alternating (syndiotactic) I_b, or regular (isotactic) I_c fashion.

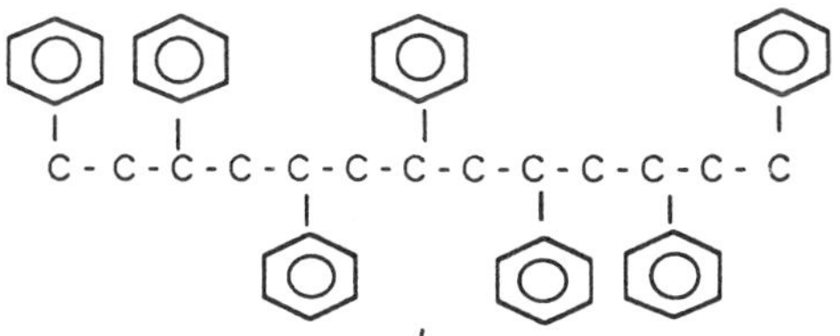

I_a

Poly(styrene) (atactic)

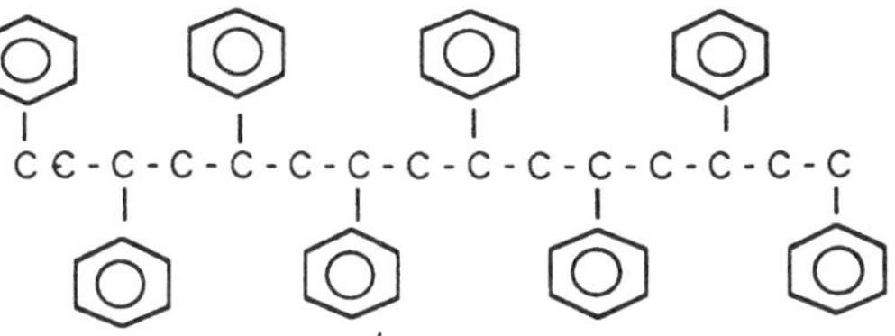

I_b

Poly(styrene) (syndiotactic)

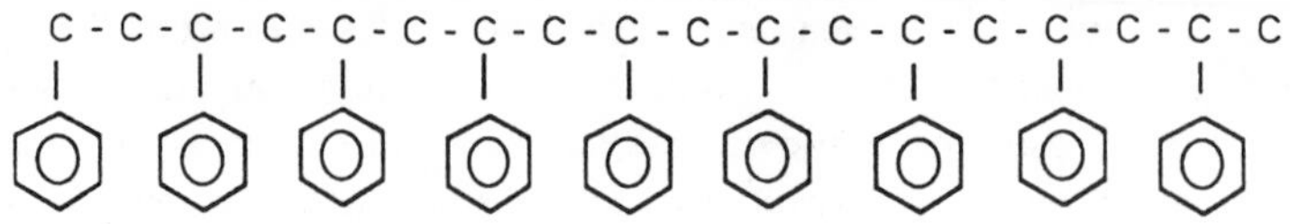

C - C - C - C - C - C - C - C - C - C - C - C - C - C - C - C - C - C

$/_c$

Poly(styrene) (isotactic)

Alternating copolymers such as poly(acetonitrile-styrene), II, also belong to this class as do block copolymers represented generally by III.

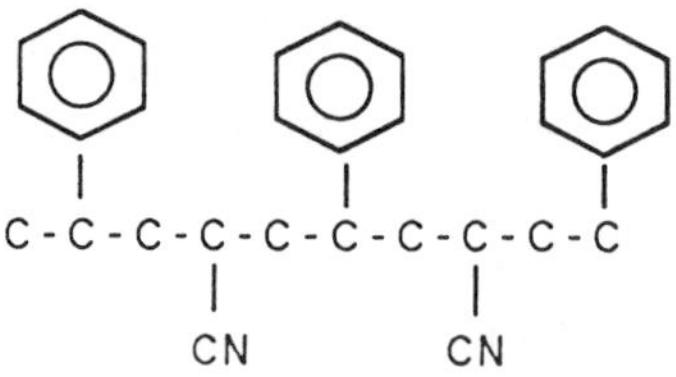

II

C - C - C - C - C - C - C - C - C - C

CN   CN

III

In type B polymers the structural constraints of the polymer chain tend to confine the chromophores in spatial positions such that they can be expected to exhibit strong mutual interactions. These may depend strongly upon the relative orientation of the interacting chromophores, and the orientations themselves will usually be dependent upon the conformation of the polymer chain.

The isolated chromophores in a polymer type A are in general adventitious, being due to end groups resulting from the polymerization process; commercial additives such as plasticizers, antioxidants, and pigments; products of thermal oxidation; or occasionally probes added deliberately. Typical examples of type A chromophores are the ketonic species present in thermally oxidized poly(propylene) (IV and V).

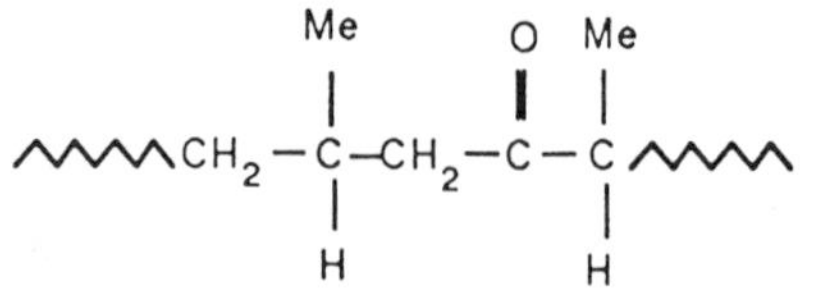

IV 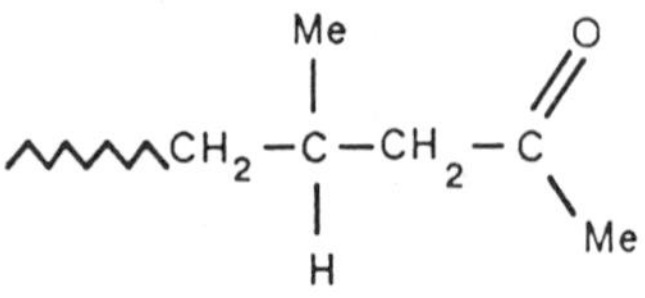   V

In the case of probe molecules added to polymeric systems, the molecule containing the chromophore may not of necessity be attached to a polymer chain. Thus, many studies have been carried out, for example, using a luminescent molecule such as VI, which is free to orientate itself along the axis of polymer chains with which it is in contact.

2,2'-(vinylidenedi-p-phenylene) bisbenzoxazole

Finally one should include in this category the large number of examples of the use of polymer matrices in which to study the luminescence of added free small molecules, particularly over a temperature range up to ambient and above where conventional solvent glasses are not available. In such studies the polymer matrix, poly(methylmethacrylate) being a favorite, is usually assumed to be inert, an assumption that certainly does not enjoy universal validity.

C. AIMS OF STUDIES ON EXCITED STATES OF SYNTHETIC POLYMERS

The many time-resolved studies carried out on excited states in synthetic polymers have been motivated by a wide spectrum of scientific and technological aims. Some of the more obvious are categorized below.

1. Fundamental Interests

These would include studies on the nature of photoemission from polymers of type B, in which interchromophoric interactions are of special interest. One should also include here fundamental studies on the photoluminescence of small molecules in polymeric environments.

2. Use of Probe Molecules

The excited state properties, principally fluorescence, of probe molecules can be used to great effect to study molecular motion, order, and energy migration in synthetic polymer systems. Thus, as an example, measurement of the relative intensities of fluorescence of a probe molecule polarized parallel to and perpendicular to the plane of linearly polarized exciting radiation as a function of orientation of a solid sample yields information concerning the ordering of polymer chains. In a mobile system, similar time-resolved polarization studies yield information on the rotational relaxation of the probe molecule. Such measurements will now be discussed with emphasis upon the techniques.

II. TIME-RESOLVED FLUORESCENCE TECHNIQUES

The great sensitivity of fluorescence spectral, intensity, decay, and anisotropy measurements has led to their widespread use in synthetic polymer systems, where interpretations of results are based upon order, molecular motion, and electronic energy migration.[2] Time-resolved methods down to picosecond time-resolution using a variety of detection methods but mainly that of time-correlated single-photon counting can in principle probe these processes in much finer detail than steady-state techniques, but the complexity of most synthetic polymers poses severe problems in terms of interpretation of results.

We now discuss some of the techniques and results of time-resolved studies on synthetic polymer systems which highlight studies of order, motion, energy migration, and heterogeneity. We begin with the time-correlated single-photon counting technique, which has probably been the most widely used method of studying fluorescence down to the 10-ps time domain.

A. TIME-CORRELATED SINGLE-PHOTON COUNTING (TCSPC)

The basic principles of TCSPC have been the subjects of many reviews.[3-10] The method relies on the basic concept that the probability distribution for emission of a single photon

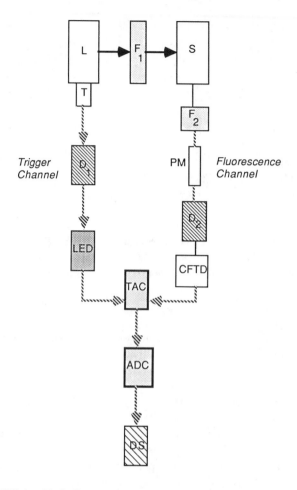

FIGURE 3.   Block diagram of a conventional single photon counting apparatus —— optical signal; ＼＼＼ electronic signal; L, excitation source; T, trigger (antenna, fiber optic, and photomultiplier tube etc.); F_1 and F_2, wavelength selection for excitation and emission; D_1 and D_2, delay lines; LED, leading edge discriminator; PM, photomultiplier; CFTD, constant fraction timing discriminator; TAC, time-to-amplitude converter; ADC, analogue-to-digital converter; and DS, data store (multichannel analyzer or computer).

following excitation gives the actual intensity against time distribution of all photons emitted, thus by sampling the time of single photon emission following a large number of excitation pulses, the probability distribution is created.

The experiment is carried out as follows, with reference to Figure 3. A trigger T, which could be a photomultiplier, an antenna pick-up, or a logical synchronizing pulse from the electronics pulsing the excitation source, generates an electrical pulse at a time exactly correlated with the time of generation of the optical pulse. The trigger pulse is routed, usually through a discriminator, to start input of a the time-to-amplitude converter (TAC) which initiates linear voltage ramp. In the meantime the optical pulse excited the sample, which subsequently fluoresces. An aperture is adjusted so that, at most, one photon is detected for each 50 exciting events. The signal resulting from this photon stops the voltage ramp in the TAC which puts out a pulse, the amplitude of which is proportional to the final ramp voltage, and hence to the time difference between start and stop pulses. The TAC output pulse is given a numerical value in the analogue-to-digital converter and a count is stored in the data storage device in an address corresponding to that number. Excitation and data storage are repeated in this way until the histogram of number of counts against address number in the storage device has enough data so that it represents, to some required precision, the fluorescence decay curve of the sample. If deconvolution is necessary, the time profile of the excitation pulse is collected in the same way by replacing the sample by a light scatterer.

B. CONVOLUTION

If the flash of light that excited the sample were infinitely narrow and if the response of the detection system were infinitely fast the observed decay curve would represent the true decay, or δ-pulse, of the sample, $G(t)$.

The form of the observed decay, $I(t)$, when the excitation function, $E(t)$, is not a δ-function can be deduced from the theory of impulse functions and leads to the convolution concept. An advantage of picosecond laser excitation is that, for all but the shortest decays, the excitation function much more clearly approximates a δ-function than for other sources. However, the response time of the detection system must be considered in convolution.

Suppose that $H(t)$ is the δ-pulse response of the detection system and $P(t)$ the measured time profile of the pump pulse, i.e., the instrument response function. $P(t)$ is a convolution of $E(t)$ and $H(t)$.

$$P(t) = E(t) \otimes H(t) \tag{10}$$

and thus

$$I(t) = P(t) \otimes G(t) = \int_0^t P(t')G(t - t')dt' \tag{11}$$

where $I(t)$ represents the δ-function response of the sample distorted by convolution with both the pump pulse and the detector response, i.e., the measured decay curve. Thus the convolution integral can be solved for $G(t)$, if $I(t)$ and $P(t)$ measured under the same conditions of instrumental distortion are known.

In the normal experiment, when the number of counts in the decay curve have reached a suitable precision, the number of counts in each channel in the MCA, $I(t_i)$ follows a Poisson distribution with a SD G_i given by Equation 12. Note that it is necessary to correct the number of counts for background noise due to photomultiplier dark noise and leaked light; in the following it is assumed such corrections have been made.

$$G_i = I(t_i) \tag{12}$$

We must also remember that Equation 11 is an approximation since the decay curve is a histogram rather than a continuous distribution; however, it is easily shown that this is a valid procedure. We should further note that there can be experimental distortions to Equation 11. These can be due to

1. Scattered excitation radiation
2. Wavelength dependence of the photomultiplier tube
3. rf interference
4. Impurity fluorescence
5. Pulse pile-up errors

1. Scattered Excitation Radiation

Scattered light should be eliminated experimentally, as far as possible, but that remaining can be dealt with by replacing Equation 11 by 13,

$$I(t) = P(t) \otimes G(t) + C \cdot P(t) \qquad (13)$$

where C = component of measured decay comprised of scattered light.

2. Wavelength Dependence

The largest assumption in the deconvolution procedure is the use of a measured instrument response function $P_E(t)$ that is recorded at the excitation wavelength, whereas the theory requires that the instrument response function is recorded at the emission wavelength $P_e(t)$. Under normal conditions, $P_E(t)$ is not identical to $P_e(t)$ and hence corrections are required.

There is an often-used mathematical treatment, the so-called shift routine, which is based upon the assumption that a wavelength-dependent PM response results merely in a time shift of the measured curve, no change in *shape* of the response function occurring. This may be an oversimplification, but there is some experimental evidence in support.

The correction to be applied is

$$I_e(t) = P_E(t + \delta) \otimes G(t) \qquad (14)$$

where δ represents the zero time-shift. It can be shown[13] that parameters derived from this time routine vary depending upon which part of the decay curve (whole or from the peak) is analyzed, thus care must be taken in the use of the method. A method has been developed by several groups which use a reference compound to achieve this correction.[11,12] In this, the experimental result is deconvolved against that obtained with a standard, single-exponentially decaying compound, and the method has been shown to overcome wavelength dependence effects. Microchannel plate photomultipliers[30a] and even some conventional photomultipliers[30b] have demonstrated a negligibly small wavelength dependence, providing a far more satisfactory solution to the problems.

The major motivation behind using the microchannel plate photomultipliers is the dramatic reduction in the transit time spread. This has led to instrument responses functions as short as 70 ps being observed. With such a high temporal resolution, the effect of differing path lengths through dispersion monochromators is now easily seen. Corrections can be made[30c] although alternative methods of dispersing the fluorescence emission are more favorable.

3. rf Interference

There is no reliable method of correcting mathematically for this, which will give rise to oscillations in the decay curve.

4. Impurity Fluorescence

This can bedevil experiments, resulting in complex decays where simple kinetics should prevail. The only answer is rigorous experimental control or purity.

5. Pulse Pile-Up Errors

In order to ensure that the single emitted photon is randomly selected from those emitted, the number of observed photons must be restricted, otherwise the observed decays will be distributed in favor of early events. The ratio of excitation events N_D to emission counts, N_E ($= F_D$) will depend upon the level of distortion deemed acceptable. It is generally said that a value of F_D as high as 0.05 is tolerable since at this count rate distortions in the early channels are negligible, while they amount to only a few percent in the late channels. However, empirical tests in our laboratory indicate that significant distortions are still present when F_D is as high as 0.05. In fact most workers keep F_D lower than 0.01 or 0.02. As far as pile-up distortions are concerned, the lower the value of F_D the better; therefore, if collection times are short (e.g., as in instruments with a cavity-dumped laser as excitation source) an F_D of 0.005 or even 0.002 is more usual. There are mathematical treatments for pulse pile-up[9] but it is better to adjust F_D to avoid them.

C. DATA ANALYSIS
This is discussed at length elsewhere,[9] but the following options are available.

1. Straight Line Fitting
This is valid only if the width of the instrument response function is negligible compared with the lifetime being measured.

2. Phase-Plane Plot[14,15]
The phase-plan plot had the (now irrelevant) advantage of not requiring a computer, but is cumbersome.

3. The Method of Moments
The methods of moments are widely used by physical biochemists.[16-19]

4. Laplace Transforms
This suffers also from the required integration of truncated data, although corrections are available.[20]

5. Modulating Functions
A modulating function is any function which, together with its first derivative, goes to zero at time t = 0 and at some time t = T. Valeur and Moirez[21] reported that modulating functions of form in which m and n are

$$\phi(t) = t^m(T - t)^n \tag{15}$$

integers are most suitable for the application of the technique to fluorescence decay convolution. Again, no account is taken of Poisson noise but unlike the previous two techniques this method does not require a cutoff correction.

6. Fourier Transforms
A number of methods based on the fast Fourier transform have been developed in recent years for application to multiexponential decays[22] and deconvolution.[23,24] All are plagued, however, by the well-known accentuation of random (Poisson) noise upon Fourier transformation of the data. Perhaps the most successful is the method of Wild et al.[23] in which the (single exponential) decay parameters are determined in Fourier space. This method is ideally suited to the excitation source employed by its proposers (a mode-locked Ar^+ ion laser with pulse repetition rate 115 MHz). A further development by Andre,[24] in which spurious oscillations are avoided by extrapolating the function g(v) by the Fourier transform of a single exponential, shows some promise, especially for decays in which the function form G(t) is unknown.

7. Least Square Fitting
This method still represents the most acceptable technique of data extraction.[9] With this technique the data points with the highest number of counts are more heavily weighted; moreover, any section of the decay curve may be excluded from the analysis, a feature especially useful if distortions are present in the data. If the sample decay is a single exponential, G(t) is given by

$$G(t) = a_1 \exp(-t/a_2) \tag{16}$$

and in order to linearize the fitting function, Equation 16 is expanded to first order in a Taylor's expansion as a function of the parameters a_1 and a_2. A linear least-squares search is then carried out to find values of the parameter increments δa_1 and δa_2 that minimize the reduced χ^2_ν, given by

$$\chi^2_\nu = \frac{\sum_{i=n_1}^{n_2} w_i[Y(t_i) - I(t_i)]^2}{n_2 - n_1 + 1 - p} \tag{17}$$

where w_i, the weighting factor, is the reciprocal of the number of counts $Y(t_i)$ in channel i, n_1 and n_2 are the first and last channels of the section of the decay to be analyzed, and p is the number of fitting parameters (two for a single exponential fit). The search for the minimum in χ^2_ν is performed according to Marquardt's technique.[25]

When the minimum in χ^2_ν has been reached it is vitally important to have reliable criteria by which the fit can be judged. The actual value should be close to 1; values of χ^2_ν much less than 1 are symptomatic of poor statistics whereas values much in excess of 1 indicate a poor fit. If all distorted data are to be rejected we would accept results for which χ^2_ν is less than 1.2, whereas if some level of distortion must be tolerated, fits with values of χ^2_ν less than 1.4 may be acceptable if they are justified by some other criteria. Since acceptable values of χ^2_ν are sometimes obtained for poor fits, it is usual to inspect a plot of the weighted residuals for nonrandom fluctuations. The weighted residual in channel i is given by

$$r_i = \sqrt{w_i}[Y(t_i) - I(t_i)] \tag{18}$$

It is generally less difficult to detect small deviations of the fitted from the observed curve in a plot of r_i vs. channel number rather than in the more traditional visual inspection of the two curves $Y(t_i)$ and $I(t_i)$. An even more sensitive plot is that of the autocorrelation function of the weighted residuals. The correlation of the residual in channel i with the residual in channel i + j is summed over a number of channels, m and normalized, i.e.,

$$C_{rj} = \frac{\frac{1}{m} \sum_{i=n_1}^{n_1+m-1} r_i \cdot r_{i+j}}{\frac{1}{n_3} \sum_{i=n_1}^{n_2} r_i^2} \tag{19}$$

In this expression $n_3 = n_2 - n_1 + 1$, the total number of channels in the section of the decay used in the fit. An upper limit, usually $n_3/2$ is put on j so that the number of terms, $m = n_3 - j$, summed in the numerator is sufficient to given proper averaging. According to Equation 19 $C_{ro} = 1$. In a successful fit C_{rj} for $j \neq 0$ is randomly scattered about zero although, because of the finite value of m, some high-frequency low-amplitude fluctuations are generally observed. These are clearly distinguishable from the type of correlation indicative of an incorrect fitting function or of distorted data.

Judgments based on inspection of the aforementioned plots are subject to the inevitable bias associated with subjective tests. Calculation of the serial correlation coefficient or as it is more commonly known, the Durbin-Watson parameter,[26,27] DW, is more sensitive than χ^2_ν to small nonrandom oscillations in the residuals and is thus of use. DW is calculated according to the equation:

$$DW = \frac{\sum_{i=n_1+1}^{n_2} (r_i - r_{i-1})^2}{\sum_{i=n_1+1}^{n_2} r_i^2} \qquad (20)$$

Acceptable values of DW have been tabulated for up to 100 data points and 5 fitting parameters. Single exponential fits yielding values of DW >1.65 are generally successful. The corresponding values for double and triple exponential fits are 1.75 and 1.8, respectively.

The success of the technique above is illustrated in Table 1, which displays analysis by this method of simulated data obtained by convolving three components of respective A and τ (shown in Table 2 as initial values), with a real instrument response function using a cavity-dumped dye laser (see below) adding random noise to simulate the experiment, and then analyzing.[33] Recovered values of A and τ are very satisfactory for a three-component fit but a two-component, (four parameters) fit is seen to be unacceptable.

D. LASERS AND TCSPC

The requirement in lasers for TCSPC applications are high repetition rate tunable pulses of short duration. Mode-locked CW lasers fall into this category providing pulse duration of <10 ps (dye lasers), 200 ps (ion lasers), and 100 ps (Nd:YAG lasers). When used in the mode-locked mode, the repetition rate is fixed by laser cavity length and is normally of the order of 76 or 82 MHz. Such high repetition rates are too fast even for TCSPC and therefore a reduction in rate is required. Extracavity devices, such as a Pockels' cell, have been used, but suffer from an inherent large rf interference signal associated with the high voltage changes; a repetition rate which is too low (see Table 2) and inefficient use of the available pulses. The most commonly used device for controlling the pulse repetition rate is the acoustic-optic, intracavity Bragg cell, known as a cavity dumper. Such a device has been used with mode-locked argon and krypton ion lasers and synchronously pumped, mode-locked dye lasers. Depending upon the electronic drive being used, repetition rates are continuously selectable from, for example, 76 MHz to single shot. When used in unmode-locked CW laser systems, the cavity dumper can provide the same flexibility and stability in pulse output, providing pulse durations of $\sim$10 ns, which have proved useful in measuring lifetimes down to 1 ns.[41,42]

Until recently, the most commonly used system was the cavity-dumped, mode-locked dye laser synchronously pumped by a mode-locked argon ion laser. While such a system provides the required repetition rate and pulse duration, the wavelength tunability was limited to dyes that could be pumped by the 514.5-nm argon ion lasing line, 560 to 760 nm, or 280 to 380 nm if the output was frequency doubled. Mode locking of the 364-nm UV line of an argon laser has recently permitted access to dyes which lase in the 380 $\rightarrow$ 500 nm spectral region of the spectrum.[33]

However, the most practical system currently available uses a mode-locked CW Nd:YAG laser as a pump source. By frequency doubling the 1064-nm output to 532 nm or frequency tripling to 355 nm, a large range of dyes from the UV to the IR are accessible, A system currently in use in our laboratory is shown schematically in Figure 4. The intensity of such a low-powered laser source (e.g., 100 mW at 600 nm) nevertheless represent 6×10^{16} photons per second, greatly in excess of the flash lamps listed in Table 2.

The time required to collect a decay curve to a given precision is closely related, of course, to the intensity of the excitation source. An additional factor is the flash repetition rate, higher rates leading to more efficient TAC operation and shorter collection times. It is this factor, as much as their higher intensities, that makes measurement times with cavity-dumped lasers so short. The advantage is lost when pulses are selected with a Pockels' cell, which is usually operated at <100 kHz. Typical collection times with 4 MHz laser excitation

TABLE 1
Analysis of Simulated
Three Component

	Initial	Recovered
A_1	0.25	0.25
τ_1/ns	2.5	2.54
A_2	0.07	0.07
τ_2/ns	10.00	9.72
A_3	0.025	0.026
τ_3/ns	40.00	39.44

TABLE 2
Comparison of Characteristics of Lasers and Flash Lamps for TCSPC

	Laser	Lamps
Pulse width	1—2 ps (mode-locked dye) 6—10 ps (mode-locked cavity-dumped dye) 200 ps (ion laser) 10 ns (cavity-dumped only)	800 ps—5 ns
Repetition rate	80 MHz (mode-locked) < 100 KHz (Pockels cell) Single-shot-4MHz, (cavity-dumped)	~ 150 KHz max
Pulse shape and stability	Good stability, Gaussian reliable shape	Unstable over long period, erratic
Optical properties	Monochromatic and polarized	Continuum plus spectral lines, unpolarized
Wavelength range	515—560 560—630 dyes 620—730 280—315 dyes 310—365 doubled 514, 488, 458 (Argon) 532 (Nd:YAG)	UV through visible depending upon fill
Time resolution	10 ps	> 50 ps
Useful intensity (photons per pulse)	10^{10} fundamental 5×10^6 (doubled)	10^6—10^9 depending upon lamp (*total* intensity, nonwavelength selected)
Advantages	High repetition rate, stability intensity; very useful for TRES anisotropy, wavelength selected decays, and complex analysis	Inexpensive wide tuning range
Disadvantages	Very expensive to purchase and maintain	Low intensity, longer data collection times, instability

are 0.5 to 1 min for a decay curve with 30,000 counts in the channel of maximum counts. It should be noted that, as the decay time of the samples under investigation gets longer, the repetition rate of the cavity-dumper will have to be lowered in order to allow the fluorescence resulting from one flash to decay before the next excitation event, and so prevent miscorrelation in the TAC. Some loss in overall intensity usually accompanies lower dumping rates. However, for most nanosecond lifetime measurements a repetition rate of 800 kHz (1250-ns interpulse separation) will be sufficiently low. Flash lamps are capable of operating at a maximum of only 150 kHz and in routine operation are rarely driven above 100 kHz. Cavity-dumped lasers are therefore more efficient than flash lamps and relatively low repetition rate lasers (Pockels cell pulse selection) in time-correlated nanosecond decay time measurements.

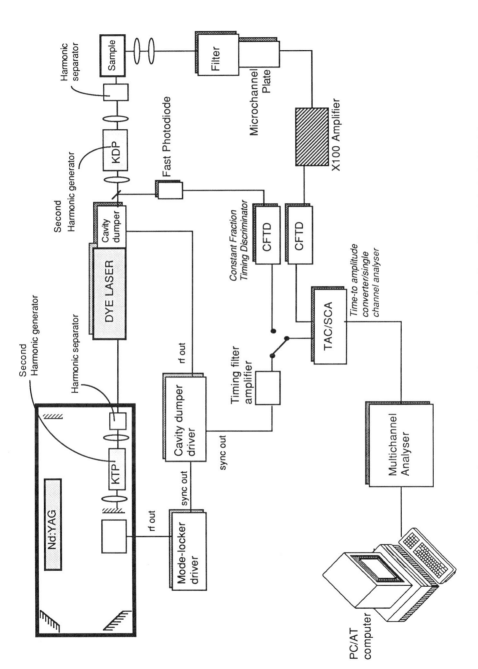

FIGURE 4.   Time-correlated single-photon counting spectrometer based on a CW mode-locked Nd:YAG laser.

The efficiency of TCSPC experiments using lasers can be improved by operating the TAC in reverse mode, i.e., by using the fluorescence photon as the start pulse, the next laser pulse as stop. The same constraints of avoidance of pulse pile-up apply, but the TAC electronics no longer required to cycle rapidly.

The ultimate time resolution possible with synchronously pumped dye laser excitation and TCSPC detection is of the order of 10 ps.[29] This will be discussed in other papers, and will thus not be pursued here.

E. FITTING FUNCTIONS

The great sensitivity and high signal-to-noise obtained with TCSPC permits some discrimination between different mathematical forms for the fitting function G(t) which would arise from different physical models. These have been discussed extensively elsewhere[2,31] but are summarized here with examples of their use. The single or dual exponential fluorescence decay encountered in simple small molecules in fluid or rigid media is not often expected in synthetic polymers. The causes of this, briefly, are heterogeneity, motion, complex formation, and energy transfer and migration.

1. Heterogeneity

For more than one simultaneously excited, noninteracting species, the decay of total fluorescence will be described in principle by Equation 21. The situation with two noninteracting species is fairly common, but as the number of species increases, interactions such as energy transfer are bound to become more probable, complicating the kinetics.

$$I(t) = \sum_i A_i e^{-t/\tau_i} \tag{21}$$

In the extreme of a large number of noninteracting sites, such as molecules adsorbed on a solid surface, in defects in molecular crystals or in some polymeric species, the decay may be better described by a distribution of decay times suitably weighted about some mean value.

A recent treatment by Albery et al.[32] gives a rate-parameter k as a distribution represented by:

$$k = k \exp(\gamma x) \tag{22}$$

Thus the decay of concentration C of a species from initial concentration C_0 is given by

$$\frac{C}{C_0} = \frac{\int_{-\infty}^{+\infty} \exp(-x^2)\exp[-\tau\exp(\gamma x)]dx}{\int_{-\infty}^{+\infty} \exp(-x^2)dx} \tag{23}$$

where

$$\tau = kt, \quad \text{and} \quad \int_{-\infty}^{+\infty} \exp(-x^2)dx = \pi^{1/2}$$

Since any sample of polymer is characterized by a distribution of molecular weights and the fluorescence of a chromophore is in principle environmentally sensitive, even for noninteracting chromophores it would not be surprising if a distribution of decay times were observed in these situations, which, however, corresponds to that observed in a free chrom-ophore in solution, and the decay should be modeled adequately by a rate constant. This is certainly not the case for interacting chromophores where the local environmental will be critical in determining the decay rate of any particular fluorophore. In a homopolymer, the principal cause of heterogeneity will be the tacticity of the polymer, isotactic, syndiotactic, and atactic polymers being expected to behave very differently. In cases where nominally atactic polymers consist of isotactic and syndiotactic sequences, the decay may in favorable simple cases be interpretable in terms of a summation of exponential decays of two kinetically distinct species. For a wide distribution of sites, a kinetic model recognizing this heterogeneity may be more appropriate, although this yields information of limited usefulness.

In copolymers, heterogeneity of the environment of a chromophore by virtue of com-position becomes of overriding concern. In our earlier work on copolymers of vinyl naphthalene[33-40] and polystyrene,[41,42] the models adopted to explain results deliberately em-phasized heterogeneity at the expense of, say, energy migration. The analysis of results was based upon the fitting of experimental decay curves to a series of *triple* exponential functions discussed below. Caution must necessarily be exercized in this procedure. Ware and co-workers[43-46] have demonstrated that single-photon counting decay data fit to even *two* com-ponent decay laws, and yielding a satisfactory fitting parameters can in fact conceal under-lying *sets* of lifetimes with a variety of intensity distribution functions of remarkable diversity. Discrimination can be made *only* when large numbers of counts per channel are made (5×10^4 to 5×10^5 in the peak). The capacity for delusion in interpretation of polymer decay data is thus very high. With this proviso, an example of results on an excimer forming polymer can be given.

In many synthetic polymers, particularly the vinyl aromatic type, excimer formation is significant (Figure 5). In the simplest case of excimer formation in free molecules in fluid solution, the decays of uncomplexed chromophore (monomer) and excimer are of the form

$$I_M(t) = A_1 e^{-\lambda_1 t} + A_1 e^{-\lambda_2 t} \tag{24}$$

$$I_E(t) = A_3(e^{-\lambda_1 t} - e^{-\lambda_2 t}) \tag{25}$$

In synthetic polymers, the interpretation is necessarily more difficult. The form of Equations 24 and 25 requires that the kinetics of formation and decay of complexes are modeled adequately by rate constants and that they take place in a homogeneous medium. If, as in synthetic polymers, the population of excimer trap sites may occur through energy migration or rotational diffusion, a rate constant may not be an adequate representation of the process, some time-dependent parameter being required. Heterogeneity may also play an important role. In our earlier work the fluorescence decay of excimer-forming polymers was modeled adequately by a scheme based upon simple excimer kinetics to which had been added terms to account for the occurrence in copolymers of monomer sites which, by their isolation, could not form excimers. For polymers which contain isotactic and syndiotactic sequences, or rather, are made up of meso and racemic triads, the kinetics may be similarly a superimposition of simple schemes appropriate for the different sequences.

The earlier simple interpretation of fluorescence decay kinetics of vinyl aromatic poly-mers has been questioned recently on the basis of a reinvestigation[48] in which it was shown that, although earlier work, based upon analysis in the monomer decay region, was self consistent, analysis in the excimer region has resulted in some inconsistencies. Figures 6 and 7 given fluorescence decay for a vinyl naphthalene-methyl methacrylate copolymer analyzed in the monomer (Figure 6) and excimer (Figure 7) regions. It can be seen that empirical triple exponential decays fit the data, but the question is whether or not these

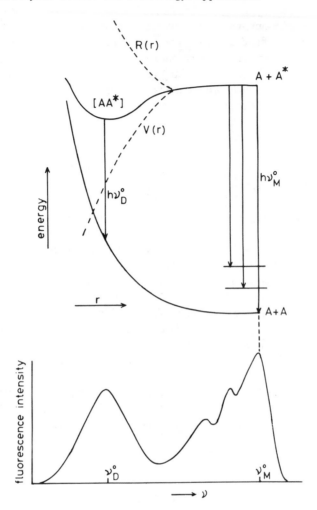

FIGURE 5. Excimer formation in aromatic molecule A. Excimer emission from AA* is red-shifted by virtue of stabilization of excited state, repulsion in ground state and structureless because of dissociation in ground state.

parameters have any physical significance. Results on four polymer samples of differing composition are summarized in Table 3 with the previously proposed model.[33-40] The monomer data are therefore fully compatible with the previously proposed model, Scheme 2. The full test of the model, however, requires that data obtained in the excimer region should also be compatible with the model, and the following points relating to Table 3 should be made.

1.   The excimer decays recorded on a 400-ns time range could not be analyzed in terms of a single, double, or triple exponential decay function. The goodness of fit parameters clearly indicated that the double exponential function did not describe the decay curves. Triple exponential analyses of the decay curves typically recovered two lifetimes which were approximately equal; the goodness of fit parameters indicated that the triple exponential function was inappropriate. The failure of the triple exponential decay scheme can be attributed to the relatively low time resolution conditions under which the decay curves were recorded and because the two shortest lifetimes were associated with growing in components. The shortest lifetime, evaluated by analyzing excimer

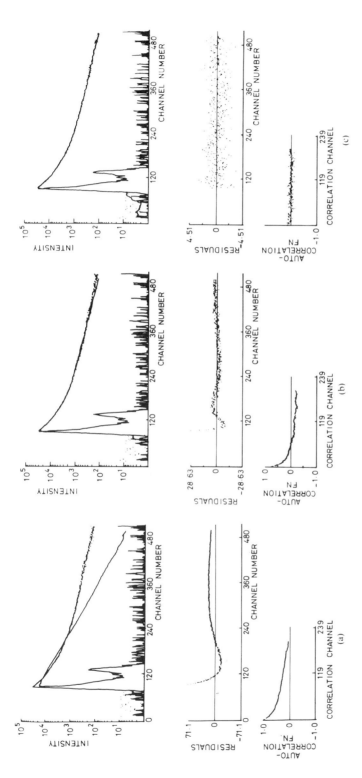

FIGURE 6. (a) Single exponential analysis of a decay curve of polymer sample b recorded at 330 nm (λ_{EX} = 298 nm). (b) Double exponential analysis of a decay curve of polymer sample b recorded at 330 nm (λ_{EX} = 298 nm). (c) Triple exponential analysis of a decay curve of polymer sample b recorded at 330 nm (λ_{EX} = 298 nm).

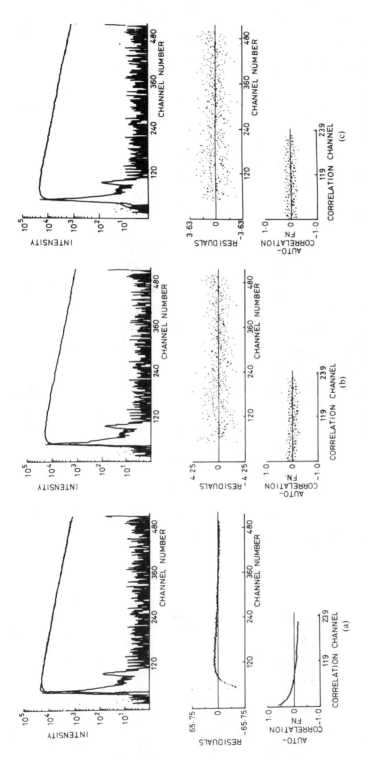

FIGURE 7. (a) Single exponential analysis of a decay curve of polymer sample b recorded at 500 nm (λ_{EX} = 298 nm). (b) Double exponential analysis of a decay curve of polymer sample b recorded at 500 nm (λ_{EX} = 298 nm). (c) Triple exponential analysis of a decay curve of polymer sample b recorded at 500 nm (λ_{EX} = 298 nm).

TABLE 3
Summary of the Lifetime Data Derived from Triple Exponential Analyses of the Fluorescence Decay Curves Recorded at 330 and 480 nm for Polymer Samples A, B, C, and D of Differing Composition Given in Reference 48

Polymer sample	Emission Wavelength (nm)	A_1	τ_1 (ns)	A_2	τ_2 (ns)	A_3	τ_3 (ns)
A	330	0.18 ± 0.01	3.4 ± 0.4	0.76 ± 0.01	13.2 ± 0.2	0.06 ± 0.01	36 ± 2
	480	−0.23 ± 0.09	4.0 ± 0.9	−0.3 ± 0.2	4.7 ± 0.2	1.00	45 ± 1
B	330	0.42 ± 0.01	2.3 ± 0.2	0.49 ± 0.01	9.1 ± 0.2	0.088 ± 0.009	39 ± 2
	480	−0.21 ± 0.05	1.0 ± 0.3	−0.46 ± 0.02	5.6 ± 0.1	1.00	41.6 ± 0.2
C	330	0.58 ± 0.02	1.5 ± 0.3	0.32 ± 0.02	7.8 ± 0.8	0.09 ± 0.01	35 ± 2
	480	−0.19 ± 0.02	1.4 ± 0.4	−0.31 ± 0.01	9.3 ± 0.9	1.00	40 ± 2
D	330	0.83 ± 0.02	0.45 ± 0.06	0.12 ± 0.01	5.5 ± 0.5	0.051 ± 0.007	23 ± 2
	480					1.00	35

Note: Evaluated by tail fits.

decay curves recorded on the 200-ns time range, was 1.0 ns, and hence only approximately 1.5 times the resolution (the time increment represented by a single channel in the multichannel analyzer) of the 400-ns time range. The two shortest lifetimes are associated with growing in components which determine the rising portion of the decay profile. Consequently, the ability to resolve the two shortest lifetimes is dependent on the temporal stability of the excitation source and the wavelength dependence of the photomultiplier tube. The above factors lead to the chi-square hypersurface becoming ill defined so that the algorithm for searching the chi-square hypersurface converges on a localized minimum instead of the true deeper minimum.

2.   The preexponential factors given in Table 3 do not sum to zero. Since the excitation process does not directly produce any excimer species[1] then $[D*] = 0$ at $t = 0$.

3.   The lifetimes do not appear to correspond to the lifetimes recovered from analyses of the monomer (330 nm) decays.

It must be tempting to conclude on the basis of these results that the triple exponential scheme does not apply to excimer formation in these copolymers.

Other functional forms for the decays could in principle be fitted to the decays, based upon emphasis on energy migration and motion of the molecules.

2. Motion

Translational diffusion results in a decay law of the form of Equation 26. Itagaki et al.[16] have proposed a decay law for restricted rotational motion of similar type:

$$I(t) = \exp(-At - 2B + ^{1/2}) \tag{26}$$

They propose that in excimer-forming vinyl aromatic polymers in fluid solution, the rate-determining step in excimer formation is rotational diffusion, leading to use of a

$$I(t) = A\exp\{-(at + bt^{1/2})\} + B\exp(-ct) \tag{27}$$

term such as Equation 27 to model the fluorescence. Data from other workers data were claimed to be compatible with this model, although the normal stringent tests of acceptability were not applied in this case. In general, fluorescence anisotropy measurements are of more importance in determining motion in synthetic polymers in fluid solutions and this will be discussed in a later section.

3. Energy Migration

In cases where energy migration is a dominant feature of luminescence, as in molecular crystals, various forms of decay are expected depending upon circumstances, but relying upon usually complex solutions, to the basic rate equations where $E(t)$ is the time-dependent population of the initially excited (exciton) state, $T(t)$ the population of the trap state, k_E the decay rate constant for band states, k_T the decay rate constant for the trap states, and $k_L(t)$ the time-dependent trapping rate functions,

$$\dot{E}(t) = -[k_E + k_L(t)]E(t) \tag{28}$$

$$\dot{T}(t) = -k_T T(t) + k_L(t)E(t) \tag{29}$$

the form of which depends upon the effective transport topology. For a strictly one-dimensional transport, Fayer has given the form of $k_L(t)$ as Equation 30.

$$k_L(t) = At^{-1/2} \tag{30}$$

For quasi-one-, two-, and three-dimensional diffusional processes, other forms are appropriate.[55] Thus very extensive theoretical and picosecond experimental work on electronic excited state transport in finite volumes of randomly distributed molecules has been reported, which shows that there are significant deviations in the behavior of finite volume systems compared with the infinite volume systems considered above. The treatment is mathematically complex and the results will not be given here explicitly. Frederickson and Frank[56] have used this treatment to suggest possible forms for the decay of monomer and growth and decay of excimer fluorescence in vinyl aromatic polymers where electronic energy migration might be a dominant process. In this work the suggestion was made that complex mathematical forms resulting from this treatment might be simulated by multiple exponentials, and thus care was needed in attaching physical significance to parameters derived from this procedure. A simplification of the Frederickson and Frank approach gave as the decay law for a monomeric species in a polymer exhibiting energy migration the expression:

$$I(t) = A \exp\{-(at + bt^{1/2})\} + B \exp(-ct) \tag{31}$$

An exact solution to Equations 28 and 29 for one-dimensional and higher dimensionality cases of diffusion has been developed. In this, the mathematical result of the model was rearranged to express the time dependence of the luminescence in the form of a power series $I(t) \propto e^{-t^\alpha}$. The result could then be compared with the results of fluorescence measurements on a poly(diacetylene) sample taken using both TCSPC techniques and streak-camera detection (see below). In the theoretical case, strongly one-dimensional diffusion should be represented by value of $\alpha = 0.45$. The experimental result, Figure 8, shows this is obeyed exactly.

4. Relaxation

Relaxation in the form of excimer or complex formation has been discussed above. Other forms, with appropriate models, include solvent relaxation[58] and vibrational relaxation. These models are of course appropriate for application to the kinetics of decay of excited electronic states monitored by techniques other than fluorescence (see Section IV).

F. TIME-RESOLVED EMISSION SPECTRA

The changes in dispersed fluorescence as a function of time can provide invaluable information in complex systems, including the identification of a number of emitting species and their temporal or kinetic relationship. A rapid method of providing uncorrected (convolved) spectra, exists with the TCSPC method using upper and lower voltage discriminators on the TAC to provide a time gate[9] (Figure 9). An example of the use of this technique to yield data on synthetic polymers is given in Figure 10 in which the TRES of a vinyl-naphthalene methyl methacrylate copolymer are recorded.

The kind of conclusions reached by use of this technique are typified by this example:

1. The excitation process produces only excited monomer species and so at $\Delta t = 0$ only the structured monomer fluorescence is observed (Figure 10A).
2. The intensity of the monomer fluorescence then decreases while the intensity of the excimer fluorescence increases due to the population of excimer sites (Figure 10B).
3. The time-resolved spectra recorded at delay times when no monomer emission should be present (based upon the lifetime of the monomer species) gave small, but significant, monomer fluorescence. This indicates that reverse dissociation is a significant deac-

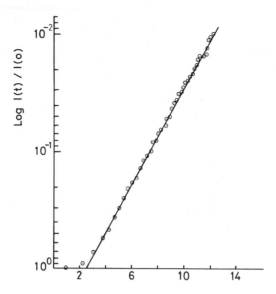

FIGURE 8.   Experimental fit (exp −tv) to PDA-1OH fluorescence at 4.2 K with α = 0.45α.

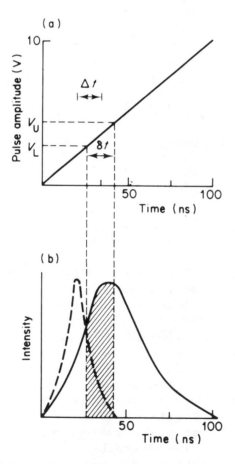

FIGURE 9.   Schematic representation of selection of time window by voltage discriminators (a) TAC voltage ramp (TAC range 100 ns). V_L, level of lower level discriminator; V_U, level of upper level discriminator; Δt, delay time; δt, gate width. (b) Hypothetical pump pulse (----) and convolved decay curve (——). Photons in time window (hatched area) selected for spectral analysis.

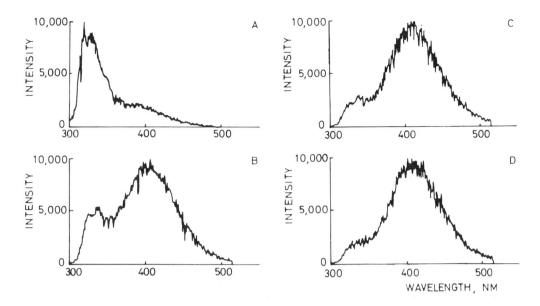

FIGURE 10. Time-resolved emission spectra of polymer sample 6. (A) $\Delta t = 0$ ns, $\delta t = 0.3$ ns; (B) $\Delta t = 21$ ns, $\delta t = 6$ ns; (C) $\Delta t = 47$ ns, $\delta t = 6$ ns; and (D) $\Delta t = 103$ ns, $\delta t = 16$ ns.

tivation process for the excimer species in these polymers. The ratio of excimer to monomer fluorescence, as a consequence, becomes approximately constant (Figure 10C).

4. The situation described above exists until complete decay of the excimer (and the associated monomer) fluorescence has occurred (Figure 10D).

It should be noted that the time-resolved spectra given in Figure 10 have been normalized to give 10,000 counts at the wavelength of maximum intensity. If a distribution of excimer sites does occur in these polymers, then the positions and perhaps the shape of the excimer spectral band should exhibit a time dependence. The time-resolved spectra of polymer sample B, see Figure 10, clearly show that the excimer spectral band does not exhibit any such time dependence, and consequently it must be concluded that only one excimer site is resolvable with these techniques in the specific case of this polymer.

The recording of TRES is seen to be a useful qualitative diagnostic in the study of synthetic polymers. There are occasions when fully corrected spectra are required, i.e., spectra free from distortions due to instrument response functions. These decay-associated spectra are constructed from decay curves by methods which have been described in full elsewhere.[59-61]

G. FLUORESCENCE ANISOTROPY

Time-resolved fluorescence anisotropy measurements can provide detailed information on the reorientation dynamics of molecules in solution. Until recently, however, this information has been limited to single rotational correlation times which are only strictly appropriate for the diffusion of spherically symmetric molecules. Improvements in instrumentation and data analysis techniques during the last decade have led to increasingly accurate measurements of fluorescence lifetimes. These capabilities also have led to parallel improvements in determinations of fluorescence anisotropies.

The advances in time-resolved techniques have fostered a reexamination of theories of the rotational motions of molecules in liquids. Models considered include the anisotropic motion of unsymmetrical fluorophores,[62,63] the internal motions of probes relative to the overall movement with respect to their surroundings, the restricted motion of molecules within membranes (e.g., wobbling within a cone),[65] and the segmental motion of synthetic macromolecules.[66,67] Analyses of these models points to experimental situations in which the anisotropy can show both multiexponential and nonexponential decay. Current experimental techniques are capable of distinguishing between these different models. It should be emphasized, however, that to accurately extract a single average rotational correlation time demands the same precision of data and analysis as fluorescence decay experiments which exhibit dual exponential decays. Multiple or nonexponential anisotropy experiments are thus near the limits of present capabilities and generally demand favorable combinations of fluorescence and rotational diffusion times.

Another key issue with regard to determinations of anisotropy decays is the wide variety of approaches to the calculations of rotational lifetimes. This is in contrast to the situation for the determination of unpolarized fluorescence decays. In the latter case, stable pulsed laser excitation sources, TCSPC, and standard procedures for deconvoluting fluorescence decays have led to the acceptance of decay times as relatively easily measured parameters by which to characterize molecular fluorescence.

The anisotropy of a system, $r(t)$, is derived from measurements of the fluorescence decays with polarizations parallel and perpendicular to the polarization of excitation:

$$r(t) = [I_{\|}(t) - I_{\perp}(t)]/[I_{\|}(t) + 2I_{\perp}(t)] = D(t)/S(t) \tag{32}$$

The different approaches to analyzing the time dependence of the anisotropy generally arise from different methods by which deconvolutions of $I_{\|}(t)$ and $I_{\perp}(t)$ are translated into a deconvoluted $r(t)$. For example, the rotational parameters can be extracted by (1) individually deconvolving $I_{\|}(t)$, (2) individually deconvolving $I_{\perp}(t)$, (3) deconvolving $D(t)$, (4) deconvolving both $D(t)$ and $S(t)$ and then reconstructing $r(t)$, (5) simultaneously fitting $I_{\|}(t)$ and $I_{\perp}(t)$, and (6) simultaneously analyzing several decay curves (global analysis),[68] etc. These and other methods have been discussed in some detail by Cross and Fleming[69] and Christensen et al.[70] At this point, there is no general agreement on which of these methods is most accurate, most efficient, least subject to typical systematic errors, etc. An important complication in such a comparison is that several previous studies have not fully appreciated that the statistical procedures applied to unpolarized fluorescence decay cannot be directly transferred to the analysis of $D(t)$, $r(t)$, etc. Whereas $I_{\|}(t)$ and $I_{\perp}(t)$ individually follow the Poisson statistics routinely employed in previous fluorescence decay measurements, deconvolution of $D(t)$ and $S(t)$ must employ properly propagated weights which are distinctively non-Poisson.[71]

The fluorescence anisotropy of the general, unsymmetric rigid rotor was first given by Belford et al.[63] and subsequently discussed by several other workers.[65,66] These treatments lead to the following expressions:

$$I_{\|}(t) = e^{-t/\tau_f}\left(1 + 2\sum_{i=1}^{5} a_i e^{-t/\tau_i}\right) \tag{33}$$

$$I_{\perp}(t) = e^{-t/\tau_f}\left(1 - \sum_{i=1}^{5} a_i e^{-t/\tau_i}\right) \tag{34}$$

$$D(t) = I_{\|}(t) - I_{\perp}(t) = 3e^{-t/\tau_f}\sum_{i=1}^{5} a_i e^{-t/\tau_i} \tag{35}$$

$$S(t) = I_{\|}(t) + 2I_{\perp}(t) = 3e^{-t/\tau_f} \tag{36}$$

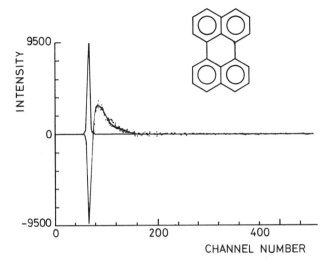

FIGURE 11.   Polarization of fluorescence, as a function of time, or per-ylene (shown on diagram) in glycerol following excitation by second har-monic argon-ion laser at 257.25 nm. Anisotropy r, is defined in terms of intensities of fluorescence parallel ($I_\parallel$) and perpendicular ($I_\perp$) to plane of polarization of laser exciting light, as $r = (I_\parallel - I_\perp)/(I_\parallel - 2I_\perp)$. Ordinate is r(t) channel number of linear in time.

$$r(t) = D(t)/S(t) = \sum_{i=1}^{5} a_i e^{-t/\tau_i} \tag{37}$$

where the τ_is are functions of the rotational diffusion coefficients around the three principal molecular axes and the a_is are functions of the direction cosines relating the absorption and emission transition dipoles to the principal rotation axes.

In a study on perylene (Figure 11) as a model standard,[70] these were further simplified by assuming that the rotational diffusion constants about the two in-plane axes are identical. This approximation reduces the number of terms in the summations from five to three and the anisotropy then can be expressed as[64]

$$r(t) = \frac{2}{5} e^{-6D_\perp t} \cdot \sum_{k=0}^{2} e^{-k^2(D_\parallel - D)t_\perp} \cdot F_k(\Theta_A \Theta_E \phi_{AE}) \tag{38}$$

where Θ_A and Θ_E are the polar angles between the absorption and emission dipoles and the unique symmetry axis, ϕ_{AE} is the difference in their azimuthal angles, and where $D_\parallel$ and $D_\perp$ refer to the rates of rotation about the unique symmetry axis perpendicular to this axis

$$F_0 = 1/4(3\cos^2\Theta_A - 1)(3\cos^2\Theta_E - 1) \tag{39}$$

$$F_1 = 3/4\sin2\Theta_E\sin2\Theta_A\cos\phi_{AE} \tag{40}$$

$$F_2 = 3/4\sin^2\Theta_E\sin^2\Theta_A\cos2\phi_{AE} \tag{41}$$

For perylene one additional simplification comes into play. Group theoretical consid-erations require that all $^1\pi\pi^*$ transitions are polarized within the molecular plane. This means that both $\Theta_A = \Theta_E = \pi/2$ and

$$r(t) = 0.10e^{-6D_\perp t} + 0.30 \cos2\phi_{AE} \cdot e^{-(2D_\perp + 4D_\parallel)t} \tag{42}$$

It should be recognized that the diffusion constants for rotation about the two in-plane symmetry axes are not rigorously equivalent; however, these rotational constants would have to be greatly different for more than two exponentials to be observed. This is due to the close interconnection between the original five exponentials which shows that $\tau_1 \sim \tau_5$ and $\tau_2 \sim \tau_3$ in Equations 33 to 37. We thus expect the fluorescence anisotropy of perylene to be well fitted by Equation 42 with the preexponential factors relating to the angle between the in-plane absorption and emission dipoles. That this is the case is illustrated in Figure 11, the results of an exacting study using laser excitation, full details of which are given elsewhere.[70]

Anisotropy parameters were derived from those obtained for D(t) and S(t) as follows:

$$D(t) = d_1 e^{-t/\tau_1} + d_2 e^{-t/\tau_2}$$

$$S(t) = s_0 e^{-t/\tau_f} \tag{43}$$

$$r(t) = d_1/s_0 \exp(-t(1/\tau_1 - 1/\tau_f)) + d_2/s_0 \exp(-t(1/\tau_2 - 1/\tau_f))$$

$$= r_1 \exp(-t/\tau_1^{rot}) + r_2 \exp(-t/\tau_2^{rot}) \tag{44}$$

The preexponential factors and rotational lifetimes given in Equation 44 were compared with those given by the two exponential model, Equation 42, to determine the rotational diffusion rates $D_{\parallel}$ and $D_{\perp}$ as well as the relative orientation of the transition dipoles in perylene.

Results showed that the fluorescence anisotropy of perylene in solutions of glycerol/water was well described by a biexponential model. For excitation at 257 nm, $r(t) = (0.77 \pm 0.006)\exp(-t(6D_{\perp})) - (0.233 \pm 0.006)\exp(2D_{\perp} + 4D_{\parallel}))$. A double exponential model should be rigorously correct for rotations in molecules with cylindrical symmetry (i.e., possessing a C_{∞} axis) and with electronic transitions polarized in the plane perpendicular to the C_{∞} axis. The first of these conditions is effectively met by the comparable rotational diffusion rates about any of the in-plane axes. The second requirement is satisfied by the $^1\pi\pi^*$ transitions monitored in this experiment. The above equation appears to describe perylene for a broad range of solvent viscosities and temperatures. Although $D_{\parallel}$ and $D_{\perp}$ depend on solvent, $D_{\parallel}/D_{\perp}$ is constant ($\sim 7/1$), indicating that the in-plane and out-of-plane rotations are equally affected by changes in environment. Both rotations must displace solvent molecules, thus blurring the distinction between sticking and slipping often applied to the two motions.

These experiments established perylene/glycerol/water as a potential standard for time-resolved anisotropy measurements which cannot be simply described by single exponential kinetics as might be found in polymers.

Improvements in time-resolved fluorescence techniques have pushed anisotropy measurements beyond the determination of single, average correlation times. Movements of macromolecules in solution have been shown to involve non- and multiexponential anisotropy decays. Considerable theoretical effort has provided models for the restricted motion of membrane probes, the internal motion of biopolymers, and the rotational motion of unsymmetrical fluorophores.

III. ALTERNATIVE METHODS OF MEASURING FLUORESCENCE DECAYS

Although the TCSPC technique has dominated fluorescence decay measurements, other techniques are available.

A. OPTICAL GATING
1. Kerr Cell

Early picosecond measurements were made using a Kerr-cell shutter and low repetition rate Nd laser. An example is given in Figure 12 in which the operation is seen to be very simple.[72] Doubled or quadrupled light from the Nd^{3+} laser excites the sample; the resultant fluorescence, however, is not seen by the photodetectors because the polarizers P_1 and P_2 are crossed. When a delayed fundamental laser pulse (at 1060 nm) arrives at the Kerr cell containing carbon disulfide, the enormous electric field of the light rotates the plane of polarization of light by 90°, opening the shutter for the duration of the delayed pulse. By making sequential measurements at different times, the intensity of fluorescence as a function of time can be recorded.

2. Up Conversion

A variation of this technique is shown in Figure 13 where the principle of up conversion is used to sample fluorescence intensity.[73,74] Here a delayed laser pulse is mixed with fluorescence in a crystal which gives an output at the sum of the laser and fluorescence frequencies. Thus fluorescence light at, for example, 900 nm (λ_1), mixed with 600 nm (λ_2) laser light gives an output at λ, given by Equation 45, which in this case is 360 nm, is easily detectable.

$$\frac{1}{\lambda} = \frac{1}{\lambda_1} + \frac{1}{\lambda_2} \tag{45}$$

B. STREAK CAMERA

Yet another means of detecting fluorescence photons is the use of a streak camera. The principle is very simple: fluorescence photons arriving at the cathode of the tube eject electrons which are accelerated towards a phosphor screen. However, en route they encounter a rapidly changing electric field, which means that electrons passing through this field at different times are deviated to different extents, thus causing a streak on the phosphorescent screen. This streak can be calibrated in terms of the arrival time of the photons causing ejection of electrons and is capable of about 10 ps time resolution.

Two versions of the streak camera system are available: the first operates the camera in the single shot mode in which a single event resulting from a single high power laser pulse can be recorded. The second operates in what is known as synchroscan mode, in which the voltage applied to the deflector plates is linearly ramped at a frequency proportional to the pulse repetition rate. In this mode, high repetition rate lasers can be used permitting averaging of the recorded signals. A typical layout of a synchroscan system is shown in Figure 14. It is based on a cavity-dumped, synchronously pumped dye laser. The optical delay enables recording of the excitation pulse and fluorescence emission to be recorded in the same time frame for ease of data analysis. A photodiode detector enables synchronization of the laser pulses with the scanning of the streak camera deflector plates. The streaked emission from the phosphor screen is detected using an optical multichannel analyzer (OMA), utilizing a Vidicon, Reticon, or CCD camera multichannel detector.

Such an experimental arrangement was used to produce fluorescence decay data on the poly(diacetylene) discussed above as exhibiting one-dimensional exciton diffusion, and shown in Figure 15.

C. FREQUENCY DOMAIN PHASE-MODULATION FLUORIMETRY

Although pulse-counting methods have tended to play a dominant role in the investigation of the time-resolved luminescence of synthetic polymers, the complementary technique, that of frequency domain phase-modulation fluorometry widely used in biochemistry, deserves

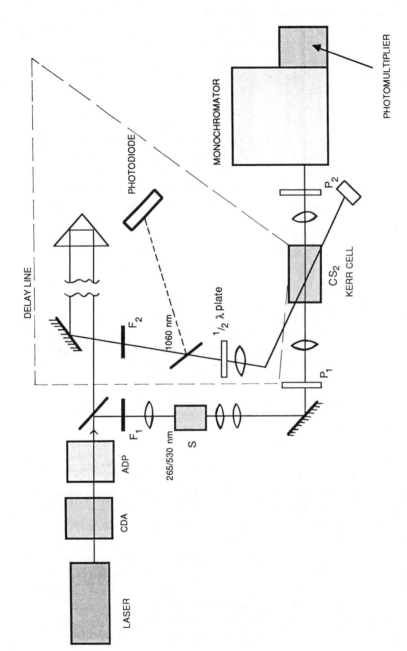

FIGURE 12. Early mode-locked arrangement for measuring picosecond fluorescence lifetimes. The laser was a Nd^{3+} in glass laser, with output at 1060 nm, doubled by CDA and quadrupled by ADP crystals to the 530- or 265-nm harmonics. The delay line, opening the CS$_2$ Kerr cell shutter is shown. (From Porter, G., Reid, E. S., and Tredwell, C. J., *Chem. Phys. Lett.*, 29, 469, 1974. With permission.)

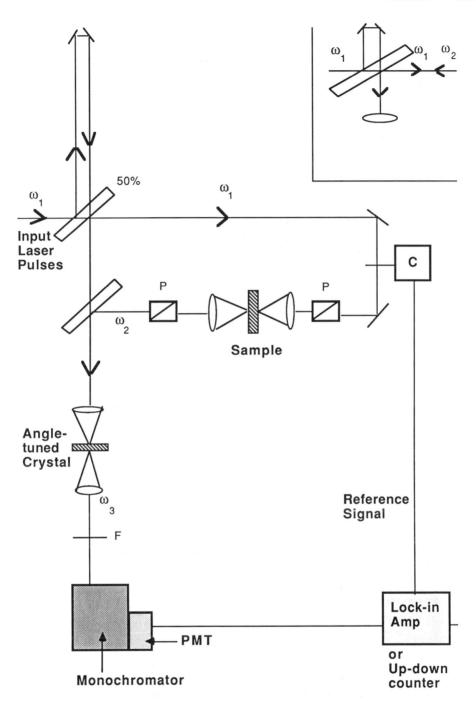

FIGURE 13.   Experimental arrangements for time-resolved fluorescence up-conversion: F, filters; P, polarizers; C, a sectored disc chopper connected to a lock-in amplifier or photon counter; ω_1, laser beam; ω_2, fluorescence. The crystal is LIO_3 (path length, 1 mm). The sample is contained in a glass cell of path length 1 mm mounted perpendicularly to the exciting beam and spun about an axis parallel to the beam, and the fluorescence is collected from the front face of the cell along the exciting beam axis. In an alternative method (inset) the sample is flowed through a cell or pumped through a nozzle to form a jet, and fluorescence is collected at 180° to the exciting beam.

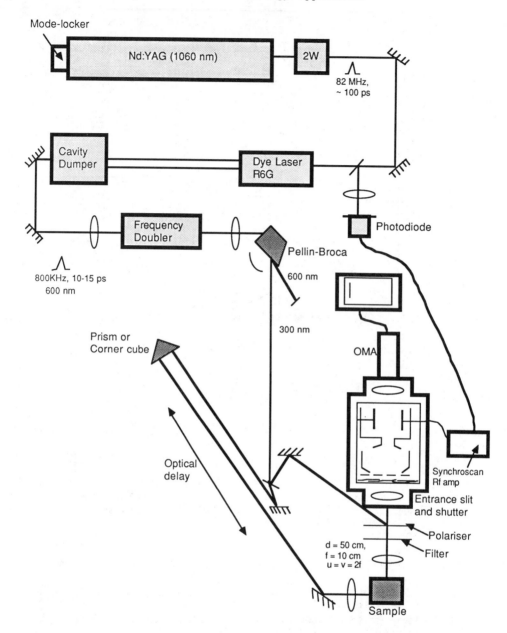

FIGURE 14.   Experimental arrangement for the study of fluorescence decay using a streak camera detection.

some coverage here. In essence, instead of employing pulsed excitation, the excitation beam is intensely modulated sinusoidally at a frequency comparable to the decay of the sample. Information concerning the decay law of the sample is obtained from the phase shift (ϕ) and the modulation (m) of the emission, both measured relative to the phase and modulation of the incident light. Whereas for pulse fluorometry, it is ideal to have a narrow excitation pulse; for phase-modulation fluorometry, it is ideal to have a wide range of modulation frequencies. As discussed above developments in laser technology have made progressively shorter pulses of light available, and hence, have stimulated growth in the field of pulse fluorometry. In contrast, most available phase-modulation fluorometers operate at only two or three modulation frequencies, which greatly limits the information content and/or resolving power. However, with the development of new modulators, and using laser excitation,

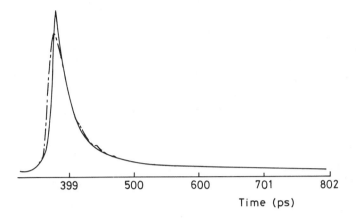

FIGURE 15. PDA-1OH luminescence decay at 4.2 K — raw data, solid line is an $\exp[-t/t_0)^{1/3}]$ fit. (From Hunt, I. G., Bloor, D., and Movaghar, B., *J. Phys. C*, 16, L623, 1983. With permission.)

phase-modulation methods have now come into their own.[76-80] An apparatus using this method is shown in Figure 16.

The output of a continuous laser is intensity modulated in an approximate sinusoidal manner using an electrooptic modulator which is placed between crossed polarizers. The modulator is biased away from the point of minimum transmission using either a DC voltage or an optical offset. The advantage of electrooptic modulators is that they provide modulation at any desired frequency up to at least 200 MHz. The modulated intensity is used to excite the sample, and the emission is observed using a photomultiplier. One photomultiplier is used as a phase reference to monitor the phase of the incident light. This photomultiplier can also be positioned in the T-format location, which is often useful for different polarized phase measurements. Measurements of the phase angle difference between the sample and reference emission is accomplished by rotation of a turret. The phase angle difference (ϕ_ω) and the relative modulation of the sample (m_ω) can be measured relative to either scattered light or a reference fluorophore. Detection of the emission is performed using the cross-correlation method. Specifically, the gain of the photomultiplier tubes (PMTs) are modulated at the frequency f + 25 Hz, where f is the modulation frequency of the incident light. The phase and modulation of the 25 Hz cross-correlation frequency is measured by a time-interval counter and a ratio digital voltameter. The two closely spaced frequency are provided by two frequency synthesizers which are each driven by the same crystal reference source (phase locked). A variety of CW lasers, including argon-ion and helium-cadmium, may be used. The analysis of data obtained using phase modulation can be achieved by the method of nonlinear least squares[76] in a manner analogous to the least-squares analysis of time-correlated, photon-counting data. Thus the measured phase and modulation data in the frequency domain are compared with values predicted by an assumed model. This method is distinct from earlier methods in that the system can be overdetermined by using many modulation frequencies and the analysis is not limited to a double-exponential decay.

The objective of the fitting procedure is twofold: (1) to test the validity of an assumed model, e.g., whether or not the decay is a single, double, or triple exponential and (2) to determine the parameters associated with the model and the uncertainties associated with the parameters.

The fitting routine finds the best possible match between the measured phase (ϕ_ω) and modulation (m_ω) values with those predicted on the basis of an assumed decay law ($\phi_{c\omega}$ and $m_{c\omega}$). The subscript $_\omega$ is an index of the modulation frequency ($\omega = 2\pi \cdot$ frequency) and

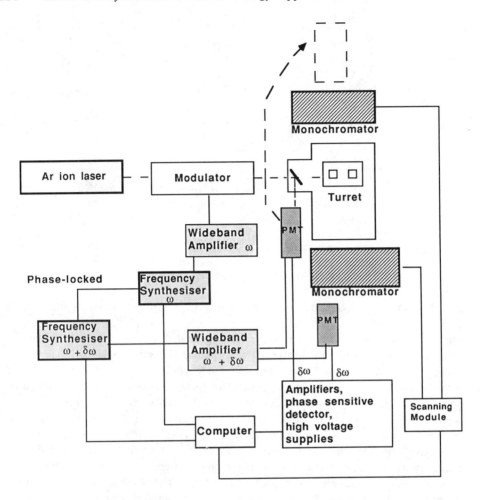

FIGURE 16. Schematic diagram of a variable-frequency phase modulation fluorometer. (From Lakowicz, J. R., Gratton, E., Cherek, M., Miliwal, B. B., and Laczko, G., *J. Biol. Chem.*, 259, 10967, 1984. With permission.)

the subscript c indicates calculates values. The impulse response function I(t), at any time t, is assumed to be represented by a sum of n exponential decays,

$$I(t) = \sum_{i=1}^{n} \alpha_i e^{-t/\tau_i} \tag{46}$$

For a mixture of noninteracting fluorophores the values of τ_i represent the individual lifetimes and the values of α_i represent the preexponential factors. The fractional steady-state intensity, f_i, of each component in the mixture is given by

$$f_i = \frac{\alpha_i \tau_i}{\sum_j \alpha_j \tau_j} \tag{47}$$

The interpretation of these parameters is more complex in the case of time-dependent solvent relaxation or other excited-state processes. The response functions may still be represented by Equation 46, except that the values of α_i and τ_i may not have physical significance.

Conveniently, one may predict the frequency-dependent values of ϕ and m for any assumed decay law. These values may be obtained from the sine, N_ω, and cosine, D_ω, transforms of the impulse response function,

$$N_\omega = \frac{\int_0^\infty I(t)\sin \omega t\, dt}{\int_0^\infty I(t)dt} \tag{48}$$

$$D_\omega = \frac{\int_0^\infty I(t)\cos \omega t\, dt}{\int_0^\infty I(t)dt} \tag{49}$$

where the subscript ω indicates the frequency ω. For a multiexponential decay these transforms are

$$N_\omega \cdot J = \sum_{i=1}^{n} \frac{\alpha_i \omega \tau_i^2}{(1 + \omega^2 \tau_i^2)} \tag{50}$$

$$D_\omega \cdot J = \sum_{i=1}^{n} \frac{\alpha_i \tau_i}{(1 + \omega^2 \tau_i^2)} \tag{51}$$

where $J = \Sigma_i \alpha_i \tau_i$. The phase and modulation values can be calculated from N_ω and D_ω and are given by

$$\phi_{c\omega} = \arctan(N_\omega/D_\omega) \tag{52}$$

$$m_{c\omega} = \sqrt{(N_\omega^2 + D_\omega^2)} \tag{53}$$

For a given sample the estimated values of α_i and τ_i are those that minimize χ^2, which is the error-weighted sum of the square deviations between the measured and calculated values. When both phase and modulation data are available χ^2 is given by

$$\chi^2 = \sum_\omega \frac{1}{\sigma_{\phi\omega}^2} (\phi_\omega - \phi_{c\omega})^2 + \sum_\omega \frac{1}{\sigma_{m\omega}^2} (m_\omega - m_{c\omega})^2 \tag{54}$$

where $\sigma_{\phi\omega}$ and $\sigma_{m\omega}$ are the estimated uncertainties in the phase and modulation data at each frequency, respectively. The choice of errors is discussed later. Minimization of χ^2 with respect to the parameters α_i and τ_i is a complex but thoroughly studied problem, as described elsewhere.

The equations described above are appropriate for phase and modulation values; measurement of the phase and modulation data is quite rapid and these data are easily measured at several emission wavelengths. Such additional measurements do not substantially increase the time needed to complete the experiment. The additional data enhance the ability to resolve closely spaced lifetimes.

Analysis of the multiple wavelength data is a straightforward extension of the procedure described above for a single emission bandpass. For a mixture of fluorophores, in which each fluorophore displays a single lifetime independent of emission wavelength, the decay law at any wavelength λ is given by the following:

$$I(\lambda, t) = \sum_{i=1}^{n} \alpha_i(\lambda) e^{-t/\tau_i} \qquad (55)$$

The preexponential factors depend on emission wavelength. The fractional steady-state intensity at each wavelength is given by

$$f_i(\lambda) = \frac{\alpha_i(\lambda)\tau_i}{\sum\limits_j \alpha_j(\lambda)\tau_j} \qquad (56)$$

The sine and cosine transforms are given by expressions comparable to Equations 48 to 51 except α_i is replaced by $\alpha_i(\lambda)$. The calculated phase and modulation values also depend on emission wavelength, i.e., $\phi_{c\omega}(\lambda)$ and $m_{c\omega}(\lambda)$. The value of χ^2 is given by

$$\chi^2 = \sum_\lambda \sum_\omega \frac{1}{\sigma_{\phi\omega}^2} [\phi_\omega(\lambda) - \phi_{c\omega}(\lambda)]^2 + \sum_\lambda \sum_\omega \frac{1}{\sigma_{m\omega}^2} [m_\omega(\lambda) - m_{c\omega}(\lambda)]^2 \qquad (57)$$

where $\phi_\omega(\lambda)$ and $m_\omega(\lambda)$ are the measured values at the indicated wavelengths. Using this procedure, the values of $\alpha_i(\lambda)$ and τ_i are varied until the minimum value of χ^2 is obtained. The τ_i values are forced to be the same at all emission wavelengths, but that the $\alpha_i(\lambda)$ values are wavelength dependent. This model is appropriate for systems in which the lifetime of each fluorophore is independent of emission wavelength, or for the case of a two-state, excited-state reaction. This model is not appropriate in instances where the decay times depend on emission wavelength, such as with solvent relaxation.

Since frequency-domain phase and modulation data are not widely available, the appearance of these data are not widely appreciated. *Simulated* data for double-exponential decays are shown in Figure 17. It was assumed that the data were measured at 16 modulation frequencies, ranging from 1 to 179.2 MHz. Measurements at this number of frequencies may be accomplished in a short time (minutes), and, unless an unreasonable number of frequencies is used, additional measurements are reported not to improve the resolution. The values in the frequency range are comparable to the upper limit of 140 MHz used in actual measurements. Random Gaussian error was added at a level comparable to the day-to-day fluctuations found in experimental data. For most simulations, a constant level of random error was independent of frequency, phase angle, or modulation. The average random error in the experimental data is $\pm 0.5°$ in phase angle and ± 0.005 in modulation. It should be stressed that 0.5% noise does not refer to 0.5% of the measured values, but rather 0.5% of the maximum possible phase and modulation values.

In the frequency domain, the effect of multiexponential decay is to increase the frequency range over which phase angles are substantially different from 0 to 90°, and over which the modulation is different from 1.0 and 0. The superiority of the dual experimental fit in Figure 17 over a single exponential fit is clear from the deviations, which for dual component decay are randomly distributed around zero, and are smaller. Adopting a three-component model did not result in a better fit.[76]

The technique may be used to record TRES and fluorescence anisotropy also. For a fluorophore in a polar and viscous environment the emission spectra shift to lower energies (longer wavelengths) with time subsequent to excitation. At any given emission wavelength the time-resolved decay of intensity is likely to be complex. Nonetheless, the intensity decay at each wavelength can be adequately represented by a sum of exponentials,

$$I(\lambda, t) = \sum_{i=1}^{n} \alpha_i(\lambda) e^{-t/\tau_i} \qquad (55)$$

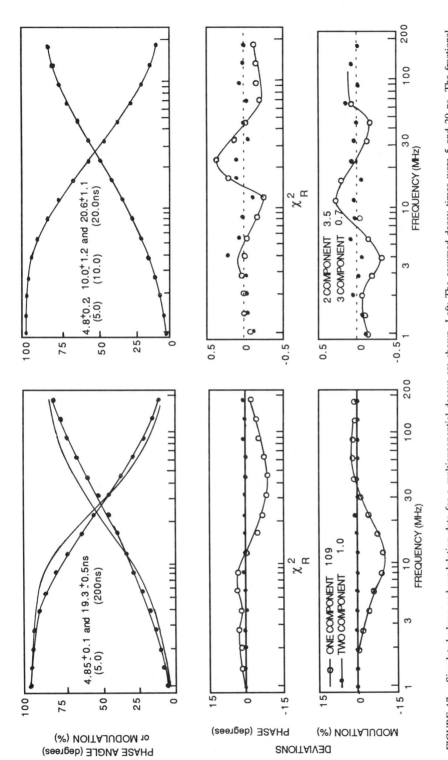

FIGURE 17. Simulated phase and modulation data for a multiexponential decays are shown. Left: The assumed decay times were 5 and 20 ns. The fractional intensities of the components were assumed to be equal ($f_1 = f_2 = 0.5$). The levels of added Gaussian noise were $\sigma_\phi = 0.5°$ and $\sigma_m = 0.005$. The dashed line (upper panel) and open circles (lower panels) indicate the best fit obtained using a single decay time. The solid lines and symbols represent the best fit using two decay times. Right: The assumed decay times were 5, 10, and 20 ns. The fractional intensities of the components were assumed to be equal ($f_1 = f_2 = f_3 = 1/3$). Gaussian noise was added at a level of 0.1% ($\sigma_\phi = 0.1°$ and $\sigma_m = 0.001$). The dashed lines and open symbols represent the best fit obtained using a two-component decay. The solid lines and symbols represent the best fit obtained using a three-component decay. (From Lakowicz, J. R., Gratton, E., Chevek, M., Miliwal, B. B., and Laczko, G., *J. Biol. Chem.*, 259, 10967, 1984. With permission.)

In this expression λ is the emission wavelength, τ_i the ith decay time, and $\alpha_i(\lambda)$ the preexponential factor of the ith component at the wavelength λ. The impulse-response function $[I,(\lambda,t)]$ at each wavelength is determined using the frequency-dependent (ω) phase (ϕ_ω) and modulation (m_ω) data at that same wavelength, minimizing χ^2 as before.

$$\chi^2 = \sum_\omega \frac{1}{\sigma^2_{\phi\omega}} (\phi_\omega - \phi_{c\omega})^2 + \sum_\omega \frac{1}{\sigma^2_{\phi\omega}} (m_\omega - m_{c\omega})^2 \tag{56}$$

In this expression the subscript $_c$ indicates values of ϕ and m calculated for an assumed decay law and $\sigma_{\phi\omega}$ and $\sigma_{m\omega}$ are the estimated uncertainties in the measured phase (ϕ_ω) and modulation (m_ω) values. ϕ_ω and m_ω refer to these values at each wavelength. The calculated values are obtained from

$$\phi_{c\omega}(\lambda) = \arctan(N_\omega(\lambda)/D_\omega(\lambda)) \tag{57}$$

$$m_{c\omega}(\lambda) = [N_\omega(\lambda)^2 + D_\omega(\lambda)^2]^{1/2} \tag{58}$$

where,

$$N_\omega(\lambda) \cdot J(\lambda) = \sum_i \frac{\alpha_i(\lambda)\omega\tau_i^2}{1 + \omega^2\tau_i^2} \tag{59}$$

$$D_\omega(\lambda) \cdot J(\lambda) = \sum_i \frac{\alpha_i(\lambda)\tau_i}{1 + \omega^2\tau_i^2} \tag{60}$$

and

$$J(\lambda) = \sum_i \alpha_i(\lambda)\tau_i \tag{61}$$

To calculate the time-resolved emission spectra the impulse functions at each wavelength are adjusted in magnitude such that the total area under the decay corresponds to the intensity of the steady-state spectrum at the same wavelength. Specifically, time-resolved emission spectra are calculated from

$$I'(\lambda, t) = \sum \alpha_i'(\lambda)e^{-t/\tau_i} \tag{62}$$

where,

$$\alpha_i'(\lambda) = \frac{\alpha_i(\lambda)F(\lambda)}{\sum_j \alpha_j(\lambda)F(\lambda)} \tag{63}$$

and $F(\lambda)$ is the steady-state intensity at λ. The time-resolved centers of gravity are calculated using

$$\nu(t) = \frac{\sum_\lambda I'(\lambda, t)\lambda^{-1}(10,000)}{\sum_\lambda I'(\lambda, t)} \tag{64}$$

The factor 10,000 is needed to change wavelengths (nanometers) to wave numbers (kK).

This expression is appropriate for equally spaced wavelengths and corrects for the dependence of slit width (in centimeters) on wavelength when the resolution in nanometers is constant. For measurement of decay of anisotropy, the measured quantities are the phase difference ($\Delta = \phi_\perp - \phi_\parallel$) between the perpendicular ($\phi_\perp$) and parallel ($\phi_\parallel$) polarized components of the emission. For an assumed decay law these values can be calculated from

$$\tan\Delta c = \frac{D_\parallel N_\perp - D_\perp N_\parallel}{N_\parallel N_\perp + D_\parallel D_\perp} \tag{65}$$

The simplest model is that in which anisotropy is assumed to decay with a single correlation time Φ.

$$r(t) = r_0 e^{-t/\Phi} \tag{66}$$

where r_0 is the anisotropy in the absence of rotational diffusion. Suppose the fluorescence intensity decays as a single exponential with decay time t. Then $N_\parallel$ and $D_\parallel$ are given by

$$N_\parallel = \frac{\omega}{3(\omega^2 + 1/\tau^2)} + \frac{2r_0\omega}{3[\omega^2 + (1/\tau + 1/\Phi)^2]} \tag{67}$$

$$D_\parallel = \frac{\omega}{3(\omega^2 + 1/\tau^2)} + \frac{2r_0(1/\tau + 1/\Phi)}{3[\omega^2 + (1/\tau + 1/\Phi)^2]} \tag{68}$$

The terms $N_\perp$ and $D_\perp$ are similar, except that $2r_0$ is replaced by $-r_0$. The values of the correlation times were determined by fitting the calculated (c) and measured (m) values of Δ to parameters which minimize χ^2.

$$\chi^2 = \sum_\omega \frac{1}{\sigma\Delta^2} [\Delta m - \Delta c]^2 \tag{69}$$

In this expression σ_Δ is the estimated uncertainty in each measured value of D. Thus the phase-modulation technique, using multifrequency modulation and laser excitation, is capable of all of the measurements that pulse-counting techniques are capable of. For complex decays, with closely spaced components, the frequency domain measurements may offer some advantages. The pulse-counting technique in general has the virtue of results which are more easily conceptualized.

IV. ALTERNATIVE METHODS OF STUDYING TRANSIENT SPECIES

A. TRANSIENT ABSORPTION METHODS

Monitoring of the electronic (and in some cases vibrational) absorption spectrum of transient species following pulsed excitation permits both the absorption spectrum of the transient and kinetics of formation and decay of the transient to be measured (Figure 18). Flash photolysis has been in existence more than 40 years; the use of lasers as excitation source is already 25 years old. The field has been reviewed very thoroughly recently,[81] and thus will not be covered again in detail here. Nanosecond measurements using doubled, tripled, and quadrupled Nd^{3+} lasers, ruby lasers, and (more recently) excimer lasers are now routine using techniques typified by Figure 19. Here we highlight picosecond measurements for which experimental techniques are much more exacting.

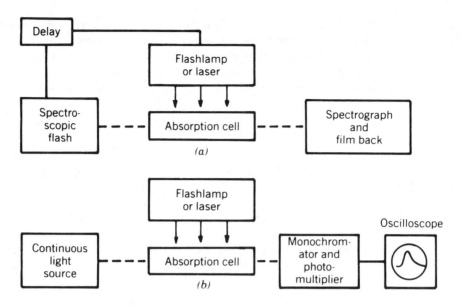

FIGURE 18.   Schematic diagrams illustrating principles of flash photolysis and flash spectroscopy (a) and kinetic spectrometry (b). (From West, M. A., *Investigation of Rates and Mechanisms of Reactions*, Vol. 6, Bernasconi, C. F., Ed., Wiley-Interscience, New York, 1986, 391. With permission.)

A laser system which allows picosecond transient absorption measurements to be made is shown in Figure 20.[82] It is perhaps worthwhile describing this in some detail. Unlike the case where fluorescence is being observed, where sensitive detection methods mean that quite weak lasers can be used to excite samples for pump and probe methods, very high intensities are required as well as short pulse durations. This is achieved as follows: light from a synchronously pumped dye laser in the form of picosecond pulses is amplified in four stages, A1 to A4 with a Nd/YAG laser pumping the amplifying dyes. This yields pulses of around 1 ps duration, probe energy >1 mJ, wavelength 595 to 605 nm, at a pulse repetition rate of 10 Hz. Any transient species produced in the sample chamber, SC, must be interrogated with a delayed pulse of white light. This is produced by continuum generation, i.e., white light produced by passage of an intense laser pulse through a liquid, in this case water. The time delay between pump and white light probe can be any time between 0.1 ps and 12 ns. Light absorbed from the white light continuum is measured by comparison of beams passed through sample cell and reference cells, analyzed with a spectroscope (monochromator), and detected using a vidicon, a detector array in the focal plane of the monochromator which allows simultaneous recording of intensities at a whole range of wavelengths. Thus the complete absorption spectrum is analyzed at every laser shot and averaging over many shots is usually performed. There are few examples of such studies in synthetic polymers, and thus as an example of the use of the technique. Figure 21 shows results of some elegant pump and probe experiments which have been carried out on the photosynthetic units from pea chloroplasts. The spectra show clearly transient absorption changing on a time scale of tens of picoseconds. Briefly two main spectral features can be observed in Figure 21, one at 690 and one at 700 nm. The one at 690 nm is dominant in early times and decays with a lifetime of approximately 15 to 20 ps. This feature undergoes a blue shift as it decays, finally being centered at 675 nm. The second main spectral feature, centered at 700 nm, is very much narrower than the 670 nm signal and occurs only in the spectra of samples with chlorophyll P700 chemically reduced. The signal at 690 nm can be attributed to the excitation of antenna chlorophyll to the singlet state (see Figure 21), while the signal at 700 nm can be attributed to the photooxidation of P700 molecules. Such

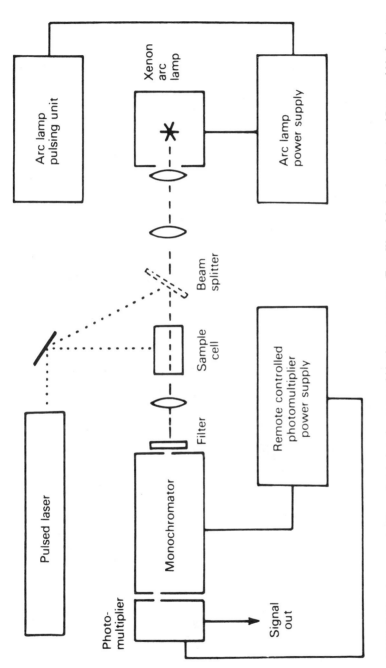

FIGURE 19.   Arrangement of laser photolysis and nanosecond kinetic spectrometry. (From West, M. A., *Investigation of Rates and Mechanisms of Reactions*, Vol. 6, Bernasconi, C. F., Ed., Wiley-Interscience, New York, 1986, 391. With permission.)

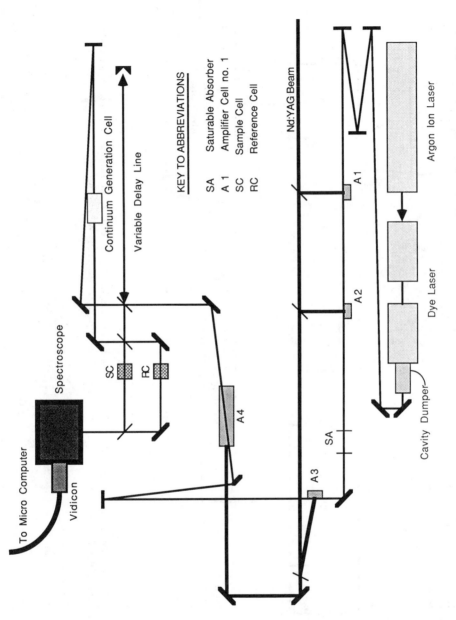

FIGURE 20. Picosecond transient absorption spectrometer, utilizing the four-stage amplification of a synchronously pumped dye laser.

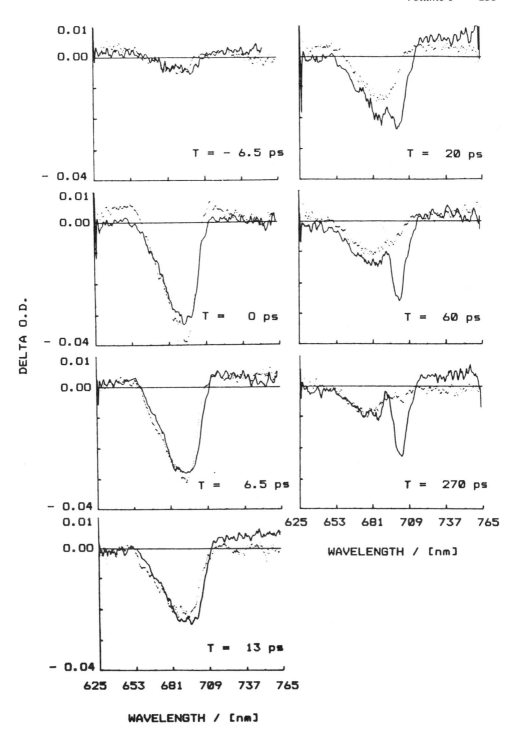

FIGURE 21. Transient absorption spectra of photosystem one reaction centers for pea chloroplasts. ——— shows spectra with chorophyll P700 chemically reduced, (----) with P700 chemically oxidized. The delay times in picoseconds are shown.

experiments provide an important means of studying the very fast chemical processes which occur very early in photosynthesis.

Transient absorption measurements require that the sample under investigation be transparent and nonscattering. For many applications, particularly those in the field of polymer science, this condition may not be met. A technique has been developed which permits recording of the diffuse reflectance spectra of transients produced upon laser flash photolysis.[83] An apparatus for the time-domain picosecond is shown in Figure 22. Here picosecond laser pulses, after three-stage dye amplification at 10 Hz to 500 μJ per pulse are frequency doubled, the residual fundamental (probe) light and frequency-doubled (pump) light are dispersed by a prism. The probe light is focused into a 4-cm pathlength cell containing water. The resultant picosecond continuum is recollimated and directed via a computer-controlled variable optical delay line onto the sample. The pump light is directed through a frequency programmable chopper operating at 35 Hz and synchronized with the dye amplifier. This removes alternate pump shots. The pump beam is then directed onto the sample as collinearly with the probe beam as possible. Small fractions of the pump and of the probe beams are deflected using beam splitters and their intensities monitored as references.

Detection is carried out with the aid of Hamamatsu photodiodies and necessary circuitry. Neutral density filters are placed in the pump reference beam as necessary. The probe reference, and probe light diffusely reflected off the sample, are collected by means of two fiber optics and directed into the input slit of a monochromator. The two beams are then detected by two photodiodes positioned closely against the exit monochromator slit. The monochromator enables the probe wavelength of interest to be selected within the confined of the spectral distribution produced by the probe fundamental in the continuum generation cell. The probe reference is passed through the same monochromator so that the beams remain faithful to each other; no optical or electronic cross-talk occurs between the two signals. The signals from the three detectors are fed to three separate gated integrators and boxcar averagers which are triggered at 10 Hz synchronized with the laser amplifier. The signals are integrated over their duration and recovered as an analogue signal and read by an A/D converter/computer interface which stores and transfers digitized signals to the computer. A typical data collection sequence would be as follows: the computer sets the optical delay to a preprogrammed position; if emission is to be measured, the pump shutter is opened. A sequence of 200 shots is recorded and the data stored. Both shutters are then opened and a second sequence of 200 shots is collected, transferred, and stored. The shutters are closed and the variable delay moves to the next programmed position. Both sequences are sorted by the computer into pump on and pump off sets. Hence, the first sequence gives the background signals (no light on detectors) and, if measured, the emission from the sample with corresponding pump reference signals. The latter data are fitted by a least squares polynomial curve fitting routine to enable a value for the emission intensity, E to be interpolated for any known pump intensity, P.

$$E(P) = \sum_{i=1}^{n+1} C_i p_i \qquad (70)$$

Where C_i is the ith coefficient of the calculated polynomial of order n, the second sequence gives sample detector intensities, D, and the probe reference intensities, N, with both pump on and pump off. A straight-line fit is performed on the pump off data to give the relationship between D_{off} and N.

$$D_{off} = \alpha N + \beta \qquad (71)$$

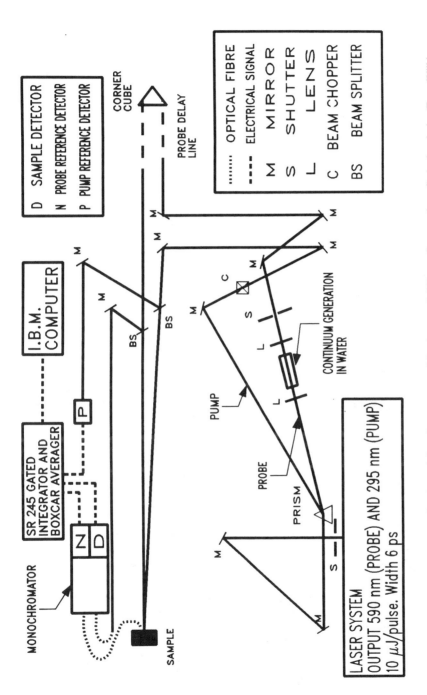

FIGURE 22. Schematic diagram of optics and detection system used in picosecond diffuse reflectance laser flash photolysis. (From Wilkinson, F., Willsher, C. J., Leicester, P. A., Barr, J. R. M., and Smith, M. J. C., *J. Chem. Soc. Chem. Commun.*, 1216, 1986. With permission.)

The quantity required is the relative change in reflectance which is equal to the relative change in the sample detector signals, D.

$$\frac{\Delta R}{R} = \frac{\Delta D}{D} = \frac{D_{off} - D_{on}}{D_{off}} \qquad (72)$$

for pump intensity P.

If emission is to be accounted for, this must be calculated using Equation 70 and the pump intensity P. D_{off} must be calculated using Equation 71 taking the value for N corresponding to D_{on}.

Hence,

$$\frac{\Delta R}{R} = \frac{\alpha N + \beta - (D_{on} - E(P) - Pg)}{\alpha N + \beta - Dg} \qquad (73)$$

where Pg and Dg are the background signals for the pump reference detector and sample detector, respectively. When the relative change in reflectance is proportional to pump energy the value obtained by using Equation 73 is normalized to P and all values are averaged. For nonlinear dependence of $\Delta R/R$ and P, the polynomial is calculated for $\Delta R/R$ upon P, and a typical value of P chosen to yield the actual absorption change and the same value of P used for a set of probe delays.

A typical result, corrected for emission, obtained with a highly scattering microcrystalline medium is shown in Figure 23.

B. TRANSIENT HOLOGRAPHIC GRATING TECHNIQUE

Another detection technique has been used to great advantage in the study of synthetic polymers, which is a variation on a transient absorption method, utilizing a transient holographic grating to study the time dependence of the polarization of absorbing molecules.[85,86] These time-resolved optical experiments rely on a short pulse of exciting plane-polarized light from a laser to photoselect chromophores which have their transition dipoles orientated in the same direction as the polarization of the exciting radiation, giving a nonrandom orientational distribution of excited state dipoles which, in the case of polymers, will randomize in time due to motions of the polymer chains to which the chromophores are attached. The precise way in which the orientated distribution randomizes depends upon the detailed character of the molecular motions taking place and is described by the orientation autocorrelation function as in time-resolved fluorescence methods. In the transient holographic grating technique a pair of polarized excitation pulses is used to create the anisotropic distribution of excited state transition dipoles. The motions of the polymer backbone are monitored by a probe pulse which enters the sample at some chosen time interval after the excitation pulses and probes the orientational distribution of the transitions dipoles at that time. By changing the time delay between the excitation and probe pulses, the orientation autocorrelation function of a transition dipole rigidly associated with a backbone bond can be determined. In transient absorption techniques, the time resolution is limited by laser pulse widths and not by the speed of electronic detectors. Fast time resolution is necessary for the experiments reported here because of the subnanosecond time scales for local motions in very flexible polymers such as polyisoprene. Figure 24 shows a block diagram of a typical apparatus used in these experiments.

The frequency doubled output of a Q-switched, mode-locked Nd:YAG laser synchronously pumps two dye lasers. One of the dye lasers (DL1) is operated at a wavelength of 649 nm. A cavity-dumped single pulse from this dye laser is frequency summed with a single pulse of the YAG fundamental (1064 nm) to produce the excitation wavelength of

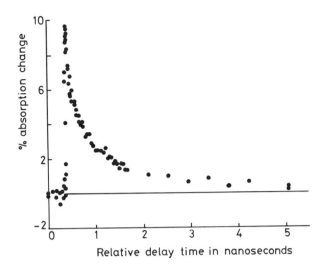

FIGURE 23. Transient absorption decay for microcrystalline 1, 5-di-phenyl-3-styryl-2-pyrazoline (emission corrected). (From Wilkinson, F., Willsher, C. J., Leicester, P. A., Barr, J. R. M., and Smith, M. J. C., *J. Chem. Soc. Chem. Commun.*, 1216, 1986. With permission.)

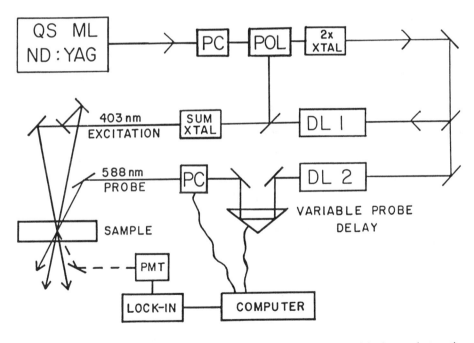

FIGURE 24. This figure shows a block diagram of a typical apparatus used in the transient grating experiments. PC: Pockels' cell; POL: polarizer with escape window; DL1, DL2: cavity-dumped dye lasers; and PMT: photomultiplier tube. (From Moog, R. S., Ediger, M. D., Boxer, S. G., and Fayer, M. D., *J. Phys. Chem.*, 86, 4694, 1982. With permission.)

403 nm. This pulse is beam split to form the grating. A second dye laser (DL2) provides the probe pulse (588 nm) which is sent down an optical delayline before entering the sample at the Bragg angle relative to the grating formed by the excitation beams. Varying the position of a retroreflector and on the optical delay line varies the timing between the excitation and probe pulses and thus provides the time base for the experiment. The diffracted

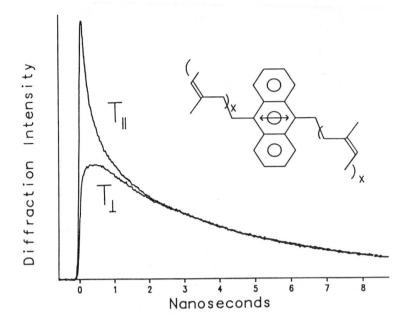

FIGURE 25.   Transient grating decays for 9,10-vis(methylene)-anthracene labeled polyisoprene in dilute hexane solution. $T_\parallel$ and $T_\perp$ are the diffraction efficiencies of the
grating for the probe beam polarized parallel and perpendicular to the excitation beams
(Equations 74 and 75). The structure of the anthracene-labeled polyisoprene is also
displayed, with the position of the transition dipole indicated by a double arrow. (From
Waldow, D. A., Hyde, P. D., Ediger, M. D., Toshiaki, K., and Ito, K., *Photophysics
of Polymers,* (ACS Symp. Ser., Vol. 358), Hoyle, C. E. and Torkelson, J. M., Eds.,
American Chemical Society, Washington, D. C., 1987, 68. With permission.)

probe beam is detected by a photomultiplier and a lock-in amplifier. A microcomputer
controls the data acquisition by varying the probe polarization and monitoring the diffracted
signal strength and the delayline position.

In the transient grating experiment, optical interference between two crossed laser pulses
creates a spatially periodic intensity pattern in an absorbing sample. This results in a spatial
grating of excited states which then diffracts a third (probe) beam brought into the sample
at some later time. The two observable experimental quantities are the intensity of the
diffracted signal for the probe beam polarized parallel ($T_\parallel(t)$) and perpendicular ($T_\perp(t)$) to
the polarization of the excitation beams. A typical result is shown in Figure 25. $T_\parallel(t)$ and
$T_\perp(t)$ for anthracene-labeled polyisoprene in dilute hexane solution is depicted. The sharp
rising edge in the data indicates the time when the excitation pulse enters the sample. At
times very soon after this, there is a large difference between the diffracted signal for the
two different probe polarizations. $T_\parallel(t)$ is larger than $T_\perp(t)$ since a larger number of excited
state transition dipoles are orientated parallel to the excitation pulse polarization. As molecular motions of the polymer occur, the excited state transition dipoles randomize their
orientations and the difference between $T_\parallel(t)$ and $T_\perp(t)$ goes to zero. Both curves continue
to decay due to the depopulation of the excited states with a lifetime of ~8 ns. Quantitatively,
the shapes of $T_\parallel(t)$ and $T_\perp(t)$ are given in terms of the orientation autocorrelation function
CF(t) and the excited state decay function K(t) as follows:

$$T_\parallel(t) = \{K(t)(1 + 2r(t))\}^2 \tag{74}$$

$$T_\perp(t) = \{K(t)(1 - r(t))\}^2 \tag{75}$$

To these equations, r(t) is the time-dependent anisotropy function which is proportional to CF(t) as

$$r(t) = r_0 CF(t) \tag{76}$$

where r_0 is the fundamental anisotropy for the transition being observed. r(t) can be obtained from the transient grating signals by the following manipulation:

$$r(t) = \frac{\sqrt{T_\parallel(t)} - \sqrt{T_\perp(t)}}{\sqrt{T_\parallel(t)} + 2\sqrt{T_\perp(t)}} \tag{77}$$

The experimental anisotropy contains information about molecular motion, but is independent of the excited state lifetime. Equation 77 indicates that the orientation autocorrelation function can be obtained directly and unambiguously, within the multiplicative constant r_0, from the results of the transient grating experiment.

By setting the time delay between the excitation and the probe beams to a given value and then alternating the polarization of the probe beam, the experimental anisotropy r(t) for time t can be obtained using Equation 77. In this mode of data acquisition, the experimental anisotropy is obtained at a relatively small number of time points but with quite good precision at each point. The results of such an experiment on a dilute solution of labeled polyisoprene in hexane are shown in Figure 26. The smooth curve running through the data points is the best fit theoretical curve using the Hall-Helfand model[90] (see below). In general, the fitting of the anisotropy involved three adjustable parameters: r_0 and the two parameters in each correlation function.

Models are required to test the variation in anisotropy with time in these polymer samples. The second order orientation autocorrelation function measured in this experiment is given by:

$$CF(t) = <P_2(\hat{\mu}(t) \cdot \hat{\mu}(O))> \tag{78}$$

In this expression, P_2 is the second Legendre polynomial and $\mu(t)$ is a unit vector with the same orientation as the transition dipole at time t. The brackets indicate an ensemble average overall transition dipoles in the sample. The correlation function has a value of one at very short times when the orientation of $\mu(t)$ has not changed from its initial orientation. At long times, the correlation function decays to zero because all memory of the initial orientation is lost. At intermediate times, the shape of the correlation function provides detailed information about the types of motions taking place. Table 4 shows the three theoretical models for the correlation function used to compare with experiment.[85,90] All describe backbone motions of the polymer.

C. TIME-RESOLVED RAMAN METHODS

Transient absorption (and reflectance) spectra of complex polyatomic molecules are, in principle, broad and featureless and do not yield structural formation. Vibrational spectra are capable of yielding structural details and recently, time-resolved resonance Raman (TR^3) spectroscopy has been used to advantage to study transients in photochemical systems. The basis of the method is summarized in Figure 27, and an apparatus employing an excimer laser photolyzing flash and an excimer laser-pumped dye laser for resonant enhancement of Raman scattering is shown in Figure 28.[84]

A typical result of the use of this system, showing the time dependence of the Raman spectrum of the radical anion of the sulfonated anthraquinone is shown in Figure 29. This

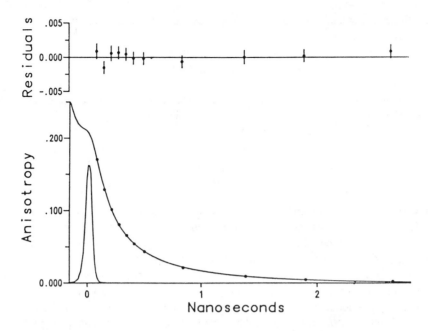

FIGURE 26.   Time-dependent anisotropy for anthracene-labeled polyisoprene in dilute hexane solution. The experimental anisotropy was obtained by setting the delay between the
excitation and probe pulses to a given position and then varying the polarization of the probe
beam. In the bottom portion of the figure, the smooth curve through the data is the best fit
to the Hall-Helfand model (τ_1 = 236 ps, τ_2 = 909 ps, and r(0) = 0.250). Unweighted
residuals for the best fit to this model are shown along with the experimental error bars in
the top portion of the figure. Note that the residuals are shown on an expanded scale ($10\times$).
The instrument response function is indicated at the left. (From Waldow, D. A., Hyde,
P. D., Ediger, M. D., Toshiaki, K., and Ito, K., *Photophysics of Polymers*, (ACS Symposium
Ser., Vol. 358), Hoyle, C. E. and Torkelson, J. M., Eds., American Chemical Society,
Washington, D. C., 1987, 68. With permission.)

TABLE 4
Correlation Functions

Function	Ref.
$Exp(-t/\tau_1)exp(-t/\tau_2)I_0(t/\tau_1)$	87
$0.5\sqrt{\pi/t}\{1/\sqrt{\tau_2} - 1\sqrt{\tau_1}\} - 1\{erfc\sqrt{t/\tau_1}$	88
$- erfc\sqrt{t/\tau_2}\}$	
$Exp(-t/\tau_1)exp(-t/\tau_2)\{I_0(t/\tau_1) + I_1(t/\tau_1)\}$	89

example is in the nanosecond time domain, but picosecond studies can also now be under-
taken, particularly with the use of new, high repetition-rate copper vapor lasers.

V. CONCLUSION

The description of techniques above which utilize pulsed lasers on the nanosecond and
subnanosecond time scale is not exhaustive, but has attempted to single out techniques of
special interest to polymer scientists.

ACKNOWLEDGMENTS

We are grateful to the Science and Engineering Research Council, U.K. and the U.S.
Army European Research Office for financial support during the preparation of this manuscript.

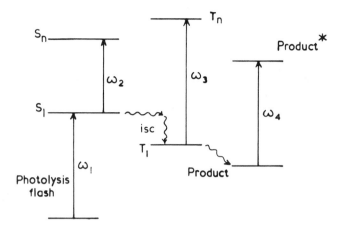

FIGURE 27. Basis of TR^3 measurement. Pump laser ω_1 excites singlet state of compound. As probe laser scans t through different resonances ω_2, ω_3, ω_4, the nonlinear resonant enhancement of Raman scattering from, in this case, the S_1 state, the T_1 state and a product can be selectively enhanced. Spectra derived are much more clearly separated than absorption spectra.

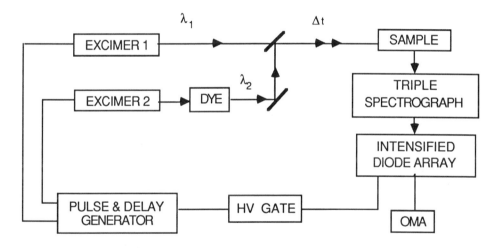

FIGURE 28. Experimental system based upon excimer laser excitation and excimer-pumped dye laser probing for TR^3 measurements.

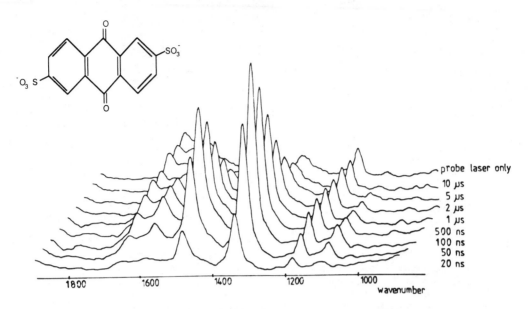

FIGURE 29.   Time dependence of spectrum observed on photolysis of anthraquinone-2,6-disulfonate (5×10^{-3} mol dm^{-3}) in the presence of NaNO$_2$ (0.1 mol dm^{-3}). Pump laser, 337 nm and probe laser, 480 nm. The spectrum corresponds to that of the radical anion.

REFERENCES

1. **Somersall, A. C. and Guillet, J. E.,** Photoluminescence of synthetic polymers, *J. Macromol. Sci. Rev. Macromol. Chem.,* C13, 135, 1975.
2. **Phillips, D.,** Complex Decay of Fluorescence in Synthetic Polymers, in *Photophysics of Polymers,* (ACS Symp. Series 358), Hoyle, C. E. and Torkelson, J. M., Eds., American Chemical Society, Washington, D.C., 1987, 308.
3. **Ware, W. R.,** Transient luminescence measurements, in *Creation and Detection of the Excited State,* Vol. 1, ed., Lamola, A. A., Ed., Marcel Dekker, New York, 1971, 213,
4. **Yguerabide, J.,** Nanosecond fluorescence spectroscopy, *Methods Enzymol.,* 26, 498, 1972.
5. **Badea, M. G. and Brand, L.,** Time-resolved fluorescence measurements, *Methods Enzymol.,* 61, 378, 1979.
6. **Knight, A. E. W. and Selinger, B. K.,** Single-photon decay spectroscopy, *Aust. J. Chem.,* 26, 1, 1973.
7. **Isenberg, I.,** Time-decay fluorimetry by photon counting, in *Biochemical Fluorescence Concepts,* Vol. 1, Chen, R. F. and Edelhoch, H., Eds., Marcel-Dekker, New York, 1984, 43.
8. **Pultney, S. K.,** Single-photon detection and timing: experiment and techniques, *Adv. Electron. Electron Phys.,* 31, 39, 1972.
9. **O'Connor, D. V. and Phillips, D.,** *Time-Correlated Single-Photon Counting,* Academic Press, London, 1984.
10. **Phillips, D., Drake, R. C., O'Connor, D. V., and Christensen, R. L.,** Time-correlated single-photon counting (TCSPC) using laser excitation, *Anal. Instrum.,* 14, 267, 1985.
11. **Lofroth, J. E.,** Deconvolution of single-photon counting data with a reference method and global analysis, *Eur. Biophys. J.,* 13, 45, 1985.
12. **Vos, K., van Hoek, A., and Visser, A. J. W. G.,** Application of reference convolution method to tryptophan fluorescence in proteins, *Eur. J. Biochem.,* 165, 55, 1987.
13. **Holzapfel, C.,** On statistics of time-to-amplitude converter systems in photon counting devices, *Rev. Sci. Instrum.,* 45, 894, 1974.
14. **Demas, J. N. and Adamson, A. W.,** Evaluation of photoluminescence lifetimes, *J. Phys. Chem.,* 75, 2463, 1971.
15. **Jezequel, J. Y., Bouchy, M., and Andre, J. C,** Estimation of fast fluorescence lifetimes with a single-photon counting apparatus and the phase plane method, *Anal. Chem.,* 54, 2199, 1982.

16. **Isenberg, I.,** On the theory of fluorescence decay experiments. I. Non-random distortions, *J. Chem. Phys.,* 59, 5696, 1973.

17. **Isenberg, I.,** On the theory of fluorescence decay experiments. II. Statistics, *J. Chem. Phys.,* 59, 5708, 1973.

18. **Small, E. W. and Isenberg, I.,** On moment index replacement, *J. Chem. Phys.,* 66, 3347, 1977.

19. **O'Connor, D. V., Ware, W. R., and Andre, J. C.,** Deconvolution of fluorescence decay curves. A critical comparison of techniques, *J. Phys. Chem.,* 83, 1333, 1979.

20. **Ameloot, M. and Hendricks, H.,** A non-iterative Laplace deconvolution method with the power of combine various experiments, in *Deconvolution and Reconvolution of Analytical Signals,* Bouchy, M., Ed., ENSIC-INPL, Nancy, France, 1982, 277.

21. **Valeur, B. and Moriez, J.,** Analyse des courbes de decroissance multiexponentialles par la methode des fonctions modulatrices — application a la fluorescence, *J. Chim. Phys. Physicochim. Biol.,* 70, 500, 1973.

22a. **Provencher, S. W.,** An eigenfunction expansion method for the analysis of exponential decay curves, *J. Chem. Phys.,* 64, 2772, 1976.

22b. **Provencher, S. W.,** A Fourier method for the analysis of exponential decay curves, *Biophys. J.,* 16, 27, 1976.

23. **Wild, U. P., Holzwarth, A. R., and Good, H.P.,** Measurement and analysis of fluorescence decay curves, *Rev. Sci. Instrum.,* 38, 1621, 1977.

24. **Andre, J. C., Viovy, J. L., Bouchy, M., Vincent, L. M., and Valeur, B.,** Use of regularisation operators together with lagrange multipliers in numerical deconvolution of fluorescence decay curves, in *Deconvolution and Reconvolution of Analytical Signals,* Bouchy, M, Ed., ENSIC-INPL, Nancy, France, 1982, 223.

25. **Bevington, P. R.,** *Data Reduction and Error Analysis for the Physical Sciences,* McGraw-Hill, New York, 1969, 235.

26. **Durbin, J. and Watson, G. S.,** Testing for serial correlation and least squares regression. I, *Biometrika,* 37, 409, 1950.

27. **Durbin, J. and Watson, G. S.,** Testing for serial correlation in least squares regression. II, *Biometrika,* 38, 159, 1951.

28. **Schnechenburger, M., Frenz, M., Tsuchiya, Y., Denzer, U., and Schleinkofer, L.,** Picosecond fluorescence microscopy for measuring chlorophyll and porphyrin components in conifers and cultured cells, *Lasers in the Life Sciences,* 4, 299, 1987.

29. **Felker, P. M. and Zewail, A. H.,** Direct picosecond time-resolution of dissipative intramolecular vibrational energy redistribution (IVR) in isolated molecules, *Chem. Phys. Lett.,* 108, 303, 1984.

30a. **Yamazaki, I., Tamai, N., Kume, M., Tsuchiya, H., and Oba, K.,** Microchannel-plate applicability to the time-correlated photon-counting method, *Rev. Sci. Instrum.,* 56, 1187, 1985.

30b. **Spears, K. G., Kramer, L. E., and Hoffland, L. D.,** Sub-nanosecond time-correlated photon-counting with tunable lasers, *Rev. Sci. Instrum.,* 49, 255, 1978.

30c. **James, D. R., Demmer, D. R. M., Venall, R. E., and Steer, R. P.,** Performance characteristics of a small side-window photomultiplier in laser single-photon fluorescence decay measurements, *Rev. Sci. Instrum.,* 54, 1148, 1983.

31. **Phillips, D., and Soutar, I.,** *Analysis of Fluorescence Decay Data from Synthetic Polymers: Heterogeneity, Motion and Migration,* Winnik, M. A., Ed., (NATO ASI Ser. C. Vol. 182), D. Reidel, Dordrecht, Netherlands, 1985, 129.

32. **Albery, W. J., Barlett, P. N., Wilde, C. P., and Darwent, J. R,** A general model for dispersed kinetics in heterogeneous systems, *J. Am. Chem. Soc.,* 107, 1854, 1985.

33. **Roberts, A. J., O'Connor, D. V., and Phillips, D.,** Multicomponent fluorescence decay in vinyl aromatic polymers, in *Luminescence from Biological and Synthetic Macromolecules,* Morowtz, H. and Steinberg, I. Z., Eds., New York Academy of Sciences, New York, 1981, 109.

34. **Roberts, A. J., Soutar, I., and Phillips, D.,** Time-resolved fluorescence spectroscopic study of excimer formation in poly(acenaphthalene) and an alternating acenaphthalene/maleic anhydride copolymer, *J. Polym. Sci. Polym. Lett. Ed.,* 18, 123, 1980.

35. **Phillips, D., Roberts, A. J., and Soutar, I.,** Transient decay studies of photophysical processes in aromatic polymers. I. Multiexponential fluorescence decays in copolymers of 1-vinylnaphthalene and methylmethacrylate, *J. Polym. Sci. Polym. Phys.,* 18, 2401, 1980.

36. **Phillips, D., Roberts, A. J., and Soutar, I.,** Transient decay studies of photophysical processes in aromatic polymers. II. Investigation of intramolecular excimer formation in copolymers of 1-vinylnaphthalene and methyl methacrylate, *Polymer,* 22, 293, 1981.

37. **Phillips, D., Roberts, A. J., and Soutar, I.,** Transient decay studies of photophysical processes in aromatic polymers. III. Concentration dependence of excimer formation in copolymers of acenaphthalene and methylmethacrylate, *Eur. Polym. J.,* 17, 101, 1981.

38. **Phillips, D., Roberts, A. J., and Soutar, I.,** Transient decay studies of photophysical processes in aromatic polymers. IV. Intramolecular excimer formation in homopolymers of 1-vinylnaphthalene, 2-vinylnaphthalene and 1-naphthylmethacrylate, *Polymer,* 22, 427, 1981.

39. **Phillips, D., Roberts, A. J., and Soutar, I.,** Transient decay studies of photophysical processes in aromatic polymers. V. Temperature dependence of excimer formation and dissociation in copolymers of 1-vinyl-naphthalene and 1-naphthylmethacrylate, *J. Polym. Sci., Polym. Phys.,* 20, 411, 1982.

40. **Phillips, D., Roberts, A. J., and Soutar, I.,** Intramolecular energy transfer, migration and trapping in polystyrene, *Macromolecules,* 16, 1593, 1983.

41. **Phillips, D., Roberts, A. J., Rumbles, G., and Soutar, I.,** Transient decay studies of photophysical processes in aromatic polymers. VII. Studies of the molecular weight dependence of intramolecular excimer formation in polystyrene and styrene-butadiene block copolymers, *Macromolecules,* 16, 1597, 1983.

42. **Rumbles, G.,** Photophysics of Synthetic Polymers, Ph.D. thesis, University of London, 1984.

43. **James, D. R. and Ware, W. R.,** A fallacy in the interpretation of fluorescence decay parameters, *Chem. Phys. Lett.,* 120, 455, 1985.

44. **James, D. R. and Ware, W. R.,** Recovery of underlying distributions of lifetimes from fluorescence decay data, *Chem. Phys. Lett.,* 126, 7, 1986.

45. **Siemiarczak, A. and Ware, W. R.,** Complex excited state relaxation in p-(9-anthryl)-N, N'-dimethylaniline derivatives evidenced by fluorescence lifetimes distributions, *J. Phys. Chem.,* 91, 3677, 1987.

46. **Wagner, B. D., James, D. R., and Ware, W. R.,** Fluorescence lifetime distributions in homotryptophan derivatives, *Chem. Phys. Lett.,* 138, 181, 1987.

47. **Vandendriessche, J., Goedeweeck, R., Collart, P., and de Schryver, F. C.,** Fluorescence probing of the local dynamics of polymers: a model approach, in *Photophysical and Photochemical Tools in Polymer Science,* Winnik, M. A., Ed., (NATO ASI Ser. C), D. Reidel, Dortrecht, Netherlands, 1986, 225.

48. **Phillips, D.,** Time-resolved fluorescence of excimer-forming polymers in solution, *Br. Polym. J.,* 19, 135, 1987.

49. **Nemzek, T. L. and Ware, W. R.,** Kinetics of diffusion-controlled reactions: transient effects in fluorescence quenching, *J. Chem. Phys.,* 62, 477, 1975.

50. **Itagaki, H., Horiè, K., and Mita, I.,** An explanation of the existence of three decay constants in a singlet monomer-excimer system, *Macromolecules,* 16, 1395, 1983.

51. **Fayer, M. D.,** Exciton coherence, in *Spectroscopy and Excitation Dynamics of Condensed Molecular Systems,* Agranovich, V. M. and Hochstrasser, R. M., Eds., North-Holland, Amsterdam, 1983, 185.

52. **Hunt, I. G., Bloor, D., and Movaghar, B.,** One dimensional recombination of charge carriers, *J. Phys. C,* 16, L623, 1983.

53. **Gochanour, C. R., Anderson, H. C., and Fayer, M. D.,** Electronic excited state transport in solution, *J. Chem. Phys.,* 70, 4254, 1979.

54. **Ediger, M. D. and Fayer, M. D.,** Electronic excited state transport among molecules distributed randomly in a finite volume, *J. Chem. Phys.,* 78, 2518, 1983.

55. **Miller, D. R. J., Pierre, M., and Fayer, M. D.,** Electronic excited state transport and trapping in disordered systems: picosecond frequency mixing, transient grating and probe pulse experiments, *J. Chem. Phys.,* 78, 5138, 1983.

56. **Frederickson, G. H. and Frank, C. W.,** Non-exponential transient behavior in fluorescent polymeric systems, *Macromolecules,* 16, 572, 1983.

57. **Rughooputh, S. D. D. V., Bloor, D., Phillips, D., and Movaghar, B.,** One-dimensional exciton diffusion in a conjugated polymer, *Phys. Rev. B.,* 35, 8103, 1987.

58. **Stubbs, C. D., Meech, S. R., Lee, A. G., and Phillips, D.,** Solvent relaxation of dansyl probes in lipid bilayer systems, *Biochem. Biophys. Acta,* 815, 351, 1985.

59. **Meech, S. R., O'Connor, D. V., Roberts, A. J., and Phillips, D.,** On the construction of nanosecond time-resolved emission spectra, *Photochem. Photobiol.,* 33, 159, 1981.

60. **Knutsen, J. R., Walbridge, D. G., and Brand, L.,** Decay-associated fluorescence spectra and the heterogeneous emission of alcohol dehydrogenase, *Biochemistry,* 21, 4671, 1982.

61. **Knutsen, J. R., Walbridge, D. G., Davenport, L., Pritt, M. D., and Brand, L.,** Deconvolution of decay associated spectra, in *Deconvolution and Reconvolution of Analytical Signals,* Bouchy, M., Ed., ENSIC-INPL, Nancy, France, 1982, 357.

62. **Tao, T.,** Time-dependent fluorescence depolarisation and brownian diffusion coefficients of macromolecules, *Biopolymers,* 8, 609, 1969.

63. **Belford, G. G., Belford, R. L., and Weber, G.,** Dynamics of fluorescence polarisation in macromolecules, *Proc. Natl. Acad. Sci. U.S.A.,* 69, 1392, 1972.

64. **Szabo, A.,** Theory of fluorescence depolarisation in macromolecules and membranes, *J. Chem. Phys.,* 81, 150, 1984.

65. **Kinoshita, K., Kawato, S., and Ikegami, A.,** The theory of fluorescence polarisation in membranes, *Biophys. J.,* 20, 289, 1977.

66. **Valeur, B. and Monnerie, L.,** Dynamics of macromolecular chains. III. Time-dependent fluorescence polarisation studies of polystyrene in solution, *J. Polym. Sci. Polym. Phys.,* 14, 11, 1976.

67. **Dubios-Violette, E., Geny, F., Monnerie, L., and Parodi, O.,** Depolarisation de fluorescence pour un colourant engage sur une macromolecule flexible, *J. Chim. Phys. Physicochim. Biol.,* 66, 1865, 1969.

68. **Knutson, J. R., Beecham, J. M., and Brand, L.,** Simultaneous analysis of multiple fluorescence decay curves: a global approach, *Chem. Phys. Lett.,* 102, 501, 1983.

69. **Cross, A. J. and Fleming, G. R.,** Analysis of time-resolved fluorescence anisotropy decays, *Biophys. J.,* 46, 45, 1984.

70. **Christensen, R. L., Drake, R. C., and Phillips, D.,** Time-resolved fluorescence anisotropy of perylene, *J. Phys. Chem.,* 90, 5960, 1986.

71. **Wahl, Ph.,** Analysis of fluorescence anisotropy decays by a least square method, *Biophys. Chem.,* 10, 91, 1979.

72. **Porter, G., Reid, E. S., and Tredwell, C. J.,** Time-resolved fluorescence in the picosecond region, *Chem. Phys. Lett.,* 29, 469, 1974.

73. **Beddard, G. S., Doust, T. A. M., Meech, S. R., and Phillips, D.,** Synchronously-pumped dye laser in fluorescence decay measurements of molecular motion, *J. Photochem.,* 17, 427, 1981.

74. **Doust, T. A. M., Porter, G., and Phillips, D.,** Picosecond spectroscopy: applications in biochemistry. I. Techniques, *Biochem. Soc. Trans.,* 12, 630, 1984.

75. **Wong, K. S., Hayes, W., Hattori, T., Taylor, R. A., Ryan, J. F., Kaneto, K., Yoshino, Y., and Bloor, D.,** Picosecond studies of luminescence in polythiophene and polydiacetylene, *J. Phys. C,* 18, L843, 1985.

76a. **Lakowicz, J. R., Laczko, G., Cherek, H., Gratton, E., and Limkeman, M.,** Analysis of fluorescence decay kinetics from variable frequency phase shift and modulation data, *Biophys. J.,* 46, 463, 1984.

76b. **Lakowicz, J. R., Gratton, E., Cherek, M., Miliwal, B. B., and Laczko, G.,** Determination of time-resolved fluorescence emission spectra and anisotropies of a fluorophore-protein complex using frequency-domain phase-modulation fluorimetry, *J. Biol. Chem.,* 259, 10967, 1984.

77. **Gratton, E., Limkeman, M., Lakowicz, J. R., Maliwal, B. P., Cherek, H., and Laczko, G.,** Resolution of mixtures of fluorophores using variable frequency phase and modulation data, *Biophys. J.,* 46, 479, 1984.

78. **Lakowicz, J. R. and Cherek, H.,** Resolution of an excited-state reaction using frequency domain fluorimetry, *Chem. Phys. Lett.,* 122, 380, 1985.

79. **Lakowicz, J. R. and Maliwal, B. P.,** Construction and performance of a variable-frequency phase-modulation fluorometer, *Biophys. Chem.,* 21, 61, 1985.

80. **Lakowicz, J. R., Laczko, G., Gryczynski, I., and Cherek, H.,** Measurement of sub-nanosecond anisotropy decays of protein fluorescence using frequency domain fluorimetry, *J. Biol. Chem.,* 261, 2240, 1986.

81. **West, M. A.,** Flash and laser photolysis, in *Investigation of Rates and Mechanisms of Reactions,* Vol. 6, Bernasconi, C. F., Ed., Wiley-Interscience, New York, 1986, 391.

82. **Gore, B. L., Doust, T. A. M., Giorgi, L. B., Klug, D. R., Ide, J. P., Crystall, B., and Porter, G.,** The design of a picosecond flash spectroscope and its application to photosynthesis, *J. Chem. Soc., Faraday Trans. 2,* 82, 2111, 1986.

83. **Wilkinson, F., Willsher, C. J., Leicester, P. A., Barr, J. R. M., and Smith, M. J. C.,** Picosecond diffuse reflectance laser flash photolysis, *J. Chem. Soc. Chem. Commun.,* 1216, 1986.

84. **Phillips, D., Moore, J. N., and Hester, R. E.,** Time-resolved resonance Raman spectroscopy applied to anthraquinone photochemistry, *J. Chem. Soc., Faraday Trans. 2,* 82, 2093, 1986.

85. **Hyde, P. D., Waldow, D. A., Ediger, M. D., Kitano, T., and Ito, K.,** Local segmental dynamics of polyisoprene in dilute solution: picosecond holographic grating experiments, *Macromolecules,* 19, 2533, 1986.

86. **Moog, R. S., Ediger, M. D., Boxer, S. G., and Fayer, M. D.,** Viscosity dependence of the rotational re-orientation of rhodamine B in mono- and polyalcohols. Picosecond transient grating experiments, *J. Phys. Chem.,* 86, 4694, 1982.

87. **Hall, C. K. and Helfand, E.,** Conformational state relaxation in polymers: time-correlation function, *J. Chem. Phys.,* 77, 3275, 1982.

88. **Bendler, J. T. and Yaris, R.,** A solvable method of polymer main-chain dynamics with applications to spin relaxation, *Macromolecules,* 11, 650, 1978.

89. **Viovy, J. L., Monnerie, L., and Brochon, J. C.,** Fluorescence depolarization decay study of polymer dynamics: a critical discussion of models using synchrotron data, *Macromolecules,* 16, 1845, 1983.

90. **Waldow, D. A., Hyde, P. D., Ediger, M. D., Toshiaki, K., and Ito, K.,** Time-resolved optical spectroscopy as a probe of local polymer motions, in *Photophysics of Polymers,* (ACS Symp. Ser., Vol. 358), Hoyle, C. E. and Torkelson, J. M., Eds., American Chemical Society, Washington, D. C., 1987, 68.

Chapter 4

PHOTOPHYSICAL STUDIES OF TRIPLET EXCITON PROCESSES IN SOLID POLYMERS

Richard D. Burkhart

TABLE OF CONTENTS

I. Introduction 148

II. Laser-Based Spectrometers for Delayed Luminescence of Solid Polymers 149

III. Phosphorescence Emission from Solid Polymers 151

IV. Time-Resolved Delayed Luminescence Spectroscopy 155

V. Migration of Triplet Excitons in Solid Polymers 156

VI. Optical Absorption of Transient Triplets 157

VII. Laser-Induced Production of Triplet States in Conjugated Polymers 158

VIII. Concluding Observations 158

References 159

I. INTRODUCTION

The long-lived luminescence emitted by many different types of molecules, and usually described as phosphorescence, was correctly attributed to molecular triplet states by Lewis and Kasha.[1] During the intervening 40-year period, a great deal of progress has been made in providing detailed characterizations of the triplet states of molecules including their optical absorption spectra, electric polarization, zero magnetic field splitting energies, and, in some cases, chemical reactivity. By far the large majority of this work has emphasized dilute solutions of the test molecules usually at low temperatures in glass-forming solvents. The end result is that a great deal of information is available concerning triplet states of isolated molecules.

One of the challenges faced by those who work in the area of polymer photophysics is the application of knowledge gained from these earlier studies on isolated molecules to the more complex situation in which interchromophore interactions are usually not negligible. In a typical case, the root-mean-square radius of gyration of a polymer coil might be 100 Å for a polymer containing from 500 to 1000 chromophore groups. This would correspond to a local chromophore concentration of 0.2 M to 0.4 M, i.e., a very dense chromophore population. In solid polymer films one might expect even higher densities and even stronger interchromophore interactions.

When a pure organic polymeric solid is subjected to a photoexcitation pulse, a complex sequence of photophysical events is set in motion involving the excited singlet and triplet states of the chromophores present. In this chapter, we are concerned with triplet-state species usually formed by intersystem crossing from the first excited singlet state. The chromophores which will be considered are primarily those containing carbon skeletons. Once these triplet states are formed, various relaxation modes come into play resulting in the eventual decay of the triplets back to the ground state. One of the major tasks for polymer photophysics is to identify the possible modes of energy relaxation and to assess their relative importance.

Because of the large chromophore density in the polymeric solid state, energy transfer processes must specifically be taken into account in any triplet relaxation mechanism. Furthermore, pure first-order relaxation is rarely encountered in these systems. For these reasons, the kinetics of triplet decay in pure amorphous polymer films is usually not amenable to a simple interpretation. The complexity is often compounded by the formation of excimeric species and by triplet-triplet annihilation.

The use of laser excitation to investigate the photophysical properties of triplet states in polymer systems has become quite common. The large pulse intensities available from laser sources are particularly valuable since triplet quantum yields are often low and detectability is frequently a problem, especially at elevated temperatures. Furthermore, in favorable cases, the coupling between electron spin angular momentum and orbital angular momentum is sufficiently large to produce measurable oscillator strengths for direct $T_1 \leftarrow {}^1S^0$ transitions. The use of laser excitation to excite these transitions can, in some instances, provide information that would be impossible to obtain with conventional photoexcitation techniques. In addition, two-photon processes may be used to produce excited states that cannot be easily formed by other methods. The fact that many laser sources are naturally plane polarized may also be utilized to produce a photoselected population of triplets. The emission polarization from systems prepared in this way provides useful insights into the various pathways of energy disposal. In the discussion which follows, specific examples with results and conclusions will be provided for a variety of solid polymeric systems studied using the methods of pulsed laser excitation.

Although the major focus of attention in this chapter is upon polymer samples in the pure solid state, it is clear that a great deal of our current understanding of the behavior of triplet excitation in polymers is due to investigations of dilute solutions of these species. Of special importance are dilute rigid solutions at 77 K. In the present chapter we touch upon the subject of polymer solutions only so far as it is necessary to clarify our discussion of solid-state phenomena and leave to others the task of a full discussion of triplet-state processes in solution.

II. LASER-BASED SPECTROMETERS FOR DELAYED LUMINESCENCE OF SOLID POLYMERS

Luminescence spectroscopy of solid polymer films places somewhat different demands upon instrument design than is conventionally required for fluid solutions.[2] In the latter case, it is a simple matter to view the luminescence signal at right angles to the excitation beam. For solid films right angle viewing is less easily accomplished without attendant problems of signal contamination from scattered light. With solid films it is usually best to align the laser beam, the sample, and the entrance slit of the emission monochromator along a common axis (the beam axis).

The obvious disadvantage of such a system is that constant vigilance must be maintained to prevent the excitation beam from entering the entrance slit of the monochromator. Two methods have commonly been used to safeguard against this potential hazard. One method is to use a rotating sector or electromechanical shutter which blocks the entrance slit during the laser pulse. The other method is to use a blocking mirror which has been coated specifically for reflection at the laser wavelength. In most instances the latter method is preferable and is effectively required if one hopes to conduct time-resolved experiments below the 10-ms range.

Various forms of the solid polymer have been used in emission spectroscopy. The choice will depend upon the specific information required and the facilities available for temperature control. In some instances, just the polymer powder contained in a small diameter quartz tube is sufficient. Usually, however, film samples are used which have been deposited by casting from solution.

Since signal-to-noise ratios in triplet spectroscopy are usually best for cooled samples, it is common practice to use liquid nitrogen as the cryogenic medium. Thus, an optical dewar flask provides a convenient sample mounting system for polymer films which have been coated onto a quartz rod or in a quartz tube.[3] Even though this system is simple to assemble it does suffer one disadvantage. At each excitation pulse some of the energy absorbed may degrade to thermal energy, resulting in a small amount of gas bubble formation at the sample-liquid nitrogen interface. The resulting light scattering may cause distortions in the emission signal. The problem is particularly serious in the collection of luminescence decay data.

A better system involves the formation of polymer films on a quartz plate which can be mounted at the end of a cryotip assembly. The use of indium gaskets to insure good thermal contact between the copper holder and the sample is essential. Even with this system it is impossible to avoid a minor amount of instantaneous sample heating at each excitation pulse. The best results are obtained using an indium gasket which covers the entire front face of the sample (typically 2.5 cm in diameter) except for a hole in the center just large enough for entry of the laser beam. By maximizing the amount of surface area in contact with the indium the rate of heat dissipation will also be maximized.

When irradiating solid film samples using high-energy laser systems, it is not difficult to exceed the tolerance of the film toward thermal degradation. It is therefore recommended to raise beam energies gradually to avoid decomposition. If a sample is accidentally degraded, the quartz plate which had been used as the support should be carefully cleaned before reuse. In some cases, luminescence signals have been observed from ostensibly clean quartz plates

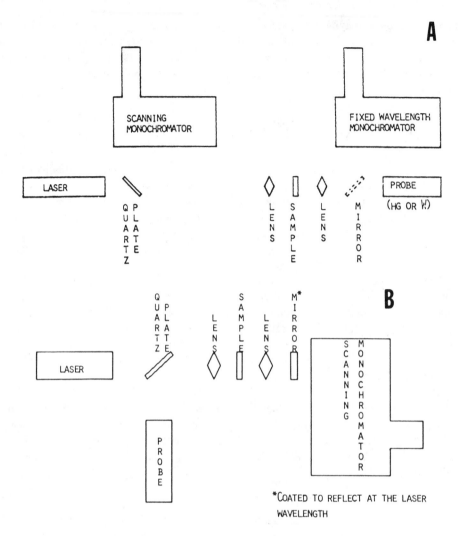

FIGURE 1.   Two arrangements of components for pulse-probe experiments on solid polymer films. (From Caldwell, N. J. and Burkhart, R. D., *Macromolecules*, 19, 1653, 1986. With permission.)

which had been involved in a degradation event. Pyrolysis of the plates at 500°C has often proved to be an effective cleaning method but it is always prudent to pretest all sample support plates for impurity luminescence.

The detection of transient triplets by optical absorption methods also presents certain instrumental problems especially for solid polymer films or for solutions in very short optical path cells. In this case, excitation and probe beams must both traverse the sample and beam steering must be carefully arranged so that the probe beam is completely contained within the region of the sample which has been photoexcited. Two different designs for the beam path in triplet-triplet absorption experiments have been successfully used in our laboratories. They are presented in Figure 1.[4] In both cases the key optical element is a quartz plate beam splitter which is inserted at a 45° angle to both excitation and probe beams. Even though only about 10% of the probe beam is reflected from the beam splitter, this is more than sufficient for the detection of the intensity of this beam by the monochromator-photomultiplier system. Conversely, about 90% of the excitation beam intensity is transmitted by the beam splitter and for most laser excitation systems this loss is not significant. The design labeled "B" in Figure 1 has the advantage that the same monochromator may be used both for transient absorption experiments and for luminescence detection.

It may be noted that alignment of the probe beam for optimum intensity becomes very important for those situations in which luminescence and probe wavelengths coincide. If the probe beam intensity can be made orders of magnitude larger than the luminescence intensity, then effectively no interference from the luminescence occurs. In any event, it is always a good precaution to determine the luminescence intensity by conducting experiments in the absence of the probe and using as the instrument sensitivity that which is normally employed for transient absorption. It may be necessary in some cases to apply numerical corrections or to generate difference spectra to account for interference from emission signals which cannot be sufficiently reduced.

Triplet-triplet absorption spectra of solid films inevitably suffer from low sensitivity due to the short optical pathlengths involved. This is a particularly difficult problem in the case of digital recording when the vertical sensitivity is on the order of eight bits. If, for example, a data recording system is set up for positive and negative displacements and has a total vertical resolution of 8 bits, then only 128 vertical counts are available for absorption or transmittance measurements. It is not unusual to encounter transmittance values on the order of 1% which would mean a difference of only one vertical count out of the 128 which are available. This is clearly an unacceptable situation for recording spectra or decay curves, but it can be improved by the use of a compensating counter current to partially offset the photocurrent produced by the probe beam. In the system described by Caldwell and Burkhart,[4] a photomultiplier power supply was used as the current compensator and a tungsten lamp supplied by a 12-V storage battery provided a stable probe beam. To assess the effectiveness of such a system let us suppose that the unattenuated probe beam provides a signal of 100 counts at a vertical sensitivity of $\pm$ 16 V and that 128 counts is the maximum number of counts available. Now suppose that current compensation is used to reduce the effective value of the photocurrent to 4 counts but this is now displayed on a $\pm$ 0.5-V scale to yield a 32-fold expansion or 128 vertical counts. If the actual transmittance is only 1%, (i.e., 1 count out of the original 100), then its apparent value will be 25% after current compensation and this would correspond to 32 vertical counts on the 0.5-V scale. It is clear that this method may only be used if a very stable probe source is available, but in that case it is quite effective.

III. PHOSPHORESCENCE EMISSION FROM SOLID POLYMERS

Some of the early experiments on the phosphorescence spectra of polymers seemed to indicate that the emission properties of the model monomeric chromophore were very similar to those of the same chromophore bonded to the polymer backbone. For example, the phosphorescence spectra of toluene and polystyrene seemed to be essentially the same.[5] In other cases, for example, *N*-isopropyl carbazole and poly(*N*-vinylcarbazole), the phosphorescence spectra were similar but not identical.[6] In all of these comparisons we are referring to dilute solutions at 77 K in rigid glassy media.

When, on the other hand, we compare the phosphorescence spectrum of a pure solid polymer film with that of the polymer in glassy solutions, then some rather distinct differences appear between the two. In Figure 2, for example, there are displayed the phosphorescence spectra of poly(*N*-vinylcarbazole) in the pure solid state and in solution. Clearly a distinct red shift and loss of structure is observed in the solid-state spectrum. There are many examples of solid-state polymer phosphorescence spectra which show this similar red shift and loss of structure compared with their phosphorescence spectra in rigid glassy solutions. Because energy transfer is usually rather efficient in solid polymers, those chromophoric groups having the lowest energy triplet states will act as energy acceptors for migrating triplet excitons. Thus, impurities which are present either as independent small molecules or as functional groups which have been accidentally bonded to the polymer chain can potentially account for some of the anomalous solid-state phosphorescence which is observed.

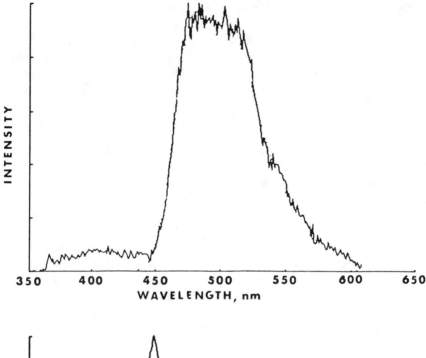

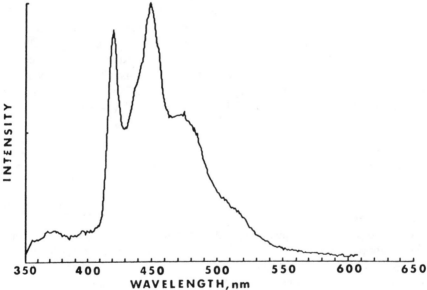

FIGURE 2.   Phosphorescence spectra of poly(*N*-vinylcarbazole) at 77 K. Upper: solid film and lower: rigid matrix of 2-methyltetrahydrofuran.

Excellent examples of impurity effects are provided by a variety of commercial polymers. For example, solid films of commercial polystyrene emit phosphorescence which is attributable to acetophenone-like impurities.[7,8] In fact, it has been suggested that light absorption by these phenyl alkyl carbonyl groups may be the initiating step of the sunlight-induced photodegradation of polystyrene.[7]

Laser photoexcitation has played a variety of roles in these studies. It has been demonstrated by laser excitation, for example, that poly(ether sulfone) is degraded photochemically and that the initiating step is probably light absorption by the diphenyl sulfone unit.[9]

These conclusions were reached, in part, by an examination of the phosphorescence spectra and phosphorescence excitation spectra of the polymer and a monomeric model compound. The results are interesting in that they demonstrate an intrinsic susceptibility of this polymer to photodegradation since it is a primary polymer functional group, rather than an impurity, which initiates photodegradation. This is not the case with polyurethanes based upon diphenylmethane 4,4′-diisocyanate. Here it was shown, again by phosphorescence spectroscopy, that the photoactive absorber is a benzophenone-like species produced by a stepwise, photoassisted oxidation process leading ultimately to the ketone species.[10] Here one finds a textbook example of the lowest energy triplet species acting as an energy acceptor and phosphorescence emitter.

Although impurity phosphorescence from solid polymers is rather common, it is not the only source of delayed luminescence having properties different from that of the isolated chromophore. In fact, at least three additional types of delayed luminescence may be identified. There are excimer phosphorescence, delayed fluorescence, and delayed excimer fluorescence. The spectroscopic characterization of these emissions and the investigation of their time-resolved features and decay kinetics have all been aided by laser techniques. We will discuss first some aspects of excimer phosphorescence in solid polymers.

There is, in fact, a certain amount of controversy surrounding the possibility of triplet excimer formation. In any event, it is safe to say that triplet excimers, if they exist, are not very well understood. A red-shifted and structureless phosphorescence from naphthalene solutions was assigned to excimer emission by Takemura and co-workers,[11] who also conducted extensive kinetic measurements on the system. Lim[12] has also assigned to a triplet excimer source the red-shifted and structureless phosphorescence observed in various disubstituted alkanes. On the other hand, Nickel and Prieto[13] have presented evidence suggesting that most, if not all, of these earlier reports claiming triplet excimer formation in fluid solutions are probably due to solvent impurities. In response to this suggestion, Locke and Lim[14] carried out a series of experiments to test the effect of the presence of biacetyl (the suspected impurity) on the triplet emission characteristics of a series of aromatic compounds. They concluded that their results could not be explained by impurity luminescence and that the original assignment of triplet excimer emission was correct.

In the case of polymeric systems there seem to be no instances of any triplet excimer formation in dilute frozen glassy matrices at 77 K. There are, however, many reports of excimer phosphorescence from solid polymer films. These emission spectra are always red shifted with respect to the phosphorescence of the corresponding isolated chromophore and their structural features are minimal. The naphthyl chromophore in poly(1-vinylnaphthalene) (P1VN)[15] and poly(2-vinylnaphthalene) (P2VN)[3] and the carbazolyl chromophore in poly(N-vinylcarbazole) (PVCA)[6,16] have been particularly well studied with respect to their potential for triplet excimer emission. Klöpffer[17] has recently reviewed many aspects of triplet excimer formation in polymers and concludes that, with respect to energetic considerations, there are several chromophore systems which should be capable of forming triplet excimers.

Because excimer fluorescence can be observed unequivocally from solutions of small molecules, it has been possible, using model systems to make deductions about the structural features of singlet excimers. For example, singlet excimer emission from 2,4-disubstituted pentanes, both in D,L-racemic and in *meso*-isomeric forms, have been studied using laser excitation.[18] Two different structural forms for the singlet excimers were identified and related to similar species thought to be responsible for excimeric fluorescences in polymers. Studies of similar sophistication have not yet appeared in connection with triplet excimer emission.

This is unfortunate since indirect evidence indicates the existence of structural variations for triplet as well as singlet excimers in solid polymer films. A rather clear example of this effect is provided by the distinct two-component phosphorescence emission signal observed

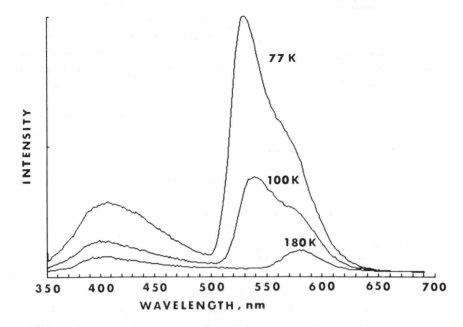

FIGURE 3.   Delayed luminescence spectra of a solid film of poly(1-vinylnaphthalene) at various temperatures. Delayed fluorescence is in the 350- to 500-nm range and phosphorescence is in the 500- to 650-nm range. (From Burkhart, R. D., Avilés, R. G., and Magrini, K., *Macromolecules*, 14, 91, 1981. With permission.)

from solid films of P1VN.[14] These spectra recorded at different temperatures are presented in Figure 3. These emission signals show the typical excimeric characteristics of a red shift relative to the isolated chromophore and a lack of structural details. It should be noted that a two-component excimer fluorescence emission from P1VN has recently been observed by Miskowski and co-workers.[19]

Excimeric phosphorescence has also been proposed to account for the emission spectra observed from solid films of PVCA at 77 K and at more elevated temperatures.[20] Time-resolved phosphorescence spectroscopy indicated the existence of two different excimeric species in PVCA.[21] Apparent two-component luminescence decay rates also supported the idea that at least two different excimeric species were the source of this phosphorescence emission. When PVCA samples prepared by cationic (PVCA-c) as well as free radical (PVCA-r) initiation were compared, significant differences in triplet spectral emission were found. The migration rates of triplet excitons, determined by luminescence quenching, were also found to be different and larger for PVCA-r than for PVCA-c. Itaya and co-workers[16] proposed that this difference was due to a larger concentration of triplet excimer trap sites in PVCA-c, which decreased the mean free path of migrating triplet excitons. This proposal is in accord with the observation that the phosphorescence emission envelope for PVCA-c is definitely red shifted compared with PVCA-r.[22]

All of these observations are consistent with the view that triplet excimers are formed in solid films of vinyl aromatic polymers. It is difficult, however, to rule out the possibility that these observations may simply be due to the presence of impurities having lowest triplet energy levels below that of PVCA. Klöpffer has pointed out that the concentration of triplet excimer forming sites is only about 10^{-4} mol/mol of basic units in most polymers.[17] To achieve impurity levels in solid polymers less than 1 part in 10,000 is difficult indeed.

Excimer-like phosphorescence from solid polymer films has been observed in so many different laboratories and under so many different conditions that it is now possible to draw up a list of the properties which impurities must have in order to account for the observations.

For example, they must be more concentrated in PVCA-c than in PVCA-r and this concentration mismatch must be the same for polymers prepared in different laboratories. Furthermore, the phosphorescence emission spectra of the impurities must be essentially structureless. In addition, the impurities could not be accidentally incorporated since the same emission spectra are observed from many different laboratories. In view of this list of requirements, it seems preferable to conclude that the red-shifted, structureless emission in solid films of PVCA is due to excimeric species.

Further information concerning the nature of triplet excimer emission from solid polymers was provided by observing phosphorescence emission spectra from solid films of a variety of different PVCA samples prepared by free radical initiation at different temperatures. A series of such polymers was examined using computer-assisted resolution of the emission band to multiple components. Attempted fits to a two-component system were unsuccessful, but for six different samples satisfactory fitting was achieved by using three components. The best fit band centers were the same for each of the three components (to ± 3 nm) for samples prepared at temperatures ranging from 0°C (photoinitiated free radical polymerization) to 125°C. The relative amplitudes of the three bands varied in a systematic way with the polymerization temperature.[23] Such orderly behavior is not expected from accidentally introduced impurities.

IV. TIME-RESOLVED DELAYED LUMINESCENCE SPECTROSCOPY

Delayed luminescence spectra, which might include both phosphorescence and delayed fluorescence emission, often consist of overlapping components having different dynamic characteristics. As an example one might point to delayed fluorescence arising from triplet-triplet annihilation which usually decays faster than the corresponding phosphorescence signal. If spectra can be recorded at different delay times after the excitation pulse, it is frequently possible to obtain simplified spectra at long delay times when all but the longest-lived component has decayed away. In such cases, time resolution may not only lead to simpler spectra but also to a satisfactory assignment of the entire emission envelope. Laser excitation is particularly effective in such cases since the pulse widths are usually sufficiently narrow to be considered a delta function at least when triplet-state processes are being monitored.

A word should be mentioned here about instrumental methods in connection with time-resolved spectroscopy. It is clear, for example, that sufficient time must elapse between excitation pulses to allow the emission from a given pulse to reach a negligible level before the next pulse is activated. A more subtle consideration involves electronic filtration of emission signals. Some data collection systems include facilities for smoothing time-dependent signals by resistance-capacitance (RC) circuits. One must be certain to verify that the RC time constant of the circuit is faster than the decay time of any emission component of the spectrum, otherwise the signal from fast components may be retained by the circuit long after the electronic excitation has decayed away. There may, in fact, be a substantial RC effect even in circuits which do not purposely incorporate electronic filtration. In cases of uncertainty about the effective RC time constant for a given circuit, the instrumental response from signals having known time characteristics may be used to remove any doubts.

One application of time-resolved spectroscopy involved the triplet excimer emission from solid films of PVCA which are quite broad and relatively structureless. They do, however, possess sufficient structural characteristics to suggest they may consist of multiple components. Early experiments on the kinetics of the phosphorescence decay of PVCA indicated that at least two emitting species were present. This suggestion was verified by time-resolved phosphorescence spectra which demonstrated that the longer wavelength component was actually the shorter lived.[20]

Time-resolved delayed luminescence spectra are especially valuable for the investigation of solid polymer films at very short time delays following excitation. These luminescence decays usually depart significantly from single exponential character. This means, among other things, that a large fraction of the total delayed emission may be concentrated into a small time regime following excitation. Under these conditions it is very difficult to obtain an accurate quantum yield for triplet formation and good estimates may only be obtained using the smallest delay times.

Effect of nonexponential decay are especially significant when dealing with delayed fluorescence signals arising from triplet-triplet annihilation. Using laser excitation and observing the emission at delay times of a few hundred microseconds, it has been found that the delayed fluorescence from PVCA solid films completely dominate the delayed emission spectrum.[24]

V. MIGRATION OF TRIPLET EXCITONS IN SOLID POLYMERS

The high density of chromophore units in the polymeric solid state guarantees the existence of significant interchromophore interactions. We have already discussed one consequence of these interactions, the formation of triplet excimers. Let us now turn to a consideration of triplet energy transfer processes which are also dependent upon the existence of relatively small interchromophore separation distances.

That triplet exciton migration takes place in polymeric systems has been demonstrated in several ways. For example, the existence of triplet-triplet annihilation implies mobility on the part of at least one partner in the annihilative process. Many investigations of the kinetics of triplet decay in molecularly doped polymers have also led to the conclusion that triplet exciton migrations occurs in these media.[25]

A precise characterization of the mobile triplet exciton in solid polymers has not, to date, been accomplished. It is tempting to assume that the mobile exciton has properties similar to those of an isolated chromophore, but such an assertion is difficult to prove and some evidence exists which suggests the contrary. A particularly interesting case is that of P1VN in which excimer phosphorescence and delayed excimer fluorescence (DEF) lifetimes are the same at 77 K.[14] Since the DEF emission arises from triplet-triplet annihilation, the identity of DEF and phosphorescence lifetimes suggests that the lifetime of the mobile exciton is significantly greater than that of the trapped exciton with which it interacts. This may be seen if we let t_t be the lifetime of the trapped exciton, t_m be the lifetime of the mobile exciton and t_{def} be the lifetime of DEF. If the trapped and mobile excitons both decay exponentially then

$$t_{def}^{-1} = t_t^{-1} + t_m^{-1} \tag{1}$$

Under these conditions and in the event that the mobile exciton resembles an isolated naphthalene chromophore, then one would expect to observe normal naphthalene phosphorescence from solid films of P1VN. The fact is that no monomeric naphthalene phosphorescence is found in these solid films.

The kinetics of delayed luminescence decay in solid films of PVCA also suggests that the mobile exciton may have properties somewhat different from that of an isolated chromophore. One indication of this possibility is associated with experiments on the activation energy of detrapping of the triplet excimer in solid films of PVAC. If one takes the sum of this activation energy and the photon energy of excimeric phosphorescence, then the result should be an energy greater than or equal to the energy of untrapped (mobile) excitons. The energy of the mobile exciton may then be compared with the triplet energy of the isolated chromophore using, for example, the phosphorescence spectrum of PVCA in a rigid glass.

On the basis of such an analysis an energy mismatch occurs such that the isolated triplet state is found to be at an energy level about 2000 cm^{-1} above that of the mobile exciton. On the other hand Masuhara and co-workers[26] have used a laser excitation optical probe method to record the triplet-triplet absorption spectrum of triplet states in a solid film of PVCA. The spectrum is very similar to that found for PVCA in rigid solution at 77 K. Thus, on the one hand, energetic considerations suggest that the mobile exciton is unlike an isolated chromophore, whereas spectroscopic considerations suggest they are similar. A satisfactory resolution of these observations has not been suggested at this time.

Laser pulse-probe studies on solid films of PVCA at ambient temperature have also revealed interesting luminescence decay behavior in the submillisecond time regime. The excimeric phosphorescence displays a rather fast (1.5 ms) lifetime component plus another component which has a lifetime of several hundred milliseconds. The lifetime of the fast component is essentially the same as that of the delayed fluorescence. It was concluded that two separate categories of mobile triplet excitons may be distinguished in these systems.[23] One type results from the initial excitation pulse through the excited singlet state plus intersystem crossing. We may use the term primary triplet excitons to describe these species. Most of the primary excitons are either trapped at excimer-forming sites or else quenched by impurities. The ones that are not trapped and remain in a mobile condition will most likely become involved in annihilative processes to produce delayed fluorescence primarily by interaction with trapped excitons. Since no monomeric type of phosphorescence is observed in these experiments, it appears that radiative decay to the ground state cannot compete with alternative modes of deactivation for these primary excitons. In addition to these primary excitons, there is also the possibility that excitons which have been trapped at excimer-forming sites may be thermally detrapped to produce secondary mobile excitons.

Evidence for reversibility in the exciton trapping process in solid films of PVCA has been derived from luminescence quenching experiments[27] and from the observation that delayed fluorescence intensities exhibit intensity extrema between 77 K and ambient temperature.[28] The mechanism for detrapping has apparently not been studied in any detail but it presumably would involve a thermally assisted intramolecular internal rotation about a backbone carbon-carbon bond, or sequence of bonds, which would result in the disruption of the excimer bond. The activation energy for this process in PVCA films has been estimated at 1000 cm^{-1}.[20]

VI. OPTICAL ABSORPTION OF TRANSIENT TRIPLETS

Very few examples may be found in the literature of triplet-triplet absorption experiments on polymeric systems. An early example of such a study was that by Bensasson and co-workers[29] on P2VN. The transient spectra for the polymeric system and that of monomeric naphthalene were similar but not identical. In general, the absorption bands are broader for the polymer and somewhat red shifted. This is the sort of behavior expected for a system consisting of a high-chromophore density in which inhomogeneous broadening would affect both emission and absorption band shapes. Laser pulse-probe methods are particularly valuable in this type of work since detectability is frequently a problem and small delay times usually produce better results due to the rapid and nonexponential triplet decays often encountered in polymer solids. A recent laser pulse-probe study was conducted on an alternating copolymer of 2-vinylnaphthalene and methylmethacrylate.[30] Since the chromophore density is reduced for the alternating copolymer it was thought that the triplet-triplet absorption bands might be similar to those found for the isolated naphthalene chromophores, but such is not the case. In fact, there is very little difference in band shape between the copolymer and the homopolymer.

Laser pulse-probe studies have also been carried out by Masuhara and co-workers to obtain transient triplet absorptions on PVCA itself and oligomers of this polymer. An early study indicated that little T-T absorption could be found for the high molecular weight polymer,[31] but later studies both on the solid polymer and on glassy solutions of PVCA indicated that a small but measurable absorption was present.[25] An attempt to obtain quantitative values for the oscillator strength of T-T absorptions of PVCA has been hampered by the existence of strong T-T bands in the same spectral region corresponding to optical absorption of the ground state. Therefore, methods based upon a determination of ground state depletion following an excitation pulse are not effective. The ground state depletion method has, however, been successfully employed to evaluate triplet state photophysical parameters for other carbazole-containing polymers.[32] It is unfortunate that so few quantitative photophysical parameters are available for triplet states of polymeric systems since comparative methods using reliable values for selected polymers would accelerate the compilation of a satisfactory data base. A critical mass of baseline data has not yet accumulated.

VII. LASER-INDUCED PRODUCTION OF TRIPLET STATES IN CONJUGATED POLYMERS

Polyacetylene, polydiacetylene, and related polymers have become the subjects of keen interest because of their potential use as the basic components of organic conductors or semiconductors. Recent studies of photogenerated excited states in solid polymers of this sort have resulted in the detection of states which are thought to be triplets. Both Hattori and co-workers[33] and Orenstein and co-workers[34] used optical absorption of transients produced upon photoexcitation of substituted polydiacetylenes to identify rather long-lived (0.1 ms) transients. Subsequent work by Robins and co-workers[35] using ESR detection of the excited states has confirmed that the transient absorptions are, indeed, due to triplets. In addition, it was found that the lifetime of each of the three triplet sublevels is short compared with the time for thermalization among these sublevels. There exists some disagreement as to the mechanistic route involved in the formation of these triplet states. Orenstein and co-workers suggested that they are formed as a result of interchain electron-hole pair generation. On the other hand, Winter and co-workers[36] using transient esr detection have concluded that triplet pairs are formed upon photoexcitation. Their reasoning is prompted by the observation of an initial spin polarization of 0 which rises to a maximum at somewhat <50 μs after the excitation pulse. The initially formed pair states, they argue, would have spin $S = 1$, but the individual members of each pair may decay independently to the singlet ground state producing a net polarization from the triplets which remain. The two studies using esr methods are interesting from the point of view that they are representative of only a handful of attempts to use esr methods in the detection of triplet states in polymers. They are particularly interesting within the context of the current discussion in that they combine the use of both laser excitation and esr detection.

VIII. CONCLUDING OBSERVATIONS

The use of laser excitation to probe the characteristics of triplet-state species in pure polymer solids has proved to be a valuable extension to conventional photoexcitation methods. The laser methods are particularly useful for those occasions in which time-resolved phenomena are of major concern and when monochromatic excitation is important. Based upon current experimental results it is now possible to attribute to triplet excitons in polymeric systems the characteristics of translational mobility and the ability to form excimeric species. The trapping process to form excimers has been shown to be reversible, at least for certain selected polymers, and it is possible to classify mobile triplet excitons according to whether they were formed in the initial excitation pulse (primary excitons) or formed by the reversal of the trapping process (secondary excitons). A major remaining problem concerns the detailed characterization of mobile triplet excitons. Some experimental results suggest that they are localized at a given chromophore group and have properties similar to those of the isolated species. In other instances it appears that they have a nonlocalized character. In a similar vein it is noted that no structural data have yet been presented giving details of the architecture of triplet excimers in solid polymers.

Finally, it should be mentioned that the pervasiveness of nonexponential decays of luminescence signals in solid polymer systems has made it difficult to provide quantitative characterizations of the various rate processes which are of keen interest. For example, quantitative values of quantum efficiencies of triplet formation, rate constants for radiative and radiationless decay, and rates of triplet exciton migration are very difficult to obtain. The importance of probing triplet luminescence signals and triplet-triplet absorption processes at short times after excitation is noted along with the unique effectiveness of laser excitation sources to accomplish these tasks.

REFERENCES

1. **Lewis, G. N. and Kasha, M.,** Phosphorescence and triplet state, *J. Am. Chem. Soc.,* 66, 2100, 1944.
2. **Parker, C. A.,** *Photoluminescence of Solutions,* Elsevier, New York, 1968, 128.
3. **Kim, N. and Webber, S. E.,** Photophysics of films of poly(2-vinylnaphthalene) doped with pyrene and tetracyanobenzene, *Macromolecules,* 15, 430, 1982.
4. **Caldwell, N. J. and Burkhart, R. D.,** Triplet state absorption and decay kinetics of poly(N-vinylcarbazole) and N-ethylcarbazole by a laser excitation-optical probe method, *Macromolecules,* 19, 1653, 1986.
5. **Vala, M. T., Haebig, J., and Rice, S. A.,** Experimental study of luminescence and excitation trapping in vinyl polymers, paracyclophanes, and related compounds, *J. Chem. Phys.,* 43, 886, 1965.
6. **Klöpffer, W. and Fischer, D.,** Triplet energy transfer in solid solutions and films of poly(vinylcarbazole), *J. Polym. Sci.,* Symp. No. 40, 43, 1973.
7. **George, G. A.,** The phosphorescence spectrum and photodegradation of polystyrene films, *J. Appl. Polym. Sci.,* 18, 419, 1974.
8. **Klöpffer, W.,** Luminescent groups in very weakly degraded polystyrene, *Eur. Polym. J.,* 11, 203, 1975.
9. **Allen, N. S. and McKellar, J. F.,** Photochemical reactions in a commercial poly(ether sulfone), *J. Appl. Polym. Sci.,* 21, 1129, 1977.
10. **Allen, N. S. and McKellar, J. F.,** Photochemical reactions in an MDI-based elastomeric polyurethane, *J. Appl. Polym. Sci.,* 20, 1441, 1976.
11. **Takemura, T., Aikawa, M., Baba, H., and Shindo, Y.,** Kinetic study of triplet excimer formation in fluid solution by means of phosphorimetry, *J. Am. Chem. Soc.,* 98, 2205, 1976.
12. **Lim, E. C.,** Molecular triplet excimers, *Acc. Chem. Res.,* 20, 8, 1987.
13. **Nickel, B. and Prieto, M. F. R.,** Triplet excimer phosphorescence from aromatic compounds in liquid solutions: a critical review and new negative results, *Z. Phys. Chem.,* 150, 31, 1986.
14. **Locke, R. J. and Lim, E. C.,** Phosphorescence of naphthalene and related compounds in fluid media: excimer phosphorescence or phosphorescence from a biacetyl-like impurity?, *Chem. Phys. Lett.,* 138, 489, 1987.
15. **Burkhart, R. D., Avilés, R. G., and Magrini, K.,** Triplet luminescence properties of poly(1-vinylnaphthalene) solid films, *Macromolecules,* 14, 91, 1981.
16. **Itaya, A., Okamoto, K., and Kusabayashi, S.,** Triplet excitation energy transfer in the vinyl polymers with pendant carbazolyl groups, *Bull. Chem. Soc. Jpn.,* 49, 2037, 1976.
17. **Klöpffer, W.,** Triplet excimers. The contribution of polymer and organic solid state research to a controversial issue, *Eur. Photochem. Newslett.,* March (29), 15, 1987.
18. **Vendendriessche, J., Palmans, P., Toppet, S., Boens, N., DeSchryver, F. C., and Masuhara, H.,** Configurational and conformational aspects in the excimer formation of bis(carbazoles), *J. Am. Chem. Soc.,* 106, 8057, 1984.
19. **Miskowski, V. M., Gupta, A., Coulter, D. R., Scott, G. M., and O'Connor, D. B.,** Excitation mobility and excimer formation in solid aromatic polymers. All, none, or some, *Polym. Prepr. Am. Chem. Soc. Div. Polym. Chem.,* 28, 76, 1987.

20. **Turro, N. J., Chow, M. F., and Burkhart, R. D.,** Photophysical properties of poly(*N*-vinylcarbazole) solid films at elevated temperatures. Evidence for a third phosphorescence species, *Chem. Phys. Lett.,* 80, 146, 1981.

21. **Burkhart, R. D. and Avilés, R. G.,** Studies on the temperature dependent phosphorescence of poly (*N*-vinylcarbazole), *J. Phys. Chem.,* 83, 1987, 1979.

22. **Burkhart, R. D. and Avilés, R. G.,** Temperature dependence of delayed emission from poly (*N*-vinyl-carbazole). A comparison of polymers prepared by free radical and cationic initiation, *Macromolecules,* 12, 1078, 1979.

23. **Burkhart, R. D. and Dawood, I.,** Computer assisted resolution of phosphorescence spectra from solid films of poly (*N*-vinylcarbazole), *Macromolecules,* 19, 447, 1986.

24. **Burkhart, R. D.,** Triplet emission from poly (*N*-vinylcarbazole) solid films at ambient temperature studied in the microsecond time regime using pulsed laser excitation, *Macromolecules,* 16, 820, 1983.

25. **Turro, N. J. and Steinmetzer, H. C.,** Electronic excitation transfer in polymers. I. Demonstration of singlet-singlet, triplet-singlet and triplet-triplet transfer in a polystyrene matrix studied by a chemiexcitation method. Evidence for forbidden and for allowed long range mechanism, *J. Am. Chem. Soc.,* 96, 4677, 1974.

26. **Masuhara, H., Inoue, K., Naoto, T., and Mataga, N.,** Intrapolymer S_1-S_1 annihilation and the triplet state of polymers having carbazolyl groups in solution, *J. Chem. Soc. Jpn. Chem. Ind. Chem.,* 1, 14, 1984.

27. **Webber, S. E. and Avots-Avotin, P. F.,** Quenching of triplet excitons in poly(*N*-vinylcarbazole) by biacetyl, *J. Chem. Phys.,* 72, 3773, 1980.

28. **Burkhart R. D. and Avilés, R. G,** The kinetics of triplet processes in poly(*N*-vinylcarbazole) from 77 to 298 K, *Macromolecules,* 12, 1073, 1979.

29. **Bensasson, R. V., Ronfard-Haret, J. C., Land, E. J., and Webber, S. E.,** Lowest triplet properties of poly(2-vinylnaphthalene) in solution, *Chem. Phys. Lett.,* 68, 438, 1979.

30. **Burkhart, R. D., Haggquist, G. W., and Webber, S. E.,** Triplet photophysical properties of the alternating copolymer of 2-vinylnaphthalene with methylmethacrylate, *Macromolecules,* 20, 3012, 1987.

31. **Masuhara, H., Ohwada, S., Mataga, N., Itaya, A., Okamoto, K. I., and Kusabayashi, S.,** Laser photochemistry of poly(*N*-vinylcarbazole) in solution, *J. Phys. Chem.,* 84, 2363, 1980.

32. **Burkhart, R. D., Boileau, S., and Boivin, S.,** A laser pulse-probe study of triplet exciton processes in poly (*N*-((vinyloxy)carbonyl)carbazole) and its monomeric analogue, *J. Phys. Chem.,* 91, 2189, 1987.

33. **Hattori, T., Hayes, W., and Bloor, D.,** Photoinduced absorption and luminescence in polydiacetylenes, *J. Phys. Chem.,* 17, L881, 1984.

34. **Orenstein, J., Etemad, S., and Baker, G. L.,** Photoinduced absorption in a polydiacetylene, *J. Phys. Chem.,* 17, L297, 1984.

35. **Robins, L., Orenstein, J., and Superfine, R.,** Observation of the triplet state of a conjugated-polymer crystal, *Phys. Rev. Lett.,* 56, 1850, 1986.

36. **Winter, M., Grupp, A., Mehring, M., and Sixl, H.,** Transient esr observation of triplet-soliton pairs in a conjugated polymer single crystal, *Chem. Phys. Lett.,* 133, 482, 1987.

Chapter 5

PHOTOPHYSICAL STUDIES OF MISCIBLE AND IMMISCIBLE AMORPHOUS POLYMER BLENDS

William C. Tao and Curtis W. Frank

TABLE OF CONTENTS

I.   Introduction 162

II.   Review of Polymer Photophysics 163
        A.   Excimer Formation 163
        B.   Electronic Excitation Transport 164

III.   Morphology of Polymer Blends 166
        A.   Blends Close to Immiscibility 166
              1.   Effect of Solubility Parameter Differences 166
              2.   Effect of Concentration 168
              3.   Effect of Molecular Weight 172
              4.   Effect of Temperature 177
        B.   Miscible Blends 180
              1.   Low Concentration — Molecular Weight Effect 180
              2.   High Concentration — Random Mixing Effects 184
        C.   Phase Separation Kinetics 186
              1.   Two-Phase Model 186
              2.   Phase Diagram 187
              3.   Effect of Molecular Weight on Spinodal Decomposition 190

IV.   Summary 191

Acknowledgments 191

References 192

I. INTRODUCTION

Amorphous polymer blends have received considerable attention because of the variety of bulk properties that may be modified by physical means with relative ease.[1-5] It has been estimated that 20 to 25% of all engineering plastics produced in the U.S. today are polymer blends or alloys. The mechanical, thermodynamic, and kinetic behavior of the blend depend intimately on the degree of mixing on the molecular level between the homopolymer or copolymer components. As a consequence, existing characterization procedures for polymer blends have been improved and new techniques have been developed.

Although the mechanical properties may depend strongly on the morphology of a blend on a distance scale of 100 Å,[6] a clear understanding of the nature of polymer-polymer interactions on the molecular level has not yet been established. Existing experimental techniques exhibit widely varying sensitivities to the distance scale over which inhomogeneities are detectable. Use of one technique may lead to the conclusion that a particular blend is compatible, while a more sensitive method may permit detection of microphase separation not found by the first. Standard characterization techniques, such as measurement of glass transition temperatures by dynamic mechanical spectroscopy (DMS) and differential scanning calorimetry (DSC), are insensitive to the presence of a second polymer phase at concentrations less than about 10%, regardless of the domain size of the inhomogeneities. Neutron,[7,8] light and X-ray scattering,[9,10] pulsed NMR,[11,12] and electron microscopy[13] have the required resolving power, but these techniques are still limited to blends that have at least 1% of the minor polymer component present. For fundamental studies of the mechanism of phase separation it is desirable to be able to examine low concentration blends, however.

In view of these difficulties, there is strong motivation to search for new characterization techniques with higher sensitivities. The new technique should also have the ability to characterize both miscible and immiscible polymer blends. The former case requires a measure of the coil size of the dispersed polymer, while in the latter case such quantities as the phase size, volume fraction of each phase, and the nature of the interface between phases are desired. In addition, simplicity of sample preparation, compatibility of preparation and measurement techniques with the processing and environmental conditions encountered normally, and applicability of the characterization technique to a broad range of polymer blends should be included.

Due to their intrinsic sensitivity, fluorescence methods may ultimately prove to meet all of the desired properties for studying such small-scale phase separation and low-concentration miscibility. The most distinctive photophysical feature of an aromatic vinyl polymer, such as polystyrene (PS) or poly(2-vinylnaphthalene) (P2VN) is the occurrence of excimer fluorescence. An excimer may be formed between two identical aromatic rings in a coplanar sandwich arrangement with the ring centers at their equilibrium van der Waals separation, if one of the rings is in an electronically excited singlet state.[14,15] Typically, the excimer fluorescence lifetime is in the range of 5 to 100 ns, and the fluorescence spectrum exhibits a characteristic broad Gaussian shape at lower energy than monomer fluorescence from an isolated aromatic ring. A convenient experimental measure of excimer fluorescence is the ratio of excimer to monomer emission intensities, I_D/I_M, obtained under photostationary state conditions.

One objective of research in our laboratory has been to establish excimer fluorescence as a quantitative photophysical tool for the elucidation of morphology and dynamics in polymer blends.[16-34] All of the work described in this paper is based on xenon arc lamp excitation of the polymer, as opposed to the use of lasers, which characterizes the general thrust of this monograph. When this research was initiated, the availability of versatile coherent UV sources was quite limited. Although the situation may be changing, we have continued to use a broad band arc lamp excitation source with a double monochromator for our photostationary excimer fluorescence measurements. We have recently begun picosecond laser transient fluorescence anisotropy measurements in collaboration with Professor M. D. Fayer at Stanford University, but will not discuss this aspect here. Rather, we will concentrate on the polymer materials science that may be studied by the excimer molecular probe, given an understanding of the photophysics.

Specifically, our objective is to review the progress made recently in using excimer fluorescence as a molecular probe of thermodynamic compatibility in solid polymer blends. The fluorescent guest aromatic vinyl polymer used in most of these studies has been P2VN, while the nonfluorescent host matrices, at least under the conditions of P2VN excitation, have included PS,[8,28] poly(vinyl methyl ether) (PVME),[27,28] and a series of poly(alkyl methacrylates).[8,28] We first review the complex photophysics of excimer fluorescence and electronic excitation transport (EET). Next we discuss the use of excimer fluorescence to evaluate the Hildebrand solubility parameter of P2VN in different poly(alkyl methacrylate) host matrices. This allowed an experimental and theoretical comparison of the thermodynamic interaction parameter of the polymer blend. We then investigate the effects of guest concentration, guest and host molecular weights, and casting temperature on blend miscibility. While the P2VN guest in poly(alkyl methacrylate) host systems form blends that are close to immiscibility, the PS/PVME blend is known to be miscible in all proportions at room temperatures.[12,13] A totally miscible blend provides a system to study the pathways of EET as a function of guest concentration and molecular weight, and a framework for models to predict I_D/I_M. Furthermore, the miscible PS/PVME blend can be forced to undergo phase separation by thermal annealing at elevated temperatures. We will discuss the use of excimer fluorescence accompanied by a model for phase separation and a phase diagram to study phase separation kinetics.

II. REVIEW OF POLYMER PHOTOPHYSICS

A. EXCIMER FORMATION

The excimer to monomer photostationary state fluorescence ratio, I_D/I_M, depends upon three major factors.[16,21] The first is the electronic stability of the excimer complex as manifested through the radiative and nonradiative decay constants of the excimer and monomer. These photophysical parameters should reflect very short-range interactions between the excimer complex on the guest polymer and the local environment in the host matrix, and are not expected to vary appreciably in a given guest-host pair. The second is the number and types of intramolecular and intermolecular excimer forming sites (EFS). Finally, the process by which these sites are sampled by EET in the solid state also affects the value of I_D/I_M.

The interrelation of these features may best be seen with reference to an earlier analysis of photostationary-state kinetics and the migrational sampling process leading to intramolecular excimer fluorescence.[30] The fluorescence intensity ratio, I_D/I_M, may be expressed as

$$\frac{I_D}{I_M} = \frac{Q_D}{Q_M} \left[\frac{1 - M}{M} \right] \tag{1}$$

where Q_D is the ratio of the fluorescence decay constant to the total decay constant for the excimer and Q_M is the analogous ratio for the monomer species. The quantity M is the probability that a photon absorbed by the aromatic vinyl polymer guest will decay along a monomer pathway. It has been shown to be a function of the concentration and segregation of the guest polymer in the blend.[30] Semerak and Frank[29] have examined P2VN blends and have demonstrated that the ratio Q_D/Q_M is approximately the same in both PMMA and PS hosts using a model compound equivalent to a single dyad of P2VN. This essentially fixes the mole fraction of EFS in the blend and allows the determination of the influence of Q_D/Q_M on I_D/I_M. Moreover, Q_D/Q_M is unaffected by the host molecular weight,[29] and thus the quantity $(1-M)/M$ is the only factor that influences I_D/I_M. In principle it is possible to determine M from a knowledge of the morphology as well as the nature of the energy migration process. We consider two simple models for calculating M in later sections.

For the present we will emphasize the trap rather than the energy migration process. We begin by considering the types of EFSs and their dependence on environmental factors such as matrix or concentration effects. There are three types of EFSs. The adjacent intramolecular EFS provides information on the conformational statistics of the guest polymer. Since the interaction is intramolecular, excimer formation is independent of chain concentration. From the conformational energy maps of PS determined by Yoon et al.,[35] it is possible to show that the tt meso dyad in the aromatic vinyl polymers will give rise to the majority of the adjacent intramolecular EFS. This restriction to a single dyadic type leads to a very small population of these EFS. Furthermore, Flory[36] has shown that modest changes in the radius of gyration of a polymer chain as a function of solvent compatibility do not affect the local dyadic configuration. Thus, adjacent intramolecular EFS should be independent of the host matrix.

The intramolecular EFS formed between nonadjacent chromophores should be sensitive to chain expansion and coiling and, thus, to the thermodynamic interaction between the guest and host polymers. Since the aromatic vinyl polymers are fairly stiff, the likelihood of this occurrence is rather low. Nevertheless, an increase in this type of EFS would reflect a decrease in the radius of gyration of the polymer coil due to a thermodynamically poor solvent. This type of EFS is also independent of guest chain concentration.

The intermolecular site formed between aromatic rings on different polymer chains should be particularly sensitive to the blend thermodynamics. More specifically, the number of intermolecular sites should increase with the degree of guest chain aggregation in a phase-separated blend. The placement of guest chain segments will be random only if the guest and host polymers are molecularly compatible.

To summarize, the intramolecular adjacent EFS is expected to be independent of both concentration and host; the intramolecular nonadjacent site is independent of concentration but could depend on the host; and the intermolecular site is dependent on both concentration and the thermodynamic nature of the host matrix.

B. ELECTRONIC EXCITATION TRANSPORT

EET, also referred to as energy migration, is the other essential element of the photophysics of excimer formation in polymers. It is generally accepted that facile energy migration between identical chromophores may occur following absorption of radiation by an aromatic ring that is pendent to the chain.[37] The mechanism of exciton hopping may be considered to be a series of single-step Forster transfers between aromatic chromophores. This migration process may be viewed as a random walk of the electronic excitation. At each step in the walk, the excitation may be emitted from an isolated chromophore or monomer. Alternately, the excitation may be trapped at an EFS followed by excimer emission or deactivation. However, development of a quantitative model for the exciton hopping is a difficult task.

The problem of a random walk with emission and traps has been the subject of numerous analytical studies. Levinson[38] treated a walk on an infinite one-dimensional lattice with emission and randomly placed traps. This approach was extended to finite-length chains by Fitzgibbon[30] and has been used to study the fluorescence of isolated polymer chains. This area will be discussed later in this review. Rosenstock[39] investigated random walks in one, two, and three dimensions with traps at the boundaries, whereas Rudemo[40] examined simple three-dimensional walks with randomly distributed traps.

Much of the significant work dealing with random walks has been accomplished by

Montroll, who has applied generating function techniques to walks on periodic lattices. This field of study has been reviewed by Barber.[41] One of the more important results has been the determination of the number of distinct sites visited in n steps as n approaches infinity for one-dimensional, two-dimensional, and different types of three-dimensional lattices. In a recent computer simulation study,[42] some of the earlier analytical results have been reproduced numerically. In this same work, hopping to all possible neighbors with a distance dependence corresponding to the Forster mechanism was also treated.

Information about the separation and relative orientation of adjacent chromophores is needed to understand the photophysics of isolated aryl vinyl polymer chains. The rate constant for a single hop between two aromatic rings, W, can be related to the monomer lifetime, τ, through the Forster expression[43]

$$W\tau = \frac{3}{2} \kappa^2 \left[\frac{R_0}{R}\right]^6 \qquad (2)$$

where R_0 is the Forster radius and κ^2 is an orientation factor relating the angles between the vector connecting the centers of mass of the two rings and the emission and absorption dipoles of the donor and acceptor, respectively. R_0 depends on the specific chemical groups under consideration and can be evaluated from spectroscopic data. The expression indicates that the rate of energy migration is extremely sensitive to the average chromophore separation, R. This spatial sensitivity has promoted the use of energy transfer as a distance and orientation probe.

There are two classes of experiments that may be performed to assess the existence of EET and thereafter to utilize its spatial sensitivity to understand polymer structure. The first class of experiments involves a system in which only donor chromophores are present; there are no traps such as excimers or extrinsic dopants. Donor-donor interactions may be examined using the phenomenon of fluorescence depolarization in which initially polarized light excites suitably oriented chromophores in the ensemble. If EET occurs, the transfer of the excitation to a neighboring unexcited donor chromophore will lead to almost complete depolarization. Measurement of the transient or photostationary anisotropy will allow the extent of interaction among the donor chromophores to be determined.

An extensive series of theoretical[44-53] and experimental[54-59] studies has been performed at Stanford University in recent years in order to develop the use of transient anisotropy measurement of EET for the study of the configurational properties of polymer chains. These theories are directed toward study of polymer chains containing a small number of chromophores randomly attached along the chain contour or at the chain ends. Clearly, these are quite different from the aromatic vinyl polymer in which there is an aromatic group on each repeat unit. The anisotropy approach may be used to obtain the end-to-end distance or the radius of gyration for the isolated chain, but they have not yet been applied directly to the study of phase separated systems, the subject of the present work.

The study of polymer blend immiscibility requires the understanding of physical parameters such as phase separation kinetics, domain sizes, coil collapse vs. interpenetration, and a multicomponent phase diagram. It is quite difficult to use fluorescence depolarization as a photophysical tool to follow these dynamic processes. The ability of fluorescence depolarization to detect subtle morphological changes in polymer blends depends critically on whether the magnitude and rate of depolarization at early times could be adequately resolved for a system with many chromophores. A polarized excitation within a chromophore-rich domain, such as that formed by phase separation or local coil contraction of high molecular weight polymers, will quickly depolarize within a short distance from the point of initial excitation. Further bulk changes outside of this effective volume would not be detected.

The second class of experiments, more suitable for studying blend miscibility, is the trapping experiment in which the migrating exciton is trapped at a low energy state. A similar many-body EET treatment to that employed in the donor-donor analysis may also be employed for the sparsely labeled polymer chain. An experimentally simpler approach, and one that is more appropriate for the aromatic vinyl polymer, is to look at the integrated trap fluorescence. This is the method used in the present work, with the EFS being the trap. From the discussion on excimer formation in the previous section, it is easily seen that the three types of EFSs are quite sensitive to changes in the interactions between the polymers in a blend. To date, the multidimensional random walk models[28] and binary phase separation model[60,61] described in this paper still prove to be significant in the study of polymer blends.

For polymer solutions and blends, the presence of energy migration increases the difficulties in the interpretation of I_D/I_M. An increased probability of remote segment contact on the same chain (i.e., local coil collapse) and clustering of segments on different chains (i.e., phase separation) have the same qualitative effect on I_D/I_M for both mechanisms. This short-circuiting process effectively increases the region of space that the exciton can traverse, thus leading to more efficient sampling of EFSs. It appears, however, that the relative contributions of the number of EFSs and the rate at which they are sampled by exciton migration may be clarified by consideration of the efficiency of exciton migration in more detail.

III. MORPHOLOGY OF POLYMER BLENDS

A. BLENDS CLOSE TO IMMISCIBILITY
1. Effect of Solubility Parameter Differences

Prior to 1979, only a few systematic studies of the excimer fluorescence of blends containing aryl vinyl polymers have been reported. Although PS was the first polymer to be studied in pure films or in solution,[62-65] no studies of PS in polymer blends were made during this period. Poly(1-vinylnaphthalene) (P1VN) was the next polymer to be studied in pure films[66,67] and in solution.[63,67,68] Fox et al.[67] gave brief qualitative results of the excimer fluorescence of poly(methyl methacrylate) (PMMA) blends containing 0.01 to 40% P1VN.

At about the same time, pure films and solutions of P2VN were first studied,[67,69] followed quickly by blends of P2VN and PS.[16-18] These blends were prepared by polymerization at 393 K of styrene that contained about 0.1% P2VN[69] or by film casting from a benzene or chloroform solution containing the guest and host polymers.[16-18] The solvent cast P2VN/PS blends containing 1% or more P2VN were visually immiscible.[17] In related studies on the effect of P2VN molecular weight on I_D/I_M, Nishijima et al.[70] found anomalously high excimer fluorescence for P2VN molecular weights greater than 140,000 and attributed it to possible phase separation.

Most of the earlier studies of polymer blends emphasized the details of excimer photophysics and tacitly assumed that the fluorescent guest was miscible in the host;[62-70] the morphology of the blends was generally not verified by other experimental or theoretical means. At the same time, but in the context of polymer physics, it was generally recognized that numerous variables can influence the thermodynamics of miscibility of polymer blends prepared by casting from a common solvent. These include the concentration of the two polymers and any residual solvent, the temperature at which the blend is prepared, the polymer molecular weight and polydispersity, and the extent of dispersive, polar, or hydrogen bonding interactions between the components. The effort at Stanford was initiated in order to create the link between the photophysics and the polymer physics. To assess the effects of these molecular and processing parameter factors on the compatibility of blends required a systematic approach in examining the sensitivity of the experimental observable I_D/I_M.

In the first study, the excimer molecular probe technique was used to examine the degree of clustering and aggregation in blends formed from dispersion of 0.2 wt% P2VN in a series of poly(alkyl methacrylate) host polymer matrices.[19] The selection of the blend system was based on several criteria. In an early treatment of polymer thermodynamics the Flory-Huggins theory of polymeric solutions was applied to polymeric blends by Krause.[71] An important point is that the binary interaction parameter, χ_{12}, which characterizes the interaction energy between individual segments of the polymers in the blend, is the major component in the enthalpy of mixing. To evaluate the influence of the chemical nature of the segment-segment interaction, Hildebrand[72] introduced the solubility parameter, δ, defined as the square root of the cohesive energy density. Direct application to polymer blend compatibility suggests that maximum interpenetration of two dissimilar polymers is expected when the binary interaction parameter vanishes, i.e., when the two polymers have the same solubility parameter.[19]

The poly(alkyl methacrylate) host matrices provide for such molecular dispersion since the range of their solubility parameters brackets the solubility parameter of P2VN. Furthermore, Krause indicates that the solubility parameter scheme will work best for polymers with similar polarity and hydrogen-bonding characteristics. Blends with electron donor-acceptor interactions between the components will likely lead to a negative interaction parameter, a condition not allowed in the regular solution theory. Since in homologous series the repeat unit varies only slightly from one member to the next, the interactions between the P2VN guest and the poly(alkyl methacrylate) hosts should vary in a qualitatively predictable fashion.

Of equal importance is the requirement that the host matrices be transparent to both the incident excitation and the P2VN monomer and excimer fluorescence. Finally, the glass transition temperatures of the poly(alkyl methacrylate) hosts should be above room temperature, making it possible to "freeze in" the chain conformational distribution of the guest polymer when films are prepared by solvent casting at room temperature. This is true for all of the hosts with the exception of poly(*n*-butyl methacrylate) and poly(vinyl acetate).

The fluorescence results for similar blends of P2VN with each of the poly(alkyl methacrylate), poly(vinyl acetate), and PS are shown in Figure 1. Here I_D/I_M has been plotted against the difference in host and P2VN guest solubility parameters, with the latter calculated by a molar group additivity approach. The major observation is that the smooth curve drawn through the data and passing through a well-defined minimum in I_D/I_M is symmetric about $\Delta\delta = 0$ for the guest and host polymers. The increase in I_D/I_M reflects the increase in the local concentration of EFSs due to decrease in the host quality. Similar results have been observed in our laboratory for poly(acenaphthalene) and P1VN.[23] Analogous behavior was subsequently observed for aromatic vinyl polymers in solvent series and solvent mixtures. Soutar[73] measured I_D/I_M in P1VN and poly(1-vinylnaphthalene-co-methyl methacrylate) in mixed solvent systems of toluene/methanol and toluene/cyclohexane. A minimum in I_D/I_M was observed when the solubility parameter of the solvent matched that of the polymer guest. Li et al.[74] estimated the δ of PS in a variety of monomeric solvents by tagging the PS chain with pyrene groups at regularly spaced intervals and monitoring the pyrene excimer to monomer fluorescence ratio. Their results yielded a value of 9.1 for δ_{PS}, which agrees quite well with published data.

It is quite clear that the excimer probe is sensitive to the host matrix. However, it is unclear which type of EFS is responsible for the host sensitivity. As mentioned before, the adjacent intramolecular EFS depends upon the population of suitable local conformational states of the chain, while the intermolecular site is sensitive solely to aggregation of the guest polymer. It is generally accepted that increasing the solvent quality will lead to expansion of the coiled polymer chain.[36] The local conformational structure, however, is insensitive to the solvent medium and therefore only the long-range conformational structure of the chain is affected. As the host matrix becomes a poorer solvent for the P2VN guest, both the bending back of segments of a polymer chain and the clustering of different chains

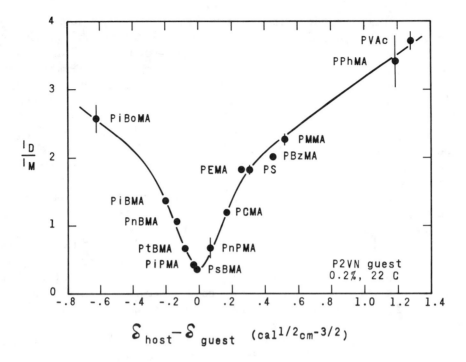

FIGURE 1. Excimer to monomer intensity ratio for 0.2 wt% P2VN blended with a series of poly(alkyl methacrylates) and solvent cast from toluene at 295 K. SDs are indicated by vertical bars where they are larger than the size of the solid circle. The results show the sensitivity of the excimer fluorescence to the enthalpic interactions between the guest and host polymers. (From Frank, C. W. and Gashgari, M. A., *Macromolecules*, 12, 163, 1979. With permission.)

would be expected to increase the nonadjacent intramolecular and intermolecular EFS and consequently to lead to an increase in I_D/I_M. As an alternative explanation, it is also possible that the higher molecular weight components of our polydisperse P2VN, which initially are compatible with the host matrix, become incompatible as the host matrix becomes a poorer solvent. This sets the stage for our next series of experiments.

The I_D/I_M vs. solubility parameter curve is directly analogous to a plot of intrinsic viscosity of polymer solutions prepared from different solvents or of the inverse swelling ratio of a cross-linked polymer in different solvents. A most significant difference, however, is that information in this case is being obtained on the molecular, in contrast to the bulk, level. For the blends prepared with 0.2 wt% P2VN in the poly(alkyl methacrylate) hosts, phase separation, as indicated by a bluish or milky tinge, was not visible until the absolute magnitude of the solubility parameter difference exceeded 0.6 $(cal/cm^3)^{1/2}$. For comparison, there were significant increases in I_D/I_M for solubility parameter differences of 0.2 to 0.4 $(cal/cm^3)^{1/2}$, as illustrated in Figure 1.

2. Effect of Concentration

Gashgari and Frank[21,24] have investigated the effect of P2VN concentration on I_D/I_M in three of the poly(alkyl methacrylate) hosts: poly(*n*-butyl methacrylate) (PnBMA), poly(ethyl methacrylate) (PEMA), and PMMA. These particular polymers were selected primarily because they exhibit a modest range of solubility parameter differences with P2VN, and they were relatively easy to prepare by solvent casting. The P2VN concentration in these blends ranges from 0.003 to 10 wt%. At infinite dilution, the value of I_D/I_M is independent of the host matrix. Regardless of host, a sharp initial rise in I_D/I_M occurs between 0.003 and 0.1 wt% P2VN followed by a transition region extending from 0.1 to 1.0 wt% P2VN.

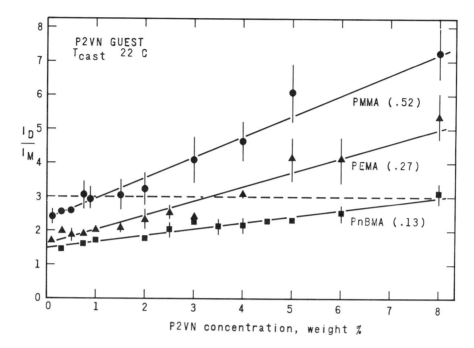

FIGURE 2.   Dependence of the observed excimer fluorescence on the P2VN guest concentration for films cast from toluene at 295 K. The numbers in parentheses beside the identifying host codes for each smoothed line through the data are the absolute values of the solubility parameter differences. (From Frank, C. W. and Gashgari, M. A., *Ann. N.Y. Acad. Sci.*, 366, 387, 1981. With permission.)

Thereafter, each blend exhibits a linear dependence of I_D/I_M on guest concentration over the range of 1.0 to 10 wt%, as shown in Figure 2. The slopes of the lines increase as the absolute value of the solubility parameter difference, shown in parentheses in Figure 2, and, hence, the binary interaction parameter increase.

The visual appearance of the films presents an interesting observation. It is not possible to state unequivocally that an optically clear film of a polymer blend is thermodynamically compatible. This is because two separate transparent layers may form, or the two polymers may have equal refractive indices, or the dispersed phase may have dimensions smaller than the wavelength of visible light. However, optically cloudy films almost certainly represent thermodynamically incompatible blends. When inspected visually, phase separation for these blends was indicated only for P2VN concentrations greater than 1 wt%. Furthermore, the demarcation between clear and cloudy films occurs approximately at $I_D/I_M = 3.0 \pm 0.5$ for each of the three host matrices. This is illustrated by the dashed line in Figure 2. Blends with I_D/I_M ratios above this value are cloudy and assuredly phase separated, whereas clear films of, as yet, unknown morphology on the molecular level are observed below the value.

In both the solubility parameter and guest concentration dependence studies, an increase in I_D/I_M consistently preceded changes in the appearance of the blends as conditions were altered toward greater immiscibility. Whereas the subjective determination of blend compatibility on the basis of optical quality is essentially bimodal, i.e., the film is either cloudy or not, the quantitative measurement of I_D/I_M allows continuous changes in the morphology of the blend to be followed. It seems reasonable to conclude that excimer fluorescence is detecting the onset of phase separation, which is occurring on a scale too small to be noticed visually.

The independence of I_D/I_M at infinite dilution on the host matrices suggests that the P2VN guest polymer coils are effectively isolated from one another, forcing excimer formation to arise totally from adjacent intramolecular interactions. The abrupt rise in I_D/I_M for

P2VN concentration between 0.003 and 0.1 wt% may be associated with an increase in the population of intermolecular and nonadjacent intramolecular EFS due to local guest chain aggregation prior to macroscopic phase separation. The linear increase in I_D/I_M for blends with P2VN concentration above 0.1 wt% shown in Figure 2 reflects the increase in intermolecular EFS and is a qualitative measure of the degree of phase separation. For blends with a poorer host matrix, cloudiness and, hence, macroscopic phase separation occur earlier, as indicated by the slopes of the linear region. A consistent interpretation of both studies is that the experimental observable I_D/I_M may be correlated qualitatively with guest chain aggregation.

Use of the excimer molecular probe to investigate blend morphology and compatibility would be reasonably straightforward if the photophysics were solely due to segmental rotation and bulk diffusive interactions between the chromophores on the polymer. Superimposed on this element of photophysics is the facile exciton migration between identical chromophores following absorption of radiation by one of the monomer units on the chain.[37] A major problem associated with the development of a quantitative model for the influence of exciton hopping on the behavior of I_D/I_M as a function of guest-host interactions involves accounting for multidimensional energy transfer. This can occur between chromophores on different chains or across loops of the same chain as well as in combination with a one-dimensional random walk along segments of a single chain. In the absence of such a comprehensive model, we can only make two limiting assumptions. The first presumes that the number of EFSs in the blend is fixed and variations in the observed I_D/I_M are attributed to changes in the efficiency of energy migration. At the other extreme, sampling of the different types of EFSs by exciton migration is assumed to be equally efficient and I_D/I_M is a function of the changing EFS trap population associated with the environmental condition.

As a first attempt to clarify the relative contributions of EFS population and energy migration efficiency to I_D/I_M, Frank and Gashgari[21] proposed a local clustering model to interpret the results of the P2VN concentration study. This model focuses on a particular chromophore on an isolated polymer chain that has just been promoted to the excited state. A reference volume V_0 surrounding this excited chromophore is defined such that it encloses all possible chromophores that can participate in energy transfer during the lifetime of the excitation. With respect to the excited ring, these chromophores inside V_0 consist of either rings from the same polymer or from different chains that have penetrated into the reference volume. In the case of high dilution and molecular dispersion of the guest and host polymer chains, energy migration will occur intramolecularly over the guest random coil and can be treated as a one-dimensional random walk. The extent to which exciton migration can sample the entire polymer coil depends on the size of the coil, the efficiency of energy migration, and the local thermodynamic condition of the surrounding host. If energy migration is slow or if the molecular weight of the coil is so large such that portions of it extend outside V_0, those portions will remain unsampled.

Changing the host matrix to a thermodynamically poorer one serves to reduce the radius of gyration of the coil and to increase the population of the nonadjacent EFS. Providing that the change in physical conditions between the guest and host maintains a compatible blend and that the coil is within V_0 such that it can be sampled by exciton migration, the ratio I_D/I_M is expected to increase. If the energy migration is not very efficient, the increased EFS will not be sampled and I_D/I_M should remain unchanged. An increase in the dimensionality of energy migration due to intramolecular cross-loop hopping[75] will result in enhanced sampling of EFS and increase in I_D/I_M.

For concentrated solid solutions and phase-separated blends, the reference volume will consist of intermolecular chromophores from chains that have penetrated V_0. The observed fluorescence behavior due to interaction between different chains within a separate domain or in the bulk could result solely from an increase in the number of intermolecular EFS and,

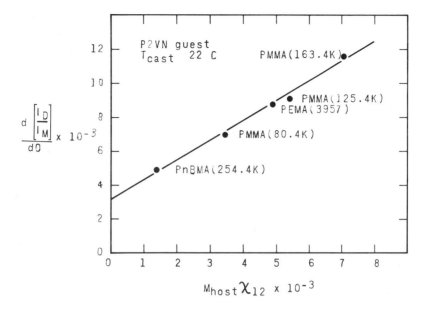

FIGURE 3. Dependence of the slope of the plot of the intermolecular excimer to monomer fluorescence intensity ratio as a function of concentration. This ratio is obtained by subtracting the value of I_D/I_M at infinite dilution from the experimentally determined value at a finite guest concentration. The results were used to generate the empirical expression in Equation 3. (From Frank, C. W. and Gashagari, M. A., *Ann. N.Y. Acad. Sci.*, 366, 387, 1981. With permission.)

therefore, I_D/I_M should increase in direct proportion to the concentration of aromatic rings within V_0. The contribution from energy migration will probably be most important at the onset of aggregation. As intermolecular contacts increase, the dimensionality of the energy migration will increase, with a consequent increase in efficiency of sampling. At high compatible concentrations, the population of EFS within the reference volume, whether intramolecular or intermolecular, increases linearly with guest concentration.

The objective of the clustering model is to describe empirically the two possible forms of intermolecular interaction giving rise to excimer fluorescence. The expression takes the form

$$\frac{I_D}{I_M} \text{ (intermolecular)} = \{A + BM_{host}\chi\}C \tag{3}$$

where C is the bulk concentration of the guest, M_{host} is the molecular weight of the host matrix, χ is the binary interaction parameter, and A and B are constants representing a collection of the photophysical rate parameters. The first term in Equation 3 denotes the intermolecular contribution from the random distribution of chromophores in the bulk, while the second term is associated with the additional contribution to intermolecular EFS due to local phase separation. The physical meaning of the expression is that the local concentration of chromophores within V_0 will be the same as in the bulk for compatible blends but will be enhanced if V_0 is located within a phase-separated domain.

The model has been applied to the P2VN concentration study and the result is illustrated in Figure 3. A plot of the slopes of the linear I_D/I_M region of the different hosts of Figure 2 against the product of the binary interaction parameter and the host molecular weight yields a straight line. Note that two additional molecular weight of PMMA are included in Figure 3. The close agreement between the different blends signifies only that the model is a good empirical representation of the data. Recently, Tao and Frank[34] have extended the cluster model to account for the effects of the molecular weights of both the host and guest polymers.

Excimer fluorescence has been shown qualitatively to be a sensitive molecular probe of polymer-polymer interactions in the amorphous state. The most significant experimental aspect of the excimer probe is that measurements of intermolecular association may be performed at extremely low concentrations. However, there is still an ambiguity associated with distinguishing between the contributions of energy migration efficiency and the number and type of EFSs. The study of energy migration must eventually involve the use of transient fluorescence to study directly the exciton migration kinetic rate constants. This has been considered but will not be discussed here.[33,34]

Nevertheless, at this point we have a skeletal basis for identifying conditions leading to possible phase separation of blends containing low concentrations (<0.5 wt%) of aryl vinyl polymers.[21] According to Equation 3, the intermolecular cluster model predicts that an increase in either the binary interaction parameter, as reflected in a larger difference between the guest-host solubility parameters, or in the bulk concentration, or in the molecular weight of the host yields a larger I_D/I_M ratio. We should note that the experiments represented by Equation 3 were for a P2VN guest of a particular molecular weight. In fact, the molecular weights of both the P2VN guest and the polymer host are important. This point will be considered in some detail in the next section. It is desirable, however, to know whether excimer fluorescence may be used solely to determine the absolute miscibility of the P2VN blends. This requires the existence of a P2VN blend that remains miscible at relatively high guest concentration such that a complementary method, e.g., differential scanning calorimetry, can yield independent information on whether or not the blend has phase separated. A further requirement is that the blend possesses a glass transition temperature well below the casting temperature so that a thermodynamically equilibrated morphology can be formed.

3. Effect of Molecular Weight

Classical Flory-Huggins theory suggests that the first problem described above can be resolved by using the experimental parameter of host molecular weight. For a given guest-host combination and, therefore, a constant binary interaction parameter, miscible blends with a high guest concentration can be prepared by lowering the host molecular weight. By lowering the host molecular weight to about 1000, a number of blends containing up to 35 wt% P2VN made with these host polymers were found to be miscible by DSC.[76-78] The I_D/I_M ratio of these compatible concentrated P2VN blends serves as a baseline for studying phase separation in these blends.

Semerak and Frank[23] investigated the dependence of I_D/I_M for P2VN guest polymer upon the molecular weight of the host. Three different P2VN guests having molecular weights 21,000, 70,000, and 265,000 were blended with eight different PS hosts having molecular weights between 2200 and 390,000 and nine PMMA hosts having molecular weights between 1100 and 350,000. In these studies, as in the previous ones, photostationary-state excitation was used and energy migration was presumed to be efficient such that the number of available EFSs is the rate-limiting feature. Therefore, an immiscible blend will have a larger value of I_D/I_M than a miscible blend of the same bulk composition.

The results of 0.3 wt% P2VN(21,000) dispersed in all of the different molecular weight PS and PMMA hosts are shown in Figure 4. The interesting features of these results are the independence of guest I_D/I_M, within experimental error, on host molecular weight for P2VN(21,000) in PS and the increase in I_D/I_M in P2VN(70,000) for PMMA host molecular weights greater than 10,000. The results for P2VN(70,000) in the PS and PMMA hosts are displayed in Figure 5. The jump in I_D/I_M is observed for both sets of host matrices, but the absolute value is approximately twice as large for PMMA compared to PS. It is of interest

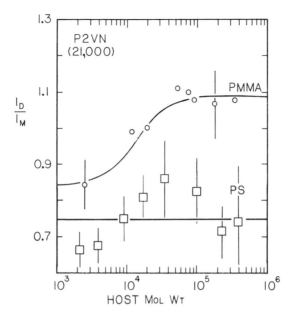

FIGURE 4.   Effect of the host molecular weight on I_D/I_M for 0.3 wt% blends of P2VN (21,000) with PMMA and PS. The uncorrected excimer and monomer intensities, measured under front-face illumination, have been determined at 398 and 337 nm, respectively. Open plotting symbols indicate that the films were optically clear. (From Semerak, S. N. and Frank, C. W., *Macromolecules*, 17, 1148, 1984. With permission.)

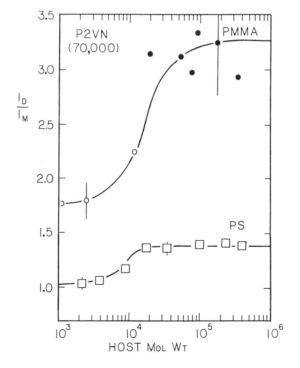

FIGURE 5.   Effect of the host molecular weight on I_D/I_M for 0.3 wt% blends of P2VN (70,000) with PMMA and PS. Open plotting symbols indicate that the films were optically clear; filled symbols indicate that the films were cloudy and obviously phase separated. (From Semerak, S. N. and Frank, C. W., *Macromolecules*, 17, 1148, 1984. With permission.)

to note that all of the PS blends remained optically clear while the PMMA blends with host molecular weight greater than 10,000 were cloudy and obviously phase separated. The solid lines simply represent smooth curves through the data points. Subsequent analysis relative to phase separation is based solely on the deviation of I_D/I_M from some lower baseline reference value.

In an immiscible blend most of the chromophores are located in the guest-rich phase; therefore, the average value of M is low and I_D/I_M is large. Thus, the fluorescence data in Figure 5 indicate that blends of P2VN(70,000) with PS(2200) are more miscible than those with PS(390,000). Similarly, the P2VN/PMMA blends are less miscible compared to the P2VN/PS blends for all host molecular weights. If optical clarity were used as the sole criterion, one would conclude, for example, that only blends with P2VN(265,000) in PS with molecular weight greater than 35,000 are immiscible.

It is not yet possible to distinguish between aggregated guest chains and those that are collapsed isolated coils solely on the basis of the observed excimer fluorescence. It is likely, however, that the 21,000 molecular weight P2VN is too low to exhibit any appreciable chain expansion as a result of host solvent effects. In light of this, and the conclusion that Q_D/Q_M is the same for P2VN in both PS and PMMA,[29] it seems reasonable to select the P2VN/PS(2100) blends as reference points since these have previously been shown to be miscible.

For a given P2VN guest molecular weight, three observations may be made:

1. The I_D/I_M ratio for any host molecular weight is always larger for PMMA than for PS.
2. The increase in I_D/I_M as a function of host molecular weight is more dramatic for PMMA than for PS.
3. Cloudy films occur in P2VN/PMMA blends with a lower host molecular weight compared to P2VN/PS blends.

All of these results are consistent with the conclusion that the P2VN/PMMA blends are less miscible than the P2VN/PS series. Moreover, it is obvious that the fluorescence method is exceedingly more sophisticated than the use of simple appearance as a diagnostic tool. Although direct comparison of the two methods could be misleading, information of this type is useful in establishing the limits of applicability of this new method.

We have been successful so far in using excimer fluorescence to predict qualitatively the physical parameters leading to blend immiscibility. These parameters, such as critical guest concentration and host molecular weight, are consistent with the values obtained by conventional methods. In order for excimer fluorescence to become a quantitative tool for analyzing phase separation, we require theoretical and experimental analysis of the blend phase diagram. Our objective is to compare theoretically calculated and experimentally generated binary interaction parameters for the P2VN/PS and P2VN/PMMA blends and to see whether the blends can be sufficiently characterized by a single value of the interaction parameter.

The thermodynamic stability of binary blends requires that conditions leading to a negative free energy of mixing and a positive second derivative of the free energy of mixing with respect to the volume fraction be simultaneously satisfied. The simplest possible theory for the thermodynamics of mixing two polymers is the Flory-Huggins lattice treatment.[79] Our use of Flory-Huggins lattice theory does not represent lack of awareness of Flory's later equation of state treatment or other recent work due, for example, to Koningsveld.[80,81] Rather, we use it because it provides a very good first-order treatment and is an excellent guide to designing experiments.

The pertinent equations required for calculation of the binodal, which is the locus of points representing the boundary between stable and metastable compositions as a function of temperature, are given by

$$\ln v_A'' = (x_A/x_B - 1)(1 - v_A'') - x_A\chi_{AB}(1 - v_A'')^2 - (x_A/x_B - 1)v_b'$$

$$+ x_A\chi_{AB}(v_B')^2 + \ln(1 - v_B') \tag{4}$$

$$\ln v_B' = (x_B/x_A - 1)(1 - v_B') - x_B\chi_{AB}(1 - v_B')^2 - (x_B/x_A - 1)v_A''$$

$$+ x_B\chi_{AB}(v_A'')^2 + \ln(1 - v_A'') \tag{5}$$

where A and B are the two polymer species, the single prime indicates the B-lean phase of the mixture, two primes indicate the A-lean phase of the mixture, v is the volume fraction, x is the degree of polymerization equal to V_m/V_r, V_m is the polymer molar volume, V_r is the reference repeat-unit molar volume (taken as the smaller of the two repeat-unit molar volumes of the polymers A and B), and χ_{AB} is the binary interaction parameter.

The experimental and theoretical comparison of the binary interaction parameter involves the following analysis. First, we determine the phase diagram for a particular P2VN blend system for a series of binary interaction parameter values. Next, we find the interaction parameter value corresponding to the experimentally determined critical host molecular weight, M*, for the blend composition in which it was determined. This quantity M* determines the upper limit of the host molecular weight for a miscible blend. Finally, the extracted binary interaction parameter can be compared to an estimate based on intrinsic viscosity measurements or molar additivity calculations.

There are several limitations of the Flory-Huggins treatment that should be considered before comparison of theory with experiment. For the more compatible P2VN/PS blend system, the thermodynamics of the solvent-cast blend should be analyzed by a ternary phase diagram. As solvent evaporates from the system, the effective glass transition temperature of the blend, T_g, must be lower than the casting temperature, T_c, in order that sufficient mobility remains to allow thermodynamic equilibrium. Further evaporation of solvent will eventually lead to an increase in T_g above T_c and a nonequilibrium blend will set in. For P2VN/PMMA blends, we also have to be concerned with the dissimilarity of the two polymers and whether the solvent may interact preferentially with one of the polymers. For the present, we will assume that the casting solvent toluene interacts equally with the P2VN guest and the different hosts and that all films prepared should have the same degree of nonequilibrium character. The latter is reasonable, given the same method of preparation with a single casting solvent. These assumptions meant that the casting solvent could be neglected in the calculation of the phase equilibrium, except that the apparent binary interaction parameter might be reduced from the nonsolvent binary parameter by an amount proportional to the amount of solvent present in the blend at the solidification point.

Other limitations of the Flory-Huggins theory include its failure to predict temperature and concentration behavior of miscible polymer pairs and the requirement of additional entropy and enthalpy factors for polymer solutions below concentrations of about 1% due to the discontinuous nature of these dilute solutions. Since this study concerns polymer blends with constant composition that were prepared at constant temperature, we expect that the Flory-Huggins lattice theory will be correct in predicting host molecular weight effects.

Selected calculated binodals for the P2VN(70,000)/PS and P2VN(70,000)/PMMA blends are presented in Figures 6 and 7, respectively. Note that these binodals are plotted as P2VN concentration vs. host molecular weight with the interaction parameter held constant, rather than the usual method of presentation giving temperature vs. concentration with the inter-action parameter held constant. The dashed line in the figures represents the bulk concentration of the guest polymer in the films. As the molecular weight of the host increases, the concentration of guest in the guest-lean phase at the binodal decreases rapidly. Note that the binodal curves for $\chi_{AB} \geq 0.05$ for the P2VN/PMMA blend system are essentially vertical

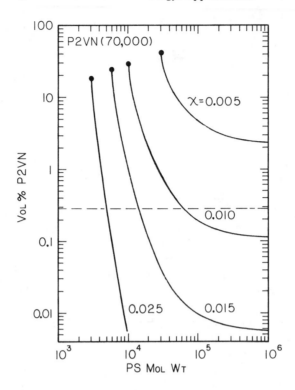

FIGURE 6.   Binodal compositions calculated for P2VN (70,000)/PS blends. Results for the P2VN lean phase are shown as volume percent P2VN vs. PS molecular weight. The interaction parameter for each curve is given and the critical point is indicated by the filled circle. The dashed line denotes 0.3 wt% P2VN. (From Semerak, S. N. and Frank, C. W., *Macromolecules*, 14, 443, 1981. With permission.)

lines over the P2VN-concentration range; hence, the location of the critical point is used in an interpolation procedure to determine the appropriate value of χ_{AB} that corresponds to the host molecular weight for which a sharp increase in I_D/I_M is observed.

We first consider the P2VN/PS blends. As noted earlier, we infer from the lack of significant change in I_D/I_M for P2VN(21,000) with the PS host molecular weight that all blends are miscible. This arises from the fact that the guest polymer is of relatively low molecular weight. An upper limit of 0.027 is established for χ_{AB} for the P2VN(21,000)/PS blends. A similar treatment of the I_D/I_M vs. host molecular weight data for P2VN(70,000) and P2VN(265,000) yielded values of 16,000 and 27,000, respectively, for the quantity M*, the critical or cloud-point host molecular weight. Subsequent interpolation with the respective phase diagram resulted in an average experimental value of $\chi_{AB} = 0.010 \pm 0.004$.

Since the experimentally determined interaction parameter is rather small, there do not seem to be any specific interactions between P2VN and PS as would be expected from their similar chemical structures. The precise value of χ_{AB} depends critically on the accuracy of the solubility parameters of the guest and host polymers because χ_{AB} is proportional to the square of the difference in two very similar quantities. Substitution of $\delta_{PS} = 9.10$ $(cal/cm^3)^{1/2}$ found from the literature, and $\delta_{P2VN} = 8.85$ $(cal/cm^3)^{1/2}$ experimentally determined, as well as the reference volume, based on PS, of 99.8 cm^3/mol into the Hildebrand binary interaction parameter expression yields $\chi_{AB} = 0.010 \pm 0.005$. The error limits arise from the uncertainties in component solubility parameters. Clearly, the agreement between the experimental and calculated interaction parameters is very good.

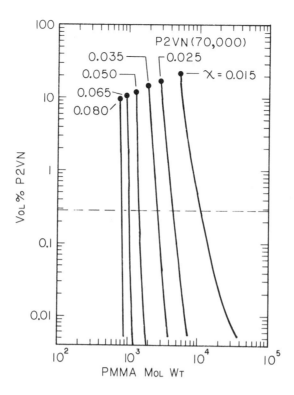

FIGURE 7. Binodal compositions calculated for P2VN (70,000)/PMMA blends. See Figure 6 for the details. (From Semerak, S. N. and Frank, C. W., *Macromolecules*, 17, 1148, 1984. With permission.)

For the P2VN/PMMA blends, a consistent value for the cloud-point host molecular weight M* < 1000 is assigned to all three blend systems with different P2VN molecular weight. Turning to the binodals, we find through interpolation an average value for χ_{AB} = 0.070 ± 0.015. Two other approaches have been developed to obtain the calculated χ_{AB} for comparison. In the first approach, due to Small,[82] the solubility parameter is determined from the sum of tabulated values for the chemical groups within the repeat unit of the polymer. The second approach is based on the expansion of a polymer chain, which is maximized in a solvent that has the same solubility parameter as the polymer. This expansion can be determined experimentally from the intrinsic viscosity of dilute polymer solutions or from swelling of a lightly cross-linked polymer sample in a variety of solvents.

Small's method produces the poorest estimates of χ_{AB}, particularly since the P2VN/ PMMA pair is incorrectly predicted to be more miscible than the P2VN/PS pair. The intrinsic viscosity method accurately predicts that the P2VN/PS pair is more miscible compared to P2VN/PMMA. The predicted value for χ_{AB} = 0.023 is quite far off compared to 0.070. This discrepancy may be removed by setting δ_{P2VN} = 8.7 $(cal/cm^3)^{1/2}$.

4. Effect of Temperature

Using excimer fluorescence from an aromatic vinyl polymer introduced as a guest in a matrix of a nonfluorescent host polymer, we have focused on three variables important to low-concentration miscibility and small-scale phase separation: enthalpic segmental inter-action between guest and host polymer,[19,21] concentration of the guest polymer in the host matrix,[24] and molecular weights of the guest and host polymers.[23,29] In all of the studies, we have been able to attribute the change in I_D/I_M to a combination of intramolecular and intermolecular aggregation of chromophores. In so doing, we have had to make several assumptions:

1.   The efficiency of electronic energy migration is assumed to be independent of host, molecular weight, and concentration.
2.   Residual solvent may be neglected in the calculation of the binodals.
3.   The effect of casting temperature on the thermodynamic equilibrium of the blend may be neglected.
4.   We have neglected the effects of temperature in the Flory-Huggins lattice treatment in the estimation of the binary interaction parameter by using I_D/I_M and cloud point information.

Although enthalpic interactions and guest concentration may be varied in a straightforward manner, an analogous examination of the effect of temperature presents considerable difficulties. Certainly, thermodynamic equilibrium may be ensured in the blend if fluorescence measurements are made sufficiently above the glass transition temperature. Unfortunately, such an experiment introduces two new factors. The first is the loss of specific knowledge about the EFS population. The formation of an excimer in the glassy state is primarily a result of competitive trapping of the migrating exciton. A secondary contribution is due to rotational sampling in which the residence time of the excitation at a particular aromatic ring is long enough for segmental motion to cause a second ring to move into the required position for excimer formation. Since the latter process depends on the local viscosity, the rate will increase with temperature, thus increasing the observed I_D/I_M. The second difficulty is that the nonradiative component of the overall excimer decay constant depends strongly on temperature[5] and, hence, I_D/I_M will change simply due to excimer complex destabilization. Both of these factors, however, bear no relation to any changes in molecular structure arising from alteration of the phase relationships.

Gashgari and Frank[24] were the first to use excimer fluorescence to investigate the temperature dependence of blend miscibility. Major emphasis was placed on two series of blends in which the guest polymer is P2VN and the host polymer is either PnBMA or PMMA. Again these particular hosts were selected because the enthalpic interactions with P2VN are sufficiently dissimilar to yield distinctly different phase relationships, and the glass transition temperatures differ widely, providing an opportunity to study nonequilibrium aspects of solvent casting. In order to eliminate the effects of temperature on the photophysical parameters while maintaining the temperature dependence of blend miscibility, the blends were prepared at elevated temperatures and quickly cooled before the fluorescence measurement was made. The intent was to freeze-in the morphology resulting from the evaporation of the solvent at the casting temperature. For the P2VN/PnBMA blends, casting at temperatures well above the T_g of the binary blend should result in thermodynamic equilibrium for the molecular structure of the blend. On the other hand, all of the P2VN/PMMA blends were prepared at casting temperatures less than or equal to T_g of the binary blend and, thus, are in nonequilibrium states.

In spite of the aforementioned difficulties in the use of visual appearance as a morphological tool, it is of interest to use the cloud point as a macroscopic threshold for phase separation and use the associated value of I_D/I_M to calculate the binodal and binary interaction parameter. All of the P2VN/PMMA blends cast at 293 K and above are visibly cloudy and phase separated for concentrations in excess of 0.5 wt%, while films prepared at lower concentrations are optically clear. The corresponding concentration separating clear and cloudy films for P2VN/PnBMA at 293 K is considerably larger at about 5%. This is consistent with PnBMA being a more thermodynamically compatible matrix for P2VN than PMMA.

If the quenching procedure is accepted as a valid means of freezing in the phase structure, then the visual as well as the fluorescence data reflect the equilibrium morphology present at the casting temperature. A most interesting correlation is observed if we denote the line separating the clear and cloudy films, in a manner done for traditional cloud point measurements, as the binodal. The cloud point curve coincides approximately with an interpolated fluorescence contour line based upon a constant value of $I_D/I_M = 3.0 \pm 0.5$. In other words, most films with I_D/I_M greater than 3.0 are cloudy while those with smaller values are clear, in the same fashion as was observed in Figures 1 and 2 for room temperature cast films. This demarcation seems to hold true for both PnBMA and PMMA hosts.

In generating the phase diagram from the solvent casting and temperature study, Gashgari and Frank[24] used a similar approach to that employed by Semerak and Frank.[23,29] As noted earlier, the most important contribution to the free energy of mixing for blends of high polymers is due to enthalpic effects represented by the binary interaction parameter. In the original Flory-Huggins treatment[83-86] χ_{12} is assumed to be independent of temperature, concentration, pressure, and molecular weight. Those assumptions are oversimplifications, however, and χ_{12} is generally considered to be dependent on temperature and concentration. One factor that was largely ignored in such approaches is the possible existence of a noncombinatorial component in the entropy of mixing. The classical combinatorial entropy contribution to the free energy of mixing is calculated assuming complete randomness of orientation of the rigid molecules.[24] The actual distribution of molecules will not be perfectly random, however, particularly if there are any specific interactions. Moreover, for nonrigid molecules there is an influence of the surrounding on the average randomness of orientation of a segment in a polymer chain relative to the orientation of the preceding segment.

Usually this noncombinatorial contribution to the entropy, χ_s, is incorporated into the binary interaction parameter as follows:[87]

$$\chi_{12}(T) = \chi_s + \frac{V_r}{RT} (\delta_1 - \delta_2)^2 \qquad (6)$$

Of prime importance to this study is the temperature dependence of χ_{12}, which is contained explicitly in the RT factor and implicitly in the solubility parameter. The latter dependence arises because in the molar group additivity approach the solubility parameter is estimated from the bulk density, which is temperature dependent. For the PMMA host matrix the enthalpic contribution to χ_{12} increases continuously with temperature over the range 173 to 373 K, while that for the PnBMA host first decreases, reaches a minimum at T_g, and then increases over the same temperature range. The increase of the enthalpic contribution with temperature above 293 K for both host matrices and the observation that the films prepared at higher casting temperatures are cloudy, whereas those cast at lower temperatures are clear support the suggestion that phase separation is occurring for blends prepared at higher casting temperatures. The modified Flory-Huggins approach should provide a good representation of the equilibrium thermodynamic state of the blends.

The binodal curves for a series of binary interaction parameters were calculated by the binodal equations of the Flory-Huggins lattice theory for different temperatures. The effective interaction parameter is given by $\chi_{12}(1 - v_s)$ where v_s is the volume fraction of residual solvent in the blend. For P2VN/PnBMA, the experimental and calculated binodals are relatively close but can be made to superimpose exactly by selecting an entropy correction term that is independent of temperature. The behavior of the PMMA host is considerably different from that of PnBMA. Although the residual solvent concentration is expected to be considerably higher than that in PnBMA, the experimental cloud point curve and calculated binodals do not even have the same curvature for any range of v_s.[24] The explanation of this behavior is that all P2VN/PMMA films represent nonequilibrium states. The calculated binodal suggests that in the absence of residual solvent the guest and host are highly incompatible and the blend would separate into two pure phases if true thermodynamic equilibrium could be reached at the elevated casting temperature.

B. MISCIBLE BLENDS
1. Low Concentration — Molecular Weight Effect

It has been shown that solvent casting at temperatures greater than T_g for the binary polymer blend, followed by rapid quenching to a temperature below T_g, can quench in the morphology characteristic of the casting temperature.[24] Assuming the morphology to represent the equilibrium state at the casting temperature, the solubility parameter approach has been shown to give a temperature-dependent χ_{12} which, when used in the Flory-Huggins theory, provides good correlation between experimental and calculated binodals. The next step in improving the quantitative aspects of excimer fluorescence as a morphological tool is to consider a polymer blend for which there is considerable information about the thermodynamics. In this way, there will be no ambiguity about whether a blend prepared under certain conditions is miscible or not. Efforts may then be directed toward developing a deeper understanding of the underlying photophysical processes. Once this is accomplished, this understanding may then be used to interpret quantitatively the fluorescence response of systems undergoing phase separation.

The mechanism of phase separation is expected to depend upon whether the single phase is metastable or unstable. In the metastable region $\partial^2 F/\partial\phi^2 > 0$ and small fluctuations in volume fraction ϕ decay with time. For phase separation to occur, a large composition fluctuation is required. Once such a nucleus is formed, it can grow by normal diffusion. During the nucleation and growth process, the composition of the growing phase remains constant as determined by the binodal curve. Within the unstable region inside the spinodal $\partial^2 F/\partial\phi^2 < 0$ and there is no thermodynamic barrier to separation. As a result, small concentration fluctuations can grow with time by a mechanism termed spinodal decomposition. A kinetic treatment of the early stages of the process has been developed by Cahn.[88]

The blend that was selected by Gelles and Frank[60,61] to accomplish this was PS/PVME. The advantage of the PS/PVME blends for photophysical studies are twofold. First, the low glass transition temperature of PVME (245 K) ensures that a low concentration blend will be in the rubbery state over a wide temperature range. As a result, the conformational population of the PS chains may be assumed to be in equilibrium. Excimer formation due to rotational sampling should be unimportant as long as the temperature is not too high above T_g. Second, the attainment of random mixing by casting from toluene permits energy migration to be studied quantitatively in the limits of low and high PS concentration. The former allows modeling of the EET as a one-dimensional random walk along an isolated PS chain while the latter permits a three-dimensional random walk model to be applied.

In the first paper of a series on energy migration in the aromatic vinyl polymers, Fitzgibbon[30] showed that the ratio I_D/I_M is given by Equation 1 where we recall that M is the overall probability that a photon absorbed by a pendant aromatic ring will lead to radiative or nonradiative decay from the monomer state. Correspondingly, 1-M is the probability that the absorbed photon will lead to decay from the excimer state. Fitzgibbon developed an analytical expression for M under conditions for which the polymer chain is isolated and the energy migration rate is much faster than then segmental rotation rate. In such a case, EET among the pendant aromatic rings may be treated as a one-dimensional random walk along the chain.

The probability M depends on three parameters: α, the probability of monomer emission both radiatively and nonradiatively at any step in the random walk, q, the trap fraction of EFSs, and L, the chain length. The following assumptions are made in the model:

1. The starting location of the excitation, or exciton, is completely random, i.e., all rings have the same absorption properties.
2. The exciton can only transfer between adjacent rings.
3. The probability that monomer decay takes place rather than transfer at a given ring, α, is constant along the chain.

4. EFSs are distributed randomly with the probability that a ring pair is an EFS equal to q.
5. The excimer does not dissociate to excited monomer, i.e., EFSs are irreversible traps.

The model successfully explains the molecular weight dependence of intramolecular excimer fluorescence observed for P2VN in 2-methyltetrahydrofuran.[30] There was a rapid increase in I_D/I_M at low molecular weight followed by a leveling at high molecular weight.

Gelles and Frank applied Fitzgibbon's one-dimensional random walk model to analogous PS fluorescence results obtained from 5 wt% PS/PVME blends. Monodisperse PS samples of different molecular weights were examined in blends with PVME over a temperature range in which no thermal deactivation of the excimer to excited monomer takes place. To simplify the analysis, the trap concentration was assumed to be independent of molecular weight, which is reasonable based on conformational statistics. Furthermore, α was assumed to be independent of molecular weight. Because different conformations result in different orientations and distances between adjacent aromatic rings, α is, in fact, not constant at each step of the walk and should be considered as an average probability. Nevertheless, since the concentrations of various conformations do not depend on molecular weight, α should be independent of chain length. Finally, Fitzgibbon and Frank[30] have found that Q_D/Q_M is also independent of molecular weight for P2VN in fluid solution. It seems reasonable to assume that this holds true for PS dispersed in PVME. Thus, for a given temperature the model has three parameters: α, q, and Q_D/Q_M.

The fluorescence results for 5 wt% PS/PVME blends are shown in Figure 8. I_D/I_M is given as a function of temperature for monodisperse PS molecular weights ranging from 2,200 to 390,000. Data at three temperatures are reported in Figure 9 showing the results of the fit to the one-dimensional model. The high viscosity of the PS/PVME blend precludes attributing the increase in the ratio to increase in rotational sampling for bonds near the ends and in the center of a chain. Similarly, the trap fraction is independent of molecular weight and chain length and is not directly responsible for the increase in I_D/I_M. The observed molecular weight dependence of the fluorescence ratio can only be explained by resorting to a photophysical model that includes energy migration.

A qualitative explanation of the energy migration model is useful to see the physical significance of the results. The walk is expected to be one-dimensional when the chains are isolated and relatively extended and the trap concentration is small. Because of the small probability of emission at each step, sites are sampled over and over. For short chains, there is smaller probability of finding a trap due to the low trap concentrations. An increase in the molecular weight will yield more distinct sites available for sampling so that the probability for finding a trap will increase.

At higher molecular weights, I_D/I_M levels off for two reasons. The first is that the length of the chain that can be sampled is limited because the lifetime of the excitation is finite. This effect is enhanced by the inefficiency of a one-dimensional walk resulting from the resampling of sites. The number of distinct sites visited in n steps for a one-dimensional walk without loss to emission or traps is proportional to $n^{1/2}$.[41] The second reason is that the trap concentration is also finite and consequently has the effect of dividing long chains into shorter segments, thus making the long chains appear photophysically similar.

The effect of temperature on I_D/I_M below 303 K is also attributed to the trap concentration and sampling lifetime. Increasing the temperature will increase the trap concentration because the tt meso conformation for excimer formation is higher in energy than the other meso dyads. This directly leads to an increase in I_D/I_M. The longer chains, however, are affected less because of the large number of sites already available for sampling. Indeed, the results show a stronger temperature dependence for the shorter chains. The non-Arrhenius temperature dependence observed is due to the increase in Q_D/Q_M with temperature and the more efficient trapping of the migrating exciton by the increased trap concentration.

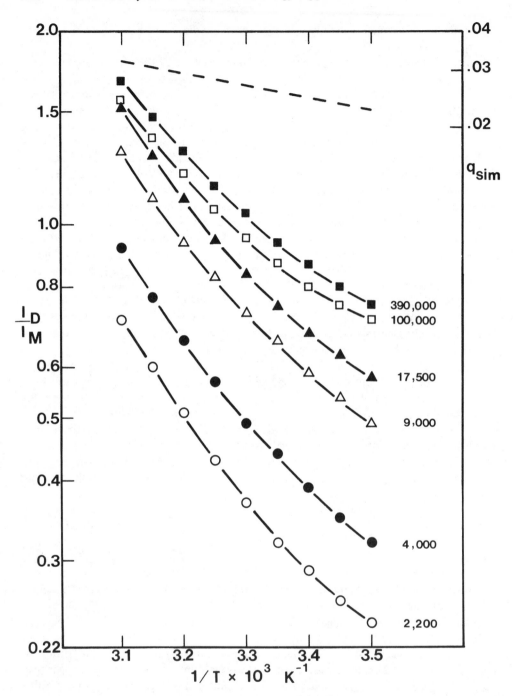

FIGURE 8.   Temperature dependence of I_D/I_M for the 5 wt% PS/PVME blend. Smooth curves are drawn through the experimental fluorescence data. The notation by each curve indicates the molecular weight of the monodisperse PS. The PVME molecular weight was 44,600 based on intrinsic viscosity measurements. The dashed line illustrates the results of a Monte Carlo simulation to determine the excimer forming site trap fraction. (From Gelles, R. and Frank, C. W., *Macromolecules*, 15, 741, 1982. With permission.)

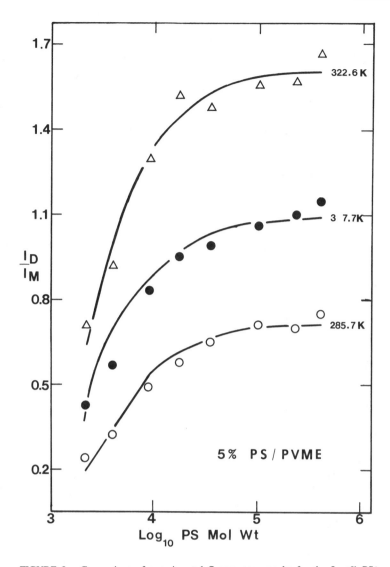

FIGURE 9. Comparison of experimental fluorescence results for the 5 wt% PS/
PVME blend with the best fit from the one-dimensional random walk model (solid
curve) for several temperatures. (From Gelles, R. and Frank, C. W., *Macromolecules*,
15, 741, 1982. With permission.)

For fluorescence data taken at and above 303 K, the one-dimensional model cannot fit
data in which the molecular weight effect persists in the high molecular weight range. At
the higher temperatures, it appears that the random walk of the excitation is no longer one-
dimensional over all or part of the PS molecular weight range. The increase in the dimensionality of the exciton migration reflects changes in the morphology of the blend. It is well
known that the PS/PVME blend system exhibits a lower critical solution temperature.[89-92]
Phase separation induced at higher temperatures would lead to intermolecular transfer of the
excitation between aggregated PS chains, imparting a three-dimensional character to the
walk. A second possibility is that there could be a local contraction of the PS coil with
increasing temperature as a result of unfavorable thermodynamic interactions with the PVME
host matrix. This would cause energy migration to occur between nonadjacent chromophores
across loops on the chain, also resulting in a more efficient walk due to the increased
dimensionality. It has been shown that decreasing the solvent quality decreases the coil

dimension of a polymer. While this may affect the conformational populations slightly, a more important factor is that it will allow more facile transfer across loops. The resulting increase in the dimensionality of the walk makes energy migration more efficient by increasing the number of distinct sites visited. In contrast to the one-dimensional case, the number of distinct sites visited in n steps is proportional to n for a three-dimensional walk on a periodic lattice. This increase in efficiency also allows the excitation to sample longer chains before emission takes place, resulting in the persistence of the molecular weight effect.

It is interesting to note that an analogous effect of thermodynamic compatibility between the guest and host matrix has also been observed in polymer solutions. A study of PS fluorescence in various solvents by Torkelson et al.[93] indicated that altering the solvent quality changed both the dependence of I_D/I_M on molecular weight and the magnitude of the ratio. In poorer solvents, the molecular weight effect persisted to higher molecular weights after the initial rapid increase. Another investigation in fluid solution of energy transfer from naphthalene groups in poly(1-naphthyl methacrylate) to vinylanthracene traps located at chain ends by Aspler and Guillet[94] showed that the efficiency of transfer increased in solvents that reduce the coil dimensions.

2. High Concentration — Random Mixing Effects

At concentrations just above the dilute-semidilute transition for an aryl vinyl polymer, energy migration should be predominately intramolecular with occasional hopping between chains when the excitation reaches a ring that is close to the exterior of a random coil. As the concentration is increased, such as in the case of phase separation or local coil collapse, the effective dimensionality can increase whenever the rings on different chains are as close as adjacent ring pairs on a single chain. Furthermore, the trap concentration will be enhanced by the formation of intermolecular EFSs. The PS/PVME blend is an ideal system for studying this change in the dimensionality of energy migration because its morphology is easily controlled by the method of preparation.

The fluorescence data for miscible blends along with Equation 1 may be used to establish a calibration curve for M, the probability of eventual nonradiative or radiative decay of the excitation by monomer fluorescence. This calibration curve incorporates changes in dimensionality of the random walk as the aromatic vinyl polymer concentration increases. Finally, these changes can be molded by incorporating different analytical expressions for the quantity M and fitted to the experimental results. The analytical expression for M in the case of a three-dimensional random walk incorporates features from three-dimensional hopping between neighboring sites on a periodic lattice and distance dependence corresponding to the Forster energy transfer mechanism for the case of no emission or trapping.[28]

The dependence of I_D/I_M on PS volume fraction of PS/PVME blends cast from toluene with PS molecular weights of 4,000 and 100,000 is shown in Figure 10. Also shown are the results for phase-separated blends cast from THF made with PS of molecular weight 100,000, to be discussed later. For blends with moderate PS concentrations, the number of distinct sites visited per step will be close to unity, provided that the walk is short. Because the effect of the excimer traps and monomer emission will be to produce walks with a smaller number of steps, this simplifying assumption is reasonable at low concentration and should work even better at higher concentrations. As a result, each step in the random walk can be considered to be independent and the probabilities for monomer decay and trapping do not change at each step. It may be shown that for deep traps, i.e., once trapped by an excimer, the excitation will be eventually lost to excimer fluorescence or deactivation and Equation 1 takes the form[28]

$$\frac{I_D}{I_M} = \frac{Q_D}{Q_M} \left[\frac{q}{1-q} \right] \frac{1}{\alpha} \tag{7}$$

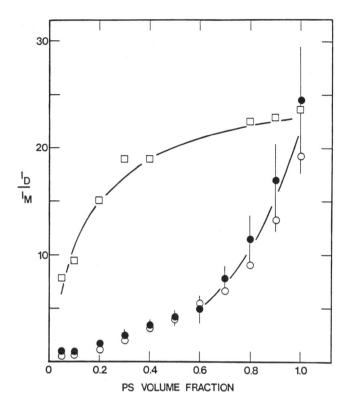

FIGURE 10. Comparison of experimental fluorescence results for the concentration dependence of the miscible PS/PVME blends prepared by casting from toluene solution with the best fit of the three-dimensional energy migration model (lower curve) and of the immiscible PS/PVME blends prepared by casting from tetrahydrofuran with the two-phase model (upper curve). (From Gelles, R. and Frank, C. W., *Macromolecules*, 15, 747, 1982. With permission.)

The quantity α should be considered to be the expected value of the fraction of times that monomer emission occurs before transfer. To fit the concentration dependence of I_D/I_M, expressions for q and α are derived in terms of the PS volume fraction ϕ. A simple lattice approach is employed in which the size of the lattice site is taken equal to the size of the PS repeat unit; the PVME segments are then broken up to fit into the same lattice.[28] Under this assumption, the separation distance between adjacent elements in the lattice will then be constant, although the number of PS rings next to any given ring will depend on concentration. It is further assumed that transfer can only take place between rings that are nearest neighbors, that the rate of transfer between two neighbors is constant with composition, and that the sum of the rates of transfer to each of the nearest neighbors equals the net rate of transfer from a given ring.

The result of this analysis of the three-dimensional spatially periodic model yields the following expression:

$$\frac{I_D}{I_M} = \frac{Q_D}{Q_M}\left[1 + \frac{Nk_e}{k_m}\phi\right]\left[\frac{q_{intra} + (N-2)\zeta\phi}{1 + [q_{intra} + (N-2)\zeta\phi]}\right] \qquad (8)$$

where ϕ is the volume fraction of PS, q_{intra} is approximately twice the probability that a given dyad is a suitable excimer forming site, N is the coordination number and ζ is the probability that two rings occupying adjacent lattice sites will be in the appropriate geometry for excimer trap formation. This probability is assumed to be the sum of the probabilities that a ring forms an excimer with each of the rings on nearest-neighbor sites. The value of Q_D/Q_M was found equal to 0.42 at room temperature from the low concentration PS/PVME study described earlier.[27] The quantity $N\phi$ is the number of nearest neighbors next to a ring to which transfer may occur. If k_e is defined as the rate constant for transfer to one nearest neighbor and k_m is the rate constant for monomer emission, then $N\phi k_e$ is the net transfer rate constant and $\alpha = k_m/(k_m + N\phi k_e)$. It was found previously that for a chain with 45% meso dyads at room temperature, the probability that a dyad is in the proper conformation to form an excimer is 0.025.[27] Therefore, $q_{intra} = 0.05$. The number of rings adjacent to a given ring that are not on the same chain is $(N - 2)\phi$. From the definition of ζ, it can be shown that the concentration of intermolecular traps is $q_{inter} = (N - 2)\zeta\phi$.

In order to fit the concentration results for toluene-cast PS/PVME blends, we have to make an assumption on the minimum PS concentration at which energy migration becomes three dimensional in nature. It was assumed that this concentration corresponds to the point at which the average chromophore separation equals the Forster transfer radius, R_0. At this separation, the probability of transfer between two rings is equal to the probability of monomer emission. If the chromophore separation is further assumed to be approximately equal to (1/ring concentration)$^{1/3}$, the ring concentration corresponding to R_0 may be obtained. From these calculations, the PS volume fraction at R_0 was found to be 0.61.

The calculation was carried out by fixing the value of $(N - 2)\zeta$ and then determining the value of NK_e/k_m for each PS volume fraction from the experimental I_D/I_M. The best results, based on minimizing the sum of squares of the residuals, were obtained for $(N - 2)\zeta = 0.31$ and $Nk_e/k_m = 52.09$. This is plotted as a solid line through the toluene-cast results in Figure 10. It is somewhat difficult to assess whether the values of the parameters used to fit the data are reasonable because it is difficult to obtain a quantitative check on the parameter $(N - 2)\zeta$. If the number of nearest neighbors, N, is taken to be 10, then the value of ζ is found to be of order 10^{-2}, which is consistent with the strict geometric requirements necessary for excimer formation. The value of k_e/k_m calculated directly from Forster transfer theory yielded a value of 2.65 compared to 5.2 from the fitted results with N = 10.

Finally, it is of interest to compare the number of steps an exciton makes at high and low PS concentrations. At very low concentrations energy migration is mainly down chain; the random walk is one-dimensional with a trap concentration $q_{intra} = 0.05$ at 300 K. If emission is ignored, the expected number of distinct sites visited is $(q_{intra})^{-1} = 20$. Since the number of steps made for the case of isolated chains is the square of the number of sites visited, the exciton travels approximately 400 steps. For high concentration PS blends with three-dimensional energy migration and no site resampling, this number is reduced to about 3. It is important to note that the nonlinear concentration dependence of I_D/I_M cannot be explained by a model that assumes that the only mechanism of excimer formation is direct absorption of a photon by a performed site. The present work shows that it can be explained by including energy migration in the analysis.

C. PHASE SEPARATION KINETICS
1. Two-Phase Model

The successful use of multidimensional energy migration to explain the fluorescence behavior for both low and high concentration PS/PVME blends certainly proved satisfying from a photophysical viewpoint. It is, however, more technologically important that excimer fluorescence may also be used quantitatively to study immiscible blends. In this section, we discuss the development of a simple two-phase morphological model that can be applied to PS/PVME blends that are phase separated by virtue of the solvent casting process. The model is extended to study the kinetics of thermally induced phase separation and to obtain a time-dependent measurement of the concentration of PS in the rich phase. Measurements of this type may be useful in verifying recent theories of spinodal decomposition in polymer blends.

In deriving an expression for I_D/I_M of a phase-separated system, we will assume that the volume fractions of PS in the rich and lean phases, ϕ_R and ϕ_L, are independent of the bulk concentration, ϕ_B. Applying the lever rule to the phase diagram, the volume fraction of the rich phase in the blend, V_R, is given by

$$V_R = \frac{\phi_B - \phi_L}{\phi_R - \phi_L} \tag{9}$$

The fraction of phenyl rings in the rich phase, X_R, which equals the probability that a photon is absorbed by a ring in the rich phase, is

$$X_R = \frac{\phi_R V_R}{\phi_R V_R + \phi_L(1 - V_R)} = \frac{\phi_R[\phi_B - \phi_L]}{\phi_R[\phi_B - \phi_L] + \phi_L[\phi_R - \phi_B]} \tag{10}$$

To simplify the analysis, two assumptions about the photophysics of a two-phase blend are made. First, we assume that the phases are large enough so that there are no boundary effects. This implies that the photophysical behavior of one phase in a phase-separated system with concentration ϕ is identical to that for a miscible blend of the same composition. The second, more critical assumption, is that there is no energy migration between the phases. This will be a good assumption when the lean phase concentration is low enough so that the PS coils are isolated and energy migration is mostly intramolecular. For concentrated PS blends, very few hops (i.e., about three) are made before the exciton is trapped due to the large number of EFS. Thus, an excitation should not be able to make enough transfers to escape the rich phase without being trapped. This, of course, pertains to concentrated phases large enough to scatter light.

If M_R and M_L are the probabilities of eventual monomer decay given that the photon is absorbed by the rich or lean phase, respectively, then the quantity I_D/I_M is given by a simple weighted average of fluorescence contributions from the two phases:

$$\frac{I_D}{I_M} = \frac{Q_D}{Q_M} \left[\frac{X_R(1 - M_R) + (1 - X_R)(1 - M_L)}{X_R M_R + (1 - X_R)M_L} \right] \tag{11}$$

In order to apply Equation 11 it is necessary to know values for Q_D/Q_M and the dependence of M on PS concentration. The ratio of the quantum yields has been estimated to be equal to 0.42 at 295 K for 5 wt% PS/PVME blends. Gelles and Frank[28] have performed fluorescence measurements for miscible blends taken at 298 K and applying Equation 1 have determined the concentration dependence of M. The significant qualitative feature is that the probability of monomer emission decreases rapidly with increasing PS concentration. An immediate test of this model is provided by the results for the phase separated PS/PVME blends cast from THF, shown also in Figure 10. The M_L and M_R values were determined for isolated PS chains and 98% PS, respectively, from the concentration dependence calibration curve. Using these values and an iterative procedure, the best fit of the experimental results was obtained for $\phi_L = 0.007$ and $\phi_R = 0.98$.

2. Phase Diagram

Interpretation of the mechanism of phase separation from fluorescence results requires that accurate values of the equilibrium binodal compositions for a particular temperature be

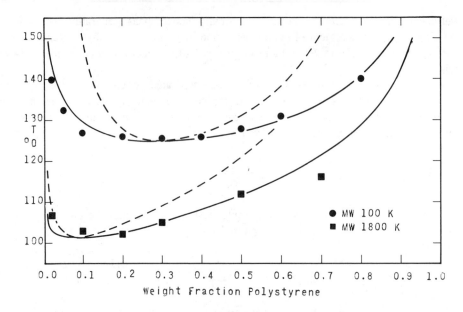

FIGURE 11.   Phase diagrams determined using a turbidimetric technique for PS/PVME blends having monodisperse PS molecular weights of 100,000 and 1,800,000 and PVME molecular weight of 44,600. The solid lines are the experimentally observed cloud point curves. The dashed lines are the spinodals calculated by assuming the cloud point curve was a true binodal and that the binary interaction parameter was concentration and temperature dependent. (From Am. Chem. Soc. Symp., Photophysics of Polymer Systems, Anaheim, CA, September 1986. With permission.)

available. These are necessary in order to calculate the volume fraction of the rich phase in the two-component system. This volume fraction is assumed to be constant during the early stages of spinodal decomposition. In an early study, Kwei et al.[91] and Nishi and Kwei[92] employed pulsed NMR and found that the Cahn-Hilliard[88] linearized nonstatistical theory of spinodal decomposition adequately described the process at short times, with the average composition of the two phases changing exponentially with time while their volume fractions remained constant. More recently, DeGennes[95] and Pincus[96] have extended the Cahn-Hilliard theory of spinodal decomposition to polymers in the melt. Their scaling relationships show that the kinetics of the early stages of the process should depend strongly on molecular weight.

Frank and Zin[97] used a turbidimetric technique to determine the cloud point curves for PS(100,000)/PVME and PS(1,800,000)/PVME blends. Cloud point temperatures for each of the PS molecular weights over the whole concentration range were also determined. The results are shown in Figure 11, where the solid points represent experimentally determined cloud point temperatures. An important extension of the experimental analysis for this phase diagram determination was the evaluation of an expression for the PS/PVME binary interaction parameter from the cloud point curves. This was done by assuming a functional form of the interaction parameter and then calculating the binodals using a free energy of mixing calculated from the Flory-Huggins configurational entropy of mixing combined with an enthalpy of mixing term based on the assumed interaction parameter. The result was that the binary interaction parameter may be represented as

$$\chi_{12}(T, C) = \frac{V_r}{RT} (-0.3067 + 0.000833T - 0.0086C) \qquad (12)$$

with temperature T in K and concentration C in weight fraction. Once an expression for the

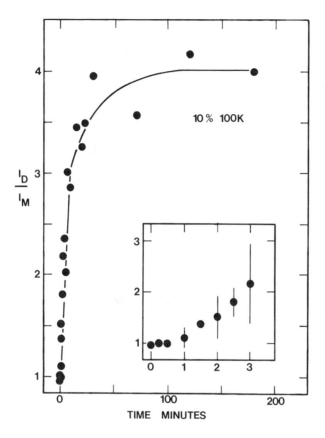

FIGURE 12. Effect of annealing time on the fluorescence of 10 wt% PS/PVME blends annealed at 423 K and then quenched to 295 K for the fluorescence measurement. The insert shows the short time behavior. (From Gelles, R. and Frank, C. W., *Macromolecules*, 16, 1448, 1983. With permission.)

interaction parameter was determined, the spinodal curves were calculated for each PS series. These curves are shown as dashed lines in Figure 11.

Two sets of experiments on phase separation kinetics have been performed. The first series of experiments by Gelles and Frank[60] was designed to demonstrate the feasibility of using excimer fluorescence to test Cahn's kinetic treatment of the early stages of spinodal decomposition. This is a linearized theory that predicts that the volume fractions of the two phases remain constant with time while the compositions change gradually. Gelles and Frank followed the excimer fluorescence from the rich phase as they thermally induced phase separation of a 10 wt% PS(100,000)/PVME blend. At an annealing temperature of 423 K, the majority of the change in I_D/I_M occurs in the first 10 min, with the ratios leveling off and becoming constant for times greater than about 50 min. Typical results are shown in Figure 12.

It is of interest to compare these data with results of microscopy studies in which an interconnected morphology was never observed for the 10 wt% films but was observed for the 50 wt% PS films.[60] This is consistent with Cahn's demonstration that the major phase must have a volume fraction greater than 0.15 to allow connectivity. A more important observation is that droplet coalescence begins to take place the same time I_D/I_M levels off. This coalescence pertains to a nucleation and growth mechanism in which the composition of the growing phase remains constant but the phase itself increases in size. This indicates that excimer fluorescence is sensitive to changes in local concentration occurring during early states of phase separation but not to the major morphological changes that occur at long times. The growth rate of the dominant concentration fluctuation, which encompasses a macroscopic diffusion coefficient, can be obtained from the linear short time portion. For 10 wt% PS(100,000)/PVME blend, the diffusion coefficient was found to be of the order of -10^{-11} cm^2/s.

3. Effect of Molecular Weight on Spinodal Decomposition

A critical feature of the modified spinodal decomposition analysis is the proposal by Pincus[96] that the growth rate of the concentration fluctuation that controls the kinetics during the early stages of decomposition is proportional to the melt reptation diffusion coefficient. As a result, the kinetics of the early region of spinodal decomposition is predicted to depend strongly upon molecular weight. The second study on phase separation kinetics of PS/PVME blends was designed to test this prediction.

For a blend of two polymers with unsymmetrical molecular weight, the time dependence of the early stages of the phase separation process is governed by the modified diffusion equation[95]

$$\frac{\partial \phi}{\partial t} = \wedge \left[\left[\frac{1}{N_A \phi_A} + \frac{1}{N_B(1 - \phi_A)} - 2\chi \right] \nabla^2 \phi - \frac{a^2}{36\phi_A(1 - \phi_A)} \nabla^4 \phi \right] \qquad (13)$$

where ϕ is the volume fraction of one of the components, t is time, $\wedge$ is the Onsager coefficient, defined at the ratio of the diffusional flux to the chemical potential gradient, N is the number of monomer units in the chain, χ is an interaction parameter, and a is a typical monomer dimension.

The solution to Equation 13 is a Fourier series that can be approximated by its maximum term[88]

$$\phi - \phi_0 = \exp\left[-t/\tau_{qmax} \right] \Sigma \left[A(q)\cos(q{\cdot}r) + B(q)\sin(q{\cdot}r) \right] \qquad (14)$$

where ϕ_0 is the initial composition, q is a wave vector, A and B are amplitudes, and r is position. The growth rate of concentration fluctuations with wave number q_{max} is the quantity $-\tau_{qmax}{}^{-1}$ and is given by

$$-\tau_q^{-1} = \wedge(q)q^2 \left[2\chi - \left[\frac{1}{N_A \phi_A} + \frac{1}{N_B(1 - \phi_A)} \right] - \frac{(aq)^2}{36\phi_A(1 - \phi_A)} \right] \qquad (15)$$

For fixed initial composition and annealing temperature, Equation 15 accompanied with the correct scaling relationships may be simplified to

$$-\tau_{qmax}^{-1} \approx \frac{1}{\phi_A N_B + \phi_B N_A} \left[\frac{T - T_s}{T_s} \right]^2 \qquad (16)$$

where T_s is the temperature at the spinodal. In order to use Equation 16, molecular weight will be defined relative to the size of a PVME repeat unit. Provided the thermodynamic driving force for phase separation is relatively constant, Equation 16 predicts that the growth rate of the dominant concentration fluctuation should decrease as the PS molecular weight increases.

The growth rate for the 10 wt% PS(100,000)/PVME blend is found experimentally to be about twice that for the 10 wt% PS(1,800,000)/PVME blend.[61] This ratio increases to 2.7 for the 50 wt% blends. Before discussing the results, it is worthwhile to look more closely at Figure 11. We note that the phase separation experiments were intended to be done under conditions such that the mechanism of phase separation was spinodal decomposition, not nucleation and growth. This may be accomplished most easily for a particular blend by selecting a system at the critical composition. At that point there would be the minimum tendency for nuclei to form during the time necessary to perform the temperature annealing experiment. As shown in the phase diagram, however, the critical composition is different for the 10 and 50 wt% PS/PVME blends. An optimum PS concentration that would minimize the extent of the metastable region that must be traversed during the temperature jump would be in the region between 0.3 to 0.4 weight fraction. Thus, the 50 wt% results should be more appropriate to test the deGennes-Pincus theory.

In fact, the predicted growth rate ratio for the 50 wt% PS/PVME blend is 3.9 compared to 2.7 found experimentally in quite reasonable agreement considering the difficulty of the quenching experiment.[97] The difference could be attributed to the slight polydispersity of the PVME in which the presence of short unentangled PVME chains that can diffuse rapidly during the early stages of phase separation is expected to weaken the molecular weight effect.

IV. SUMMARY

Applications of excimer fluorescence as a molecular level probe of chain configuration for isolated polymer chains and as a morphological tool for the study of phase separation are reviewed. In establishing the feasibility of any new experimental method for providing more detailed information about a physical phenomenon, it is necessary to compare the new approach with existing techniques. The major conclusions are summarized below.

First, the excimer probe is quite sensitive to the thermopdynamic compatibility between the guest polymer and the host matrix. The most significant experimental aspect of the excimer probe is that measurements of intermolecular association may be performed at extremely low concentrations, thus providing an advantage over classical methods such as DSC, DMS, and optical means. Increases in I_D/I_M consistently precede visible signs of phase separation of the blends.

Second, methods for the interpretation of the fluorescence ratio of aromatic vinyl polymer as a guest in solid blends have been presented which allow the experimental determination of the binary interaction parameter. Low concentration P2VN blends containing PMMA were found to be less miscible than the analogous PS blends.

Third, use of a two-phase photophysical model has demonstrated that, although casting above T_g greatly enhances the chance of the two-phase separated blends attaining thermodynamic equilibrium, equilibrium is not guaranteed unless other kinetic barriers are also absent. The casting temperature and annealing experiments have demonstrated the ability of the modified Flory-Huggins theory with a temperature-dependent binary interaction parameter to predict the thermodynamics of solvent-cast polymer blends.

Finally, excimer fluorescence of isolated polymer chains dispersed in a rigid solvent of good quality can be analyzed in terms of a one-dimensional random walk in which transfer only to nearest neighbors is possible. For sufficiently high guest polymer concentration, a three-dimensional model for singlet energy migration has been found to explain quite well the observed dependence of I_D/I_M on concentration of the guest polymer. Furthermore, the technique of excimer fluorescence can be used to study quantitatively the kinetics of phase separation in polymer blends.

ACKNOWLEDGMENTS

This work was supported by the Polymers Program of the National Science Foundation under Grant DMR84-07847.

REFERENCES

1. **Platzer, N. A. J., Ed.,** *Adv. Chem. Ser.,* 142, 1975.
2. **Klempner, D. and Frisch, K. C., Eds.,** *Polym. Sci. Technol.,* 10, 1977.
3. **Sperling, L. H., Ed.,** *Recent Advances in Polymer Blends, Grafts and Blocks,* Plenum Press, New York, 1974.
4. **Paul, D. R. and Newman, S., Ed.,** *Polymer Blends* , Vol. 1. Academic Press, New York, 1978.
5. **Olabisi, O., Robeson, L. M., and Shaw, M. T.,** *Polymer-Polymer Miscibility,* Academic Press, New York, 1979.
6. **Macknight, W. J., Karasz, F. E., and Fried, J. R.,** *Polymer Blends,* Paul, D. R. and Newman, S., Eds., Academic Press, New York, 1978, chap. 5.
7. **Kirste, R., Kruse, W. A., and Ibel, K.,** Determination of the conformation of polymers in the amorphous solid state and in concentrated solution by neutron diffraction, *Polymer,* 16, 120, 1975.
8. **Ballard, D. G. H., Rayner, M. G., and Schelten, J.,** Structure of a compatible mixture of two polymers as revealed by low-angle neutron scattering, *Polymer,* 17, 640, 1976.
9. **Hayashi, H., Hamada, R., and Nakajima, A.,** Small-angle X-ray scattering study on conformation of amorphous polymer chain in the bulk, *Macromolecules,* 9, 543, 1976.
10. **Wilkes, G. L. and Stein, R. S.,** Application of rheo-optical techniques for characterizing multicomponent-multiphase polymeric materials — a brief overview, *J. Polym. Sci., Polym. Symp.,* 60, 121, 1977.
11. **Kwei, T. K., Nishi, T., and Roberts, R. F.,** A study of compatible polymer mixtures, *Macromolecules,* 7, 667, 1974.
12. **McBrierty, V. J., Douglass, D. C., and Kwei, T. K.,** Compatibility in blends of poly(methyl methacrylate) and poly(styrene-co-acrylonitrile). II. An NMR study, *Macromolecules,* 11, 1265, 1978.
13. **Sperling, L. H.,** Interpenetrating polymer networks and related materials, *J. Polym. Sci., Macromol. Rev.,* 12, 141, 1977.
14. **Birks, J. B.,** *Photophysics of Aromatic Molecules,* Wiley-Interscience, New York, 1970, chap. 7.
15. **Klopffer, W.,** *Organic Molecular Photophysics,* Birks, J.B., Ed., John Wiley & Sons, New York, 1973, chap. 7.
16. **Frank, C. W. and Harrah, L. A.,** Excimer formation in vinyl polymers. II. Rigid solutions of poly(2-vinyl naphthalene) and polystyrene, *J. Chem. Phys.,* 61, 1526, 1974.
17. **Frank, C. W.,** Excimer formation in vinyl polymers. III. Fluid and rigid solutions of poly(4-vinyl biphenyl), *J. Chem. Phys.,* 61, 2015, 1974.
18. **Frank, C. W.,** Observation of relaxation processes near the glass transition by means of excimer fluorescence, *Macromolecules,* 8, 305, 1975.
19. **Frank, C. W. and Gashgari, M. A.,** Excimer fluorescence as a molecular probe of polymer blend compatibility. I. Blends of poly-(2-vinyl naphthalene) with poly(alkylmethacrylates), *Macromolecules,* 12, 163, 1979.
20. **Frank, C. W., Gashgari, M. A., Chutikamontham, P., and Haverly, V. J.,** Excimer fluorescence as a molecular probe of polymer blend compatibility II. The effect of concentration of blends of aromatic vinyl polymers with poly(alkylmethacrylates), in *Structure and Properties of Amorphous Polymers,* Walton, A. G., Ed., Elsevier, New York, 1980, 187.
21. **Frank, C. W. and Gashgari, M. A.,** Excimer fluorescence as a molecular probe of polymer-polymer interactions in the amorphous solid state, *Ann. N.Y. Acad. Sci.,* 366, 387, 1981.
22. **Frank, C. W.,** Excimer fluorescence: a new tool for studying polymer alloys, *Plast. Compd.,* January/February, 67, 1981.
23. **Semerak, S. N. and Frank, C. W.,** Excimer fluorescence as a molecular probe of blend miscibility. III. Effect of molecular weight of the host matrix, *Macromolecules,* 14, 443, 1981.
24. **Gashgari, M. A. and Frank, C. W.,** Excimer fluorescence as a molecular probe of blend miscibility. IV. Effect of temperature in solvent casting, *Macromolecules,* 14, 1558, 1981.
25. **Semerak, S. N. and Frank, C. W.,** Photphysics of excimer formation in aryl vinyl polymers, *Adv. Polym. Sci.,* 54, 32, 1983.
26. **Semerak, S. N. and Frank, C. W.,** Excimer fluorescence as a molecular probe of blend miscibility. V. Comparison with differential scanning calorimetry, *Adv. Chem. Ser.,* 203, 757, 1983.
27. **Gelles, R. and Frank, C. W.,** Energy migration in the aromatic vinyl polymers. II. Miscible blends of polystyrene with poly(vinyl methyl ether), *Macromolecules,* 15, 741, 1982.
28. **Gelles, R. and Frank, C. W.,** Energy migration in the aromatic vinyl polymers. III. Three dimensional migration in polystyrene/poly(vinyl methyl ether) blends, *Macromolecules,* 15, 747, 1982.
29. **Semerak, S. N. and Frank, C. W.,** Excimer fluorescence as a molecular probe of polymer blend miscibility. VI. Effect of molecular weight in blends of poly(2-vinyl naphthalene) with poly(methyl methacrylate), *Macromolecules,* 17, 1148, 1984.
30. **Fitzgibbon, P. D. and Frank, C. W.,** Energy migration in the aromatic vinyl polymers. I. A one-dimensional random walk model, *Macromolecules,* 15, 733, 1982.

31. **Thomas, J. W. and Frank, C. W.,** Energy migration in the aromatic vinyl polymers. IV. Blends of poly-(2-vinyl naphthalene) with poly-(cyclohexyl methacrylate), *Macromolecules,* 18, 1034, 1985.
32. **Fitzgibbon, P. D. and Frank, C. W.,** Excimer formation in dilute solution. I. Effect of pressure on 1,3-bis(2-naphthyl) propane and poly(2-vinyl naphthalene), *Macromolecules,* 14, 1650, 1981.
33. **Tao, W. C. and Frank, C. W.,** Energy migration in the aromatic vinyl polymers. VII. The application of a one-dimensional electronic excitation transport model for transient fluorescence of poly(2-vinylnaphthalene) in alkyl benzene solution and polystyrene blends, *J. Phys. Chem.,* 93, 776, 1989.
34. **Tao, W. C. and Frank, C. W.,** Excimer fluorescence as a molecular probe of polymer blend miscibility. VIII. Polymeric and glassy solvent host matrices, *Polymer,* 29, 1625, 1988.
35. **Yoon, D. J., Sundararajan, P. R. and Flory, P. J.,** Conformational characteristics of polystyrene, *Macromolecules,* 8, 776, 1975.
36. **Flory, P. J.,** Spatial configuration of macromolecular chains, *Science,* 188, 1268, 1975.
37. **Klopffer, W.,** Energy transfer and trapping in rigid solutions of aromatic polymers, *Spectrosc. Lett.,* 11, 863, 1978.
38. **Levinson, N.,** Emission probability in a random walk, *J. Soc. Ind. Appl. Math.,* 10, 442, 1962.
39. **Rosenstock, H. B.,** Random walks with spontaneous emission, *J. Soc. Ind. Appl. Math.,* 9, 169, 1961.
40. **Rudemo, M.,** On an absorption and emission problem for random walk, *SIAM J. Appl. Math.,* 14, 1293, 1966.
41. **Barber, M. N. and Ninham, B. W.,** *Random and Restricted Walks,* Gordon and Breach, New York, 1970.
42. **Zumofen, G. and Blumen, A.,** Random walks with variable step length on regular lattices, *Chem. Phys. Lett.,* 78, 131, 1981.
43. **Forster, T. H.,** Transfer mechanisms of electronic excitation, *Discuss. Faraday Soc.,* 27, 7, 1959.
44. **Gochanour, C. R., Andersen, H. C., and Fayer, M. D.,** Electronic excited state transport in solution, *J. Chem. Phys.,* 70, 4254, 1979.
45. **Loring, R. F., Anderson, H. C., and Fayer, M. D.,** Electronic excited state transport and trapping in solution, *J. Chem. Phys.,* 76, 2015, 1982.
46. **Fredrickson, G. H. and Frank, C. W.,** Interpretation of electronic energy transport experiments in polymeric systems, *Macromolecules,* 16, 1198, 1983.
47. **Gochanour, C. R. and Fayer, M. D.,** Electronic excited-state transport in random systems, time resolved fluorescence depolarization measurements, *J. Phys. Chem.,* 85, 1989, 1981.
48. **Frederickson, G. H. and Frank C. W.,** Nonexponential transient behavior in fluorescent polymeric systems, *Macromolecules,* 16, 572, 1983.
49. **Fredrickson, G. H., Andersen, H. C., and Frank, C. W.,** Electronic excited-state transport and trapping as a probe of intramolecular polumer structure, *J. Chem. Phys.,* 79, 3572, 1983.
50. **Fredrickson, G. H., Andersen, H. C., and Frank, C. W.,** Electronic excited-state transport on isolated polymer chains, *Macromolecules,* 16, 1456, 1983.
51. **Fredrickson, G. H., Andersen, H. C., and Frank, C. W.,** Electronic excited state transport and trapping on polymer chains, *Macromolecules,* 17, 54, 1984.
52. **Fredrickson, G. H., Andersen, H. C., and Frank, C. W.,** Electronic excitation transport as a probe of chain flexibility, *Macromolecules,* 17, 1496, 1984.
53. **Fredrickson, G. H., Andersen, H. C., and Frank, C. W.,** Macromolecular pair correlation functions from fluorescence depolarization experiments, *J. Polym. Sci., Polym. Phys.,* 23, 591, 1985.
54. **Fredrickson, G. H.,** Intermolecular correlation functions from forster energy-transfer experiments, *Macromolecules,* 19, 441, 1986.
55. **Ediger, M. D. and Fayer, M. D.,** New approach to probing polymer and polymer blend structure using electronic excitation transport, *Macromolecules,* 16, 1839, 1983.
56. **Ediger, M. D., Domingue, R. P., Peterson, K. A., and Fayer, M. D.,** Determination of the guest radius of gyration in polymer blends: time-resolved measurements of excitation transport induced fluorescence depolarization, *Macromolecules,* 18, 1182, 1985.
57. **Peterson, K. A., Zimmt, M. B., Linse, S., Domingue, R. P., and Fayer, M. D.,** Quantitative-determination of the radius of gyration of poly(methyl methacrylate) in the amorphous solid-state by time-resolved fluorescence depolarization measurements of excitation transport, *Macromolecules,* 20, 168, 1987.
58. **Peterson, K. A. and Fayer, M. D.,** Electronic excitation transport on isolated, flexible polymer chains in the amorphous solid state randomly tagged or end tagged with chromophores, *J. Chem. Phys.,* 85, 4702, 1986.
59. **Peterson, K. A., Zimmt, M. B., Linse, S., and Fayer, M. D.,** Ensemble average conformation of isolated polymer coils in solid blends using excitation transport, *ACS Symp. Ser.,* 358, 323, 1987.
60. **Gelles, R. and Frank, C. W.,** Phase separation in polystyrene/poly-(vinyl methyl ether) blends as studied by excimer fluorescence, *Macromolecules,* 15, 1486, 1982.
61. **Gelles, R. and Frank, C. W.,** Effect of molecular weight on polymer blend phase separation kinetics, *Macromolecules,* 16, 1448, 1983.

62. **Yanari, S. S., Bovey, F. A., and Lumry, R.,** Fluorescence of styrene homopolymers and copolymers, *Nature,* 200, 242, 1963.

63. **Vala, M. T., Haebig, J., and Rice S. A.,** Experimental study of luminescence and excitation trapping in vinyl polymers, paracyclophanes, and related compounds, *J. Chem. Phys.,* 43, 886, 1965.

64. **Longworth, J. W.,** Conformations and interactions of excited states. II. Polystyrene, polypeptides, and proteins, *Biopolymers,* 4, 1131, 1966.

65. **Basile, L. J.,** Effect of styrene monomer on the fluorescence properties of polystyrene, *J. Chem. Phys.,* 36, 2204, 1962.

66. **David, C., Demarteau, W., and Geuskens, G.,** Energy transfer in polymers. II. Solid poly(vinylnaphthalene)-benzophenone systems and copolymers vinylnaphthalene-vinylbenzophenone, *Eur. Polym. J.,* 6, 1397, 1970.

67. **Fox, R. B., Price, T. R., Cozzens, R. F. and MacDonald, J. R.,** Photophysical processes in polymers. IV. Excimer formation in vinylaromatic polymers and copolymers, *J. Chem. Phys.,* 57, 534, 1972.

68. **Nishijima, Y., Yamamoto, M., Mitani, K., Katayama, S., and Tanibuchi, T.,** Photoluminescence of polyvinyl/naphthalene (I) emission of excimer fluorescence, *Rep. Prog. Polym. Phys. Jpn.,* 13, 417, 1970.

69. **Harrah, L. A.,** Excimer formation in vinyl polymers. I. Temperature dependence in fluid solutions, *J. Chem. Phys.,* 56, 385, 1972.

70. **Nishijima, Y., Yamamoto, M., Katayama, S., Hirota, K., Sasaki, Y., and Tsujisaki, M.,** Excimer emission of high polymers in solution (i) fluorescence spectra of various naphthalene-containing polymers, *Rep. Prog. Polym. Phys. Jpn.,* 15, 445, 1972.

71. **Krause, S.,** Polymer compatibility, *J. Macromol. Sci., Rev. Macromol. Chem.,* C7, 251, 1972.

72. **Hildebrand, J. H. and Scott, R. L.,** *Regular Solutions,* Prentice-Hall, Englewood Cliffs, NJ, 1962.

73. **Soutar, I.,** Studies of intramolecular excimer formation in synthetic polymers, *Ann. N.Y. Acad. Sci.,* 366, 24, 1981.

74. **Li, Z. B., Winnik, M. A., and Guillet, J. E.,** A fluorescence method to determine the solubility parameters δ_H of soluble polymers at infinite dilution. Cyclization dynamics of polymers. II, *Macromolecules,* 16, 992, 1983.

75. **Guillet, J.,** Studies of energy migration and trapping in polymers containing naphthalene groups, *Polym. Prepr. Am. Chem. Soc. Div. Polym. Chem.,* 20, 395, 1979.

76. **Massa, D. J.,** Physical properties of blends of polystyrene with poly(methyl methacrylate) and styrene-(methyl methacrylate) copolymers, *Adv. Chem. Ser.,* 176, 433, 1979.

77. **Parent, R. R. and Thompson, E. V.,** Fracture surface-morphology and phase-relationships of polystyrene-poly(methyl methacrylate) systems — low-molecular-weight poly(methyl methacrylate) in polystyrene, *Adv. Chem. Ser.,* 176, 381, 1979.

78. **Parent, R. R. and Thompson, E. V.,** Fracture surface morphology and phase relationships of polystyrene/poly(methyl methacrylate) systems. I. Low-molecular-weight polystyrene in poly(methyl methacrylate), *J. Polym. Sci., Polym. Phys. Ed.,* 16, 1829, 1978.

79. **Flory, P. J.,** *Principles of Polymer Chemistry,* Cornell University Press, Ithaca, NY, 1953, chap. 12 and 13.

80. **Koningsveld, R. and Scholte, Th.,** Determination of liquid-liquid phase relations with the ultracentrifuge, *Kolloid Z. Z. Polym.,* 218, 114, 1967.

81. **Koingsveld, R. and Staverman, A. J.,** Determination of critical points in multicomponent polymer solutions, *J. Polym. Sci., C,* 16, 1775, 1967.

82. **Small, P. A.,** Some factors affecting the solubility of polymers, *J. Appl. Chem.,* 3, 71, 1953.

83. **Flory, P. J.,** Thermodynamics of high polymer solutions, *J. Chem. Phys.,* 9, 660, 1941.

84. **Flory, P. J.,** Thermodynamics of high polymer solutions, *J. Chem. Phys.,* 10, 51, 1942.

85. **Huggins, M. L.,** Solutions of long chain compounds, *J. Chem. Phys.,* 9, 440, 1941.

86. **Huggins, M. L.,** Theromodynamic properties of solutions of long-chain compounds, *Ann. N.Y. Acad. Sci.,* 43, 1, 1942.

87. **Koningsveld, R. and Kleintjens, L. A.,** Thermodynamics of polymer solutions, *J. Polym. Sci., Polym. Symp.,* 61, 221, 1977.

88. **Cahn, J. W.,** Phase separation by spinodal decomposition in isotropic systems, *J. Chem. Phys.,* 42, 93, 1965.

89. **Bank, M., Leffingwell, J. and Thies, C.,** The influence of solvent upon the compatibility of polystyrene and poly(vinyl methyl ether), *Macromolecules,* 4, 43, 1971.

90. **Bank, M., Leffingwell, J., and Thies, C.,** Thermally induced phase separation of polystyrene-poly(vinyl methyl ether) mixtures, *J. Polym. Sci.,* A-2, 10, 1097, 1972.

91. **Kwei, T. K., Nishi, T., and Roberts, R. F.,** A study of compatible polymer mixtures, *Macromolecules,* 7, 667, 1974.

92. **Nishi, T. and Kwei, T. K.,** Cloud point curves for poly(vinyl methyl ether) and monodisperse polystyrene mixtures, *Polymer,* 16, 285, 1975.

93. **Torkelson, J. M., Lipsky, S., and Tirrell, M.,** Polystyrene fluorescence: effects of molecular weight in various solvents, *Macromolecules,* 14, 1603, 1981.
94. **Aspler, J. S., Hoyle, C. E., and Guillet, J. E.,** Singlet energy transfer in a 1-naphthyl methacrylate-9-vinylanthracene copolymer, *Macromolecules,* 11, 925, 1978.
95. **de Gennes, P. G.,** Dynamics of fluctuations and spinodal decomposition in polymer blends, *J. Chem. Phys.,* 72, 4756, 1980.
96. **Pincus, P.,** Dynamics of fluctuations and spinodal decomposition in polymer blends. II. *J. Chem. Phys.,* 75, 1996, 1981.
97. **Frank, C. W. and Zin, W. C.,** Morphology in miscible and immiscible polymer blends: interplay of polymer photophysics with polymer physics, *ACS Symp. Ser.,* 358, 18, 1987.

Chapter 6

ELUCIDATION OF POLYMER COLLOID MORPHOLOGY THROUGH TIME-RESOLVED FLUORESCENCE MEASUREMENTS[1]

Mitchell A. Winnik

TABLE OF CONTENTS

I.   Introduction 198

II.   Materials 198
        A.   PIB-PMMA Particles 198
        B.   Model Compounds 199
        C.   PEHMA-PVAc Particles 199

III.   Specific Examples 200
        A.   Sorption Experiments with Anthracene 200
        B.   Fractal Analysis of Doped Particles 201
        C.   Excimer Emission from Labeled Stabilizer 202
        D.   Oxygen Quenching and the Core-Stabilizer Interface 207

IV.   Summary 208

References 209

I. INTRODUCTION

This article examines methods based upon fluorescence spectroscopy for elucidating the morphology of polymer systems in which one suspects phase separation to have occurred and for characterizing the interface between two polymer components. Phase-separated systems are quite common and they take many forms. They include simple polymer blends made up of two homopolymers, interpenetrating networks where one believes both phases are continuous, and a variety of systems composed of block- and graft copolymers, in bulk or mixed with corresponding homopolymers. Polymer interfaces are also important in coatings and in adhesion processes. The techniques we discuss are in principle applicable to all of these diverse systems.

In order to provide a framework for discussing these methods, we will examine a single type of material, a latex particle composed of a major component which is in its glassy state at room temperature, and a minor component which is in its rubbery state at room temperature. These materials were prepared as nonaqueous dispersions[2] of colloidal polymer particles in which the monomer yielding the major component was polymerized (free radical) in the presence of a small amount of the performed rubbery polymer. Grafting occurs and the final product contains a mixture of homopolymer and graft copolymer.

When these particles were prepared originally, their internal structure was unknown. Since the particles formed stable colloidal dispersions in hydrocarbon liquids,[3] one reasoned that colloidal stability was conferred by a surface covering of the rubbery polymer (the *stabilizer*) whose chains would be swollen by solvent and act as a barrier to particle-particle aggregation. Because of this, there was a tendency to refer to these materials as "core-shell" structures.

We now appreciate that these latex particles, at least the ones discussed here, have a more complex and more interesting internal structure.[4] It seems certain that the minor component, the rubbery polymer, forms a continuous network penetrating throughout the polymer interior.[5] There are also indications, not nearly so well documented, that the major component also forms a continuous phase within the particle. Thus, these materials have an interpenetrating network morphology.

Because the major component is below its T_g, the materials we examine have metastable structures generated out of kinetic convenience by a mechanism which is not at all understood. We suspect that the graft copolymer plays a crucial role in structure formation. While we are deeply interested in the mechanism whereby such structures form, our focus here is on the tools we have used to evaluate these structures. It is important to keep in mind that the morphology of the particle can be perturbed by swelling the rubbery phase, as when one redisperses the dried particles in hydrocarbon solvents, and that both in the dispersion and in the powder the internal structure in the particle might evolve through physical aging.

II. MATERIALS

A. PIB-PMMA PARTICLES

Particles of poly(methyl methacrylate) (PMMA) containing polyisobutylene (PIB) were prepared by polymerizing MMA in the presence of degraded butyl rubber.[6] The butyl rubber ($M \approx 10,000$) was a commercial product (Kalene® 8000) which by ^{1}H and ^{13}C NMR showed that it was PIB containing approximately 2% unsaturation due to isoprene units incorporated during its preparation. Particles could be prepared from reaction mixtures containing mole ratios of PIB monomer units to MMA monomer reactant varying from 4 to 20%, and these reactions yielded particles containing a monomer mole ratio IB/MMA which varied from 1 to 7%. Fluorescent dyes could be incorporated into the PMMA chains by adding a dye-containing comonomer such as *1a, 2a, 3a,* or *4a* during the polymerization step.

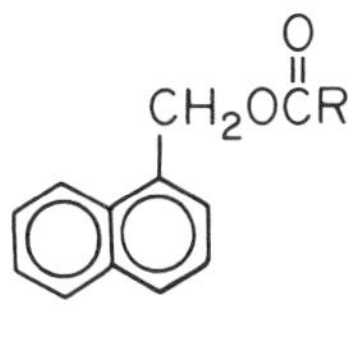

$$1 \; a,b \qquad 2 \; a,b$$

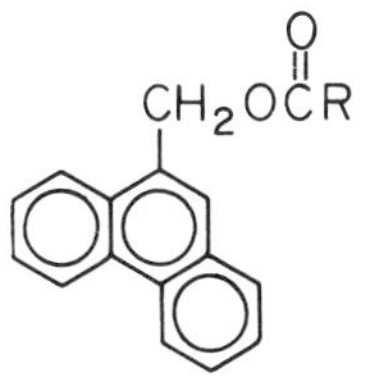

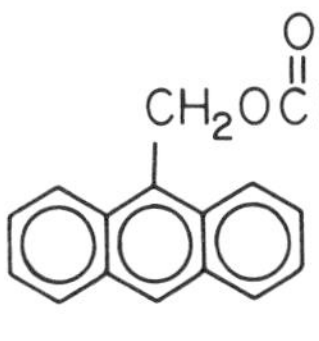

$$3 \; a, b \qquad 4 \; a, b$$

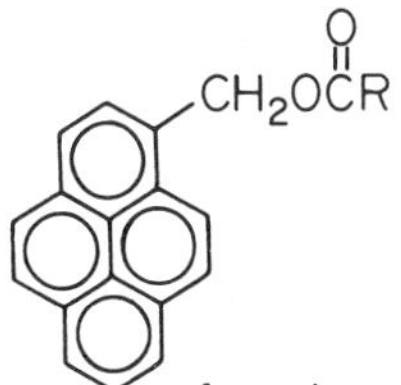

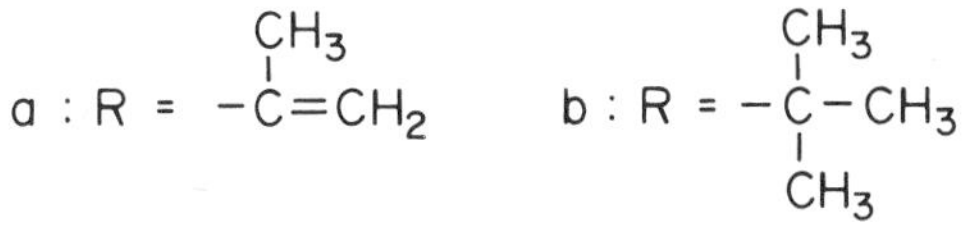

$$a : R = -\underset{\underset{CH_3}{|}}{C}=CH_2 \qquad b : R = -\underset{\underset{CH_3}{|}}{\overset{\overset{CH_3}{|}}{C}}-CH_3$$

In a typical reaction, 90 g of MMA and 10 g PIB are dissolved in 300 ml of isooctane along with a dye derivative (0.01 to 10 g). An initiator (e.g., AIBN) is added and the reaction is heated to 80°C for about 4h. Particles form with diameters of about 1 μm and with a fairly narrow distribution of particle sizes. These are centrifuged and redispersed in fresh solvent (5 c) to remove any unreacted polymer or unbound dye derivatives. The syntheses and characterization of these materials are described in detail elsewhere.[6] Because the dyes are bound to the PMMA chains, we refer to this material as core labeled.

It is useful to have stabilizer labeled particles as well. We have not yet carried out chemistry on PIB. We have, however, hydroformylated polybutadiene (PBD) and then reduced the polymer to obtain a rubber containing a small amount of $-CH_2OH$ groups.[7] Dyes such as pyrene (Py) can be attached as esters to these groups. These reactions transform PBD into an essentially saturated, isooctane-soluble polymer labeled with Py groups. When MMA is polymerized in the presence of this polymer, particles form which contain Py groups attached to the hydrocarbon polymer.

B. MODEL COMPOUNDS

Fluorescence experiments on labeled polymer systems are always interpreted by reference to the fluorescence behavior of small dye molecules containing the same chromophore. For this purpose we often use the alkylmethyl pivalate esters *1b*, *2b*, *3b*, and *4b* shown above.

C. PEHMA-PVAc PARTICLES

Poly(2-ethylhexyl)methacrylate (PEHMA) is very soluble in isooctane. Labeled derivatives can be prepared by copolymerization of EHMA with methacrylates such as *1a* to *4a*. When vinylacetate (VAc) is polymerized in the presence of the PEHMA copolymers, particles form which contain the fluorescent dyes bound to the stabilizer component. These particles are typically 0.3 μm in diameter and have a fairly narrow size distribution.[8]

Core-labeled PEHMA-PVAc particles were prepared by adding a trace of *2a* (100 ppm) to VAc plus unlabeled PEHMA in isooctane at 80°C in the presence of a free radical initiator. A representation of core-labeled particles (CLPs) and stabilizer-labeled particles (SLPs) containing the phenanthrene (Phe) group is shown below.

III. SPECIFIC EXAMPLES

A. SORPTION EXPERIMENTS WITH ANTHRACENE

A small sphere of PMMA will absorb a large molecule such as anthracene (An) very slowly. In polymer matrices well below T_g, diffusion coefficients decrease exponentially with the molecular size of the diffusant.[9] One can estimate that it would take years for a molecule the size of An to diffuse to the center of 1 μm sphere of PMMA.

If the PMMA component of a PMMA particle is labeled with naphthalene (N), it becomes possible to follow the penetration of An into the particle. N* can transfer its energy to An by the dipole coupling (Förster) mechanism.[10] This direct energy transfer process

$$N^* + An \rightarrow An^* + N$$

can occur over a distance (here, up to about 45 Å) which is substantial compared to molecular dimensions but small when compared to the particle size. Energy transfer is accompanied by quenching of the N* fluorescence. This means that as the amount of energy transfer increases, the intensity of An fluorescence (I_{An}) at 400 nm will increase and that of N fluorescence (I_N) at 340 nm will decrease. The quenched N* groups will also have a shorter fluorescence decay time. When a dispersion of particles core-labeled with N are exposed to a dilute cyclohexane solution of anthracene, the observation of energy transfer indicates that some An molecules can get close to the N groups in the particle.

The classic experimental result is shown in Figure 1 for a particle (N10) containing 10 mol% N groups in the PMMA component.[4-6] When this dispersion in cyclohexane is mixed with a dilute solution of An in cyclohexane and excited with a pulse of light at a wavelength (290 nm) where the N is excited preferentially, the fluorescence decay from these N groups (monitored at 337 nm) shows two components. At short times (Figure 1) there is a weak, rapidly decaying component leading to a much stronger long component with a decay time τ_1^N of 32 ns. When the monitoring wavelength is changed to 450 nm, where the An fluorescence decay is detected, one also observes two components; the short component ($\tau_s^A = 5$ ns) is the normal decay of the An molecules which absorb the excitation at 290 nm. The long component, with a decay time $\tau_1^A = 32$ ns, must be excited by energy transfer from N. This result establishes that energy transfer occurs in the system.

More important is the observation that upon increasing the concentration of An in the continuous medium, the decay time τ_1^N decreases.[4,5] This change occurs essentially as fast as one can prepare the samples. It indicates that An can diffuse rapidly into the particles to interact with essentially all the N groups contributing to the long component in the N fluorescence decay. This simple sorption experiment undermines the core-shell model in this system; the only way An can diffuse deep into the core of the PMMA particles in a short period of time is via a system of solvent-swollen channels which serve as a conduit. In this way the An molecules can receive energy from N groups near the PMMA-(PIB + solvent) interface. As a consequence we conclude that to form these channels much of the PIB component is buried in the particle.[5]

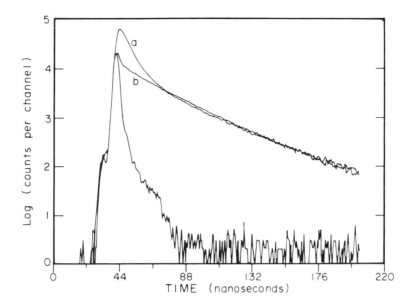

FIGURE 1.   Fluorescence decay of an N-10 dispersion in isooctane in the presence of 3×10^{-4} *M* An. Upper curve measured at $\lambda = 420$ to 480 nm (An emission); lower curve measured at 337 nm. Their respective long-time decay times are 31.6 and 32.1 ns.

B. FRACTAL ANALYSIS OF DOPED PARTICLES

If hydrocarbon-soluble substances are added to alkane dispersions of PMMA and the solvent removed, these dopants tend to remain in the PIB phase. One can take advantage of this property of the system to study the form of the PIB phase. One experiment involves doping this phase with tetraphenyl lead and carrying out X-ray scattering experiments.[11] The results of these studies are also at odds with the core-shell structure.

Other kinds of experiments become possible if the PIB phase is doped with fluorescent dyes or dye-quencher pairs. A particularly powerful experiment involves direct energy transfer measurements between donor-acceptor pairs. One adds the donor and acceptor molecules to the system one wishes to study and assumes they occupy random available sites in the accessible phase. The fluorescence decay curve of the energy donor $I_D(t)$ is very sensitive to the spatial distribution of nearby acceptor groups. This distribution depends upon the local geometry, and it is this geometry which one wishes to probe.

Förster's original theory[10] of direct energy transfer was developed for energy transfer in three dimensions, although the proper treatment of two-dimensional systems[12] (e.g., lipid bilayer membranes) has been known for some time. A major advance in the utility of the method for studying complex systems was provided by Klafter and Blumen,[13] who extended the theory to fractal geometries.

We begin with a brief review of energy transfer in three dimensions by the dipole-coupling mechanism.[10] The theory treats the case of noninteracting donors (D) in the presence of a random but macroscopically uniform distribution of acceptors (A) under conditions where mass diffusion is unimportant. The presence of A changes the unquenched, exponential decay profile of D to a form described by

$$I_D(t) = \exp[(-t/\tau_D) - P(t/\tau_D)^{1/2}] \tag{1}$$

with

$$P = \frac{4\pi^{3/2} N_A R_0 [A]}{3000} \tag{2}$$

Here [A] is the concentration of acceptor in moles per liter, N_A is Avogadro's number, and R_0 is a characteristic distance. In the Förster dipole-coupling energy transfer mechanism, the energy transfer rate and efficiency vary with interchromophore separation r as $(R_0/r)^6$. R_0 is defined as the distance at which the energy transfer rate equals $1/\tau_D$.

For example, when we prepare solutions of phenanthrene (Phen, τ_D = 58 ns) and 9-anthracenemethyl pivalate, *3b* in viscous media such as butyl rubber or rigid matrices such as PMMA, the fluorescence decay data fit Equations 1 and 2 and yield a value for R_0 = 26 Å.

On a fractal[14] of dimensionality $1 \leq \bar{d} \leq 3$, the fluorescence decay profile of D is altered by the effect of geometry on the acceptor group distribution. According ot Klafter and Blumen,[13] for a process such as Förster energy transfer which varies as r^{-6}, the donor fluorescence decay profile is given by

$$I_D(t) = \exp[(-t/\tau_D) - P(t/\tau_D)^{\bar{d}/6} \tag{3}$$

where τ_D is the unquenched donor lifetime, P is a factor proportional to the acceptor concentration, and d is the effective dimensionality.

We have applied this method to the study of the PIB phase of PMMA-PIB particles.[20] We chose phenanthrene (Phen) as the donor, in part because it gives a clean exponential decay in polymer matrices, and 9-anthrancenemethyl pivalate (*2b*) as the acceptor. Adding known quantities of Phen and *2b* to a pentane dispersion of unlabeled particles, followed by solvent removal under vacuum, leaves the donor and acceptor molecules dissolved in the PIB phase of the system. Fluorescence decay measurements give data which can be fitted to Equation 3. For a variety of acceptor concentrations, these decay curves indicate a local dimensionality of 1.82 ± 0.15 and, as predicted by Equation 4, P values that increase in proportion to the amount of *2b* added to the sample. These results are presented in Figure 2. Essentially identical results have also been obtained for the PEHMA-PVAc particles.

The picture that emerges from these direct energy transfer studies is one in which a considerable amount of the stabilizer chains is buried in the particle interior and that locally the effective dimensionality is close to two. In the experiments on the Phen + *2b* doped particles, the R_0 value is 26 Å and the maximum distance probed in the experiment is about 40 Å. Within this restricted geometry, the PIB phase is locally flat, conceivably only a macromolecular monolayer or bilayer in thickness. Figure 3 portrays a model of the particle structure.

Such a morphology could come about during particle synthesis by the coagulation of smaller (primary) particles which cluster but continue to grow because of monomer polymerization within their interior. This model would yield a compact, void-free structure containing continuous PIB and PMMA phases. We have some indications that the details of the structure vary with the PIB content for the particles. Elucidating these details remains a problem for future experiments.

C. EXCIMER EMISSION FROM LABELED STABILIZER

When PVAc particles are prepared from PEHMA copolymers containing Py groups and examined by fluorescence spectroscopy, excimer emission is observed[8,15] (Figure 4). Excimer formation[10] is well known for simple solutions of polymers labeled to a small extent (0.5 mol% or more) with Py groups. Information about the stabilizer component of nonaqueous dispersion particles is obtained by comparing excimer formation in dispersions of the stabilizer-labeled particles (SLP-Py) with that in solutions of the labeled copolymer (the stabilizer precursor, Stab-Py).

Both fluorescence spectra are shown in Figure 4. These are arbitrarily normalized at the (0-0) band of the locally excited pyrene monomer emission. One sees that anchoring the PEHMA in the particle results in less relative excimer emission as measured by the ratio I_E/I_M of excimer to monomer intensities.

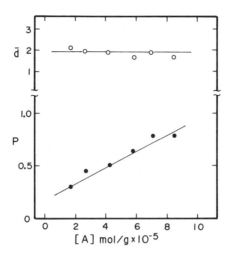

FIGURE 2.   A plot of the effective dimensionality and parameter P obtained by fitting Phe fluorescence decay profiles to Equation 3 for PIB-PMMA particles doped with 0.00 mg Phen per gram of particle and the amounts of *2b* indicated on the x-axis.

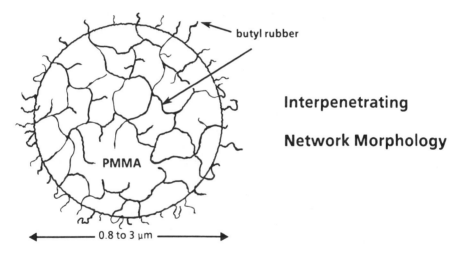

Interpenetrating

Network Morphology

Local Structure

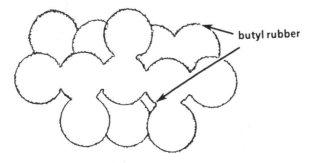

FIGURE 3.   A representation of the gross and microscopic morphology of PIB-PMMA particles as inferred from fluorescence quenching and direct energy transfer studies. The model of the local structure accomodates the result that d = 2 in the direct energy transfer experiments.

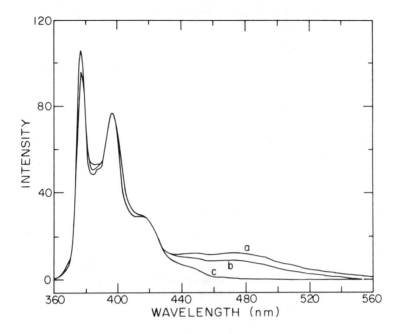

FIGURE 4.   Fluorescence spectra of (b) PEHMA-PVAc particles labeled with Py groups in the stabilizer chains (SLP-Py), (a) pyrene-labeled PEHMA (Stab-Py), and (c) the model compound *4b* (4×10^{-6} *M*) in deoxygenated cyclohexane at 23°C.

$$Py + Py \overset{h\nu}{\underset{k_M}{\rightleftharpoons}} Py^* + Py \overset{k_1}{\underset{k_{-1}}{\rightleftharpoons}} (PyPy)^* \overset{}{\underset{k_E}{\longrightarrow}}$$

Scheme I

Excimer emission is normally described[10] in terms of the simple two-state model of Scheme I. Here, upon excitation of a Py group, excimer formation occurs by diffusion with a rate described by k_1. Excimer dissociation, k_{-1}, competes with other excimer processes, radiative and nonradiative, with a total rate described by k_E to yield ground state Py. The reciprocal lifetime of Py in the absence of excimer formation, τ_0^{-1}, is equal to k_M.

Scheme I predicts that

$$\frac{I_E}{I_M} = \frac{k_{fE}}{k_{fM}} \frac{k_1}{(k_{-1} + k_E)} \tag{4}$$

where k_{fE} and k_{fM} are the radiative rate constants for excimer and exciplex, respectively. When the excimer dissociation rate is sufficiently small, one can also deduce from Scheme I that

$$k_1 = \tau^{-1} - k_M \tag{5}$$

where τ is the exponential decay time of Py*. Under our conditions $k_E \approx 10\,k_{-1}$, and $I_M(t)$ is not exponential.[15] A more complete treatment is necessary;[16] nevertheless the general arguments outlined below are valid and the conclusions drawn do not change. According to Equation 5 a decrease in I_E/I_M under conditions which do not affect k_{fE}, k_{fM}, or k_E indicates a decrease in the rate of excimer formation.

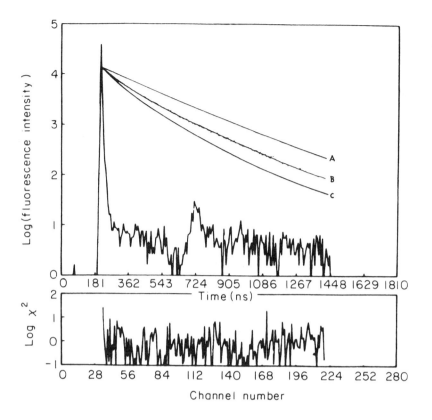

FIGURE 5.   Fluorescence decay profiles of (A) *4b* (2×10^{-6} *M*), (B) Stab-Py (1.5 mg ml^{-1}), and (C) SLP-Py particles (12.4 mg ml^{-1}) in deoxygenated cyclohexane at 23°C.

When the two fluorescence spectra in Figure 4 are compared, one observes less excimer in the particle-bound stabilizer SLP-Py than in the solution of its precursor, Stab-Py. The two factors which might lead to a smaller extent of excimer formation upon incorporation of the labeled PEHMA polymer into the particle are (1) an increase of the overall separation of Py groups (lowering their local concentration) or (2) a decrease in Py group mobility.

Two other experiments indicate that anchoring the stabilizer in the particle *increases* the local concentration of Py groups. First, the excitation spectra of both the monomer and excimer of SLP-Py are broadened with respect to Stab-Py. This result commonly indicates an increase in proximity of Py groups. The excitation spectra also show for both SLP-Py and Stab-Py that excimer and monomer derive from a common precursor.[15] When Py groups aggregate or dimerize prior to excitation by light, one sees a shift in the excimer excitation spectrum relative to that for the monomer.[17] No such indication of aggregation is found.

The second experiment involves fluorescence decay measurements. These experiments indicate *faster* excimer formation when the stabilizer is incorporated into the particle — a result apparently at odds with the fluorescence spectra in Figure 4 — and imply a higher local pyrene concentration. Comparison of Figures 4 and 5 illustrates the problem. As one sees in Figure 5, the particle-bound stabilizer undergoes faster fluorescence decay due to Py/Py* interactions implying faster excimer formation. In Figure 4, one sees that the particle-bound stabilizer gives less excimer emission, implying less excimer formation.

To explain these phenomena and resolve the apparent contradiction between the data in Figures 4 and 5, we must first quantify the fluorescence decay experiment. The time profile of $I_M(t)$ is nonexponential for both SLP-Py and Stab-Py (Figure 5). This result is to be expected, since in a statistical copolymer there is a distribution of contour separations of Py groups. Since the cyclization rate is sensitive to chain length,[16] each contour separation will have its own k_1 value. This distribution of k_1 values in the experiment is much more important than the influence of k_{-1} on the $I_M(t)$ profile. If we rewrite Equation 5 as $<k_1> = <\tau>^{-1} - k_M$, where $<k_1>$ is the mean rate constant for pyrene-pyrene interaction and $<\tau>$ the pyrene mean decay time, we see that a decrease in $<\tau>$ upon anchoring Stab-Py into the particles implies faster excimer formation.

This dilemma can be resolved[21] by recognizing that when the geometry is sufficiently constrained, not all Py/Py interactions will lead to the same excimer. Restricted overlap could enhance radiationless processes (quenching) over radiative processes. For small molecules and lightly substituted polymers in solution, the Py pairs have enough mobility to avoid this problem.

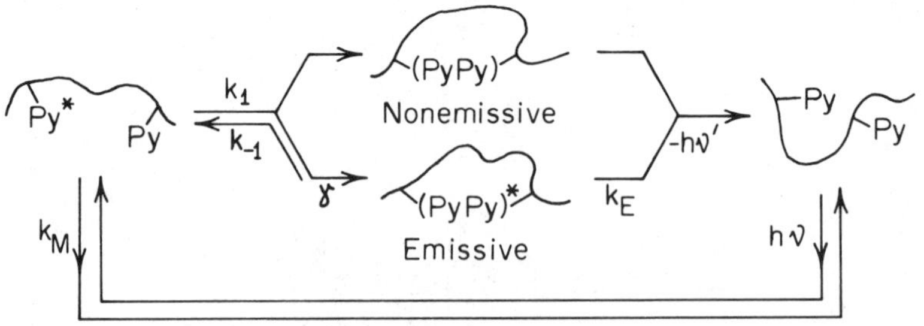

Scheme II

These factors can be accommodated in Scheme II where only a fraction β of the Py/Py interactions leads to the emissive excimer. Taking $\beta = 1$ for the copolymer Stab-Py in solution, we derive from Scheme II that

$$\frac{(I_E/I_M)_{SLP}}{(I_E/I_M)_{Stab}} = \frac{\beta_{SLP}(\langle\tau\rangle_{SLP} - k_M)}{(\langle\tau\rangle_{Stab} - k_M)} \tag{6}$$

When we examined a variety of SLP samples and PEHMA copolymers containing Py in the form of **4**, we found a unique value of $\beta = 0.25$. This result substantiates the model and furthermore implies that incorporating PEHMA into the particles so restricts the Py group motion that only 25% of the Py/Py interactions lead to the sandwich excimer geometry responsible for excimer emission. Lengthening the tether between the Py group and the polymer backbone should enhance the chain mobility. For example in 4-(1-pyrene)butyl methacrylate, three additional carbons separate the Py group from the methacrylate. We prepared new samples of PEHMA copolymer and SLPs containing Py in the form of $Ph(CH_2)_4O_2C$ groups. When the above experiments were repeated,[18] β increased to 0.48.

One advantage of carrying out experiments with labeled polymer chains is that one can dilute them with unlabeled chains to examine chain overlap and interpenetration in the particles. When we repeated the above experiments using mixtures of P(EHMA-*co*-**4**a) copolymers with unlabeled PEHMA, we obtained the curious result that I_E/I_M and $<\tau>$ for the SLP particles were *independent of the fraction of labeled PEHMA in the particle*. This result indicates little or no overlap of adjacent stabilizer chains incorporated in the particle. If the PEHMA phase in the particle were sufficiently thick, such a result would be very surprising. Chain interpenetration is the norm in three dimensions. In two dimensions, chain overlap is much more difficult to accomplish.[18] Here, too, it appears that the stabilizer chains, both those on the surface and those buried in the particle interior, are located in essentially a two-dimensional-like environment.

D. OXYGEN QUENCHING AND THE CORE-STABILIZER INTERFACE

Oxygen quenches fluorescence at short range. In simple systems quenching follows the Stern-Volmer Equation, 7. This equation is derived by assuming a uniform concentration [Q] of quencher, an excited species that decays exponentially at all quencher concentrations, and a quenching process characterized by a unique rate constant.

$$\frac{\tau_0}{\tau} = \frac{I_0}{I} = 1 + k_q \tau_0 [Q] \tag{7}$$

In polymer systems most of these assumptions break down. Dyes in polymer matrices commonly show nonexponential fluorescence decays. Furthermore Q can partition between different environments, and k_q in these environments may take different values. In our experiments we chose to label the stabilizer with Phe groups, since Phe does not easily form an excimer[10] and normally has an exponential decay profile. We prepared PVAc particles using P(EHMA-co-2a) (Stab-Phe) as a stabilizer precursor[8] and examined the effect of oxygen concentration on the fluorescence of solutions of *2b* and Stab-Phe in cyclohexane.[19] These systems were very well behaved: I(t) was exponential, I_0/I was equal to τ_0/τ, Equation 8 was followed, and the k_q values differed by about a factor of 2. Diffusion of *2b* contributes to quenching, whereas that of the much larger Stab-Phe does not.

Even in cyclohexane dispersions of the SLP-Phe particles, fluorescence quenching follows Equation 7. Here I(t) is exponential only for $[O_2] = 0$, yet $\tau_0/<\tau>$ was equal to I_0/I. The nonexponentiality of I(t) points to a (fairly narrow) distribution of O_2 concentrations and diffusivities in the particle. The value of k_q obtained here is close to that found for Stab-Phe, suggesting that the stabilizer phase of the particle dispersions are an open structure swollen with solvent and freely permeable to oxygen.[19]

In the core-labeled particles (CLP-Phe) dispersed in cyclohexane, a quite different behavior is seen.[21] The Stern-Volmer plot, Figure 6 (top), is curved suggesting that at elevated O_2 concentrations, some Phe groups are protected against quenching. That any quenching at all occurs is unexpected. In films of PVAc labeled with Phe, less than 1% quenching is observed upon exposure of the outgassed polymer to pure O_2. The $k_q[Q]$ term depends upon the O_2 concentration in the film and the local O_2 diffusion coefficient. These are both much smaller in PVAc than in cyclohexane. Some feature of the particle structure increases the oxygen permeability of part of the particle dispersion and protects only some of the Phe groups from quenching.

If a fraction f_a of the Phe groups are similarly accessible to quenching and the remainder $(1 - f_a)$ are protected, Equation 8 can be rewritten as

$$\frac{I_0}{I_0 - I} = \frac{1}{f_a \alpha k_q \tau_0 [Q]} + \frac{1}{f_a} \tag{8}$$

In such a system, a plot according to Equation 9 should be linear, with an intercept of f_a^{-1}, and a slope proportional to the product of k_q and the quencher partition coefficient α. Such plots for oxygen quenching of SLP-Phe and CLP-Phe are presented in the lower portion of Figure 6. Both plots are linear. For SLP-Phe the intercept is 1.0, indicating that all Phe groups are similarly accessible to O_2. For CLP-Phe, $f_a = 0.5$, indicating that half of the Phe are protected, but half are very much accessible to oxygen quenching.

The meaning of these data is clear. To the extent that Phe serves as a marker for the location of the PVAc component of the particle, the value of f_a indicates what fraction of PVAc has been transformed into a phase of high oxygen permeability. Here 5.6 monomer mol% PEHMA is able to transform about half of the PVAc present into a phase with different properties.

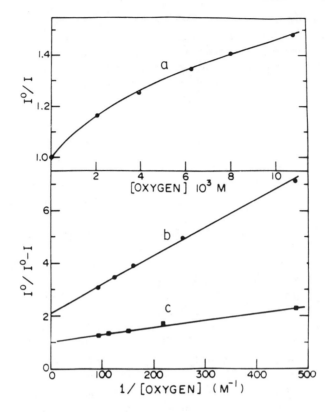

FIGURE 6.   Above: (a) Stern-Volmer plot (I_0/I vs. [O_2]) for cyclo-hexane dispersions of CLP-Phe particles. Below: Fractional quenching plot ($[I_0/(I_0 - I)]$ vs. [O_2]$^{-1}$) for CLP-Phe particles (b) and for SLP-Phe particles (c) dispersed in cyclohexane.

Some other observations are pertinent. First, preliminary experiments by reversed-phase high performance liquid chromatography indicate that nearly half the PVAc is in fact grafted to PEHMA. Second, experiments on powder samples of the dried particles indicate that the Phe groups in both SLP-Phe and CLP-Phe are much less susceptible to quenching than those in their respective cyclohexane dispersions. When SLP-Phe is equilibrated with O_2 at 1 atm, for example, the cyclohexane dispersion yields a value of $I_0/I = 4$ (80% quenching), compared to a value of 1.18 for the powder. For CLP, the I_0/I values are 1.5 (40% quenching) in the dispersion vs. 1.04 in the powder.

Taken together, these results support a microphase or interpenetrating network structure for the PVAc-PEHMA particles with a significant fraction of the stabilizer buried in the particle interior. Graft formation between PEHMA and PVAc occurs during particle for-mation. The PEHMA component of this graft copolymer forms a continuous network, locally two-dimensional, which may derive from the fact that the final particles form during synthesis by aggregation of smaller primary particles covered at least in part by the PEHMA graft copolymer. In cyclohexane, the graft component is swollen by solvent. This solvent-rich phase has a high permeability for oxygen. It is the solvent in the graft copolymer component, we believe, that is responsible for formation of the extensive interphase we detect in these oxygen quenching experiments.

IV. SUMMARY

This chapter described experiments based upon fluorescence quenching measurements carried out on colloidal-sized spherical polymer particles. When this work was begun there were models for the morphology of these materials, most prominent of which was the core-shell model. There was and still is, however, very little high resolution experimental evidence about morphology, although electron microscopy studies indicate that a number of types of latex particles prepared by the paint industry do have globally a true core-shell structure.

Our experiments on the two nonaqueous dispersion materials described here indicate that the core-shell model is not valid. Rather, these particles have an interpenetrating network structure of the type pictured in Figure 3. It appears that the gross particle morphology derives from an aggregation of small polymer particles. We infer that these primary particles are composed of the core polymer (PMMA or PVAc) and are partially covered with a monolayer of the stabilizer (PIB or PEHMA). As these fuse together to form the global structure, the rubbery component becomes trapped in the interior and forms the interconnecting network.

Since the stabilizer is present in these materials as a graft copolymer, it is firmly anchored in place. It can be swollen by hydrocarbon solvents that normally would not penetrate into PMMA or PVAc. This swelling has serious consequences on the stabilizer-core-material interface. Some facets of these changes are becoming apparent through the fluorescence quenching experiments which have been carried out.

REFERENCES

1. This paper is No. 21 in the series Fluorescence Studies of Polymer Colloids; for No. 20, see Croucher, M. D., Winnik, M. A., and Egan, L., *Colloids and Surfaces*, 31, 311, 1988.
2. **Barrett, K. E. J.,** *Dispersion Polymerization in Organic Media,* Wiley-Interscience, New York, 1975.
3. **Tadros, Th. F.,** *The Effect of Polymers on Dispersion Properties,* Academic Press, New York, 1982.
4. **Pekcan, O., Winnik, M. A., and Croucher, M. D.,** A microphase model for sterically stabilized polymer colloids: fluorescence energy transfer from naphthalene labelled dispersions, *J. Polym. Sci., Polym. Lett. Ed.,* 17, 1011, 1983.
5. **Winnik, M. A.,** Fluorescence techniques in the study of polymer colloids, *Polym. Eng. Sci.,* 24, 87, 1984.
6. **Pekcan, O., Winnik, M. A., Egan, L., and Croucher, M. D.,** Luminescence techniques in polymer colloids. I. Energy transfer studies in non-aqueous dispersions, *Macromolecules,* 16, 699, 1983.
7. **Tencer, M. and Winnik, M. A.,** unpublished results.
8. **Egan, L., Winnik, M. A., and Croucher, M. D.,** Synthesis and characterization and kinetic studies of poly(vinylacetate) non-aqueous dispersions, *J. Polym. Sci., Polym. Chem. Ed.,* 24, 1895, 1986.
9. **Berens, A. R. and Hopfinger, H. B.,** Diffusion of organic vapors at low concentrations in glassy PVC polystyrene and PMMA, *J. Membr. Sci.,* 20, 183, 1982.
10. **Birks, J. B.,** *Photophysics of Aromatic Molecules,* Wiley-Interscience, New York, 1970.
11. **Winnik, M. A., Williamson, B., and Russell, T. P.,** Small angle x-ray scattering studies of polymer colloids: non-aqueous dispersions of polyisobutylene stabilized poly(methyl methacrylate) particles, *Macromolecules,* 20, 899, 1987.
12. **Kellerer, H. and Blumen, A.,** Anisotropic excitation transfer to acceptors randomly distributed on surfaces, *Biophys. J.,* 46, 1, 1984.
13. **Klafter, J. and Blumen, A.,** Fractal behavior in trapping and reaction, *J. Chem. Phys.,* 80, 875, 1984.
14. **Mandelbrot, B. B.,** *The Fractal Geometry of Nature,* W. H. Freeman, San Francisco, 1982.
15. **Winnik, M. A., Egan, L. S., Tencer, M., and Croucher, M. D.,** Luminescence studies on sterically stabilized polymer colloid particles: pyrene excimer formation, *Polymer,* 28, 1553, 1987.
16. **Winnik, M. A.,** The end-to-end cyclization of polymer chains, *Acc. Chem. Res.,* 8, 73, 1985.
17. **Winnik, F. M., Winnik, M. A., and Tazuke, S.,** The interaction of hydroxypropylcellulose and acqueous surfactants: fluorescence probe studies and a look at pyrene labelled polymers *J. Phys. Chem.,* 91, 594, 1987.
18. **de Gennes, P. G.,** *Scaling Concepts in Polymer Physics,* Cornell University Press, Ithaca, New York, 1979.
19. **Egan, L. S., Winnik, M. A., and Croucher, M. D.,** Oxygen quenching studies of fluorophore-labelled non-aqueous dispersions of poly(vinylacetate), *Langmuir,* 4, 438, 1988.
20. **Pekcan, O., Winnik, M. A., and Croucher, M. D.,** Direct energy transfer studies on doped and labeled polymer latex particles, *Phys. Rev. Lett.,* 61, 641, 1988.
21. **Egan, L. S.,** Luminescence Studies of Fluorescently Labeled Polymer Colloid Particles, Ph.D. thesis, University of Toronto, 1986.

Chapter 7

STUDIES OF THE DEPOLARIZATION OF POLYMER LUMINESCENCE: LASERS OR NOT LASERS?

J. L. Viovy, L. Monnerie, V. Veissier, and M. Fofana

TABLE OF CONTENTS

I.   Introduction 212

II.   Techniques 212
        A.   Ultraviolet Absorption and Luminescence 212
        B.   Fluorescence Anisotropy Decay 212
        C.   Apparatus 213
        D.   Holographic Transient Polarization Grating...........................215
        E.   Light Sources..216
        F.   Data Analysis 217

III.   Probes and Labels 217
        A.   Need for an Extrinsic Luminescent Site 217
        B.   Probes 218
        C.   Side-Chain Labeling..218
        D.   Main-Chain Labeling..218
        E.   Flexible Oligomer Probes 220

IV.   Studies of Local Dynamics 221
        A.   Chain Motions: Theoretical Models 221
        B.   Discussion of Models-of-Motion Using Labeled Chains 224
        C.   Use of Flexible Probes 226
        D.   Rigid Probes in a Polymer Melt or Elastomer 230

V.   Conclusions 231

Acknowledgments 233

References 234

I. INTRODUCTION

Polymer materials have numerous specific mechanical properties and various spectacular ones are still steadily discovered thanks to progress in material engineering and synthesis. These achievements are made possible by improvements of the understanding of the dynamics on a molecular scale, which in turn rely on progress in spectroscopic techniques. In the wide range of such techniques, fluorescence anisotropy decay (FAD) plays a specific role. On a positive side, it is the only technique able to provide an accurate continuous sampling of the single-chain orientation autocorrelation function (OACF). This advantage is limited by the fact that an extrinsic fluorescent species must be introduced into the polymer in one way or another. In this chapter, we focus on the recent progress permitted by high flux sources such as lasers and synchrotrons. The instrumental aspects of the most accurate methods for studying the OACF of luminescent molecules, namely single photon fluorescence anisotropy decay and holographic transient polarization grating (HTPG) are presented in Section II, and the relevance of different light sources is discussed. Section III deals with the choice of probes and labels, which is an essential aspect of this method. Some typical examples of the use of polarization decay are given in Section IV, and finally we discuss the specific role that lasers should play in this field in the next future.

II. TECHNIQUES

A. ULTRAVIOLET ABSORPTION AND LUMINESCENCE

Electronic transitions in molecules can be very complex,[1,2] but luminescence depolarization studies of polymer dynamics generally use rather simple well-defined transitions,[3,4] summarized in Figure 1. Light absorption brings the molecule in a singlet excited state which depends on the absorbed wavelength, then intramolecular radiationless energy conversion brings it back to the lowest vibrational level of the first singlet excited state (S_1). The molecule remains in this state for a time varying between 10^{-10} and 10^{-7} s, depending on the molecule, temperature, and environment. It can return to the fundamental S_0 state by various nonradiative processes or return to one of the vibrational excited states of the S_0 state by emitting a fluorescence photon. Conversion to a long-lived triplet state, which may lead to phosphorescence or to delayed fluorescence, is not considered here. In simple cases, the fluorescence intensity decay after the withdrawal of the excitation is exponential. The fluorescence lifetime directly imposes the time scale on which information can be gained using fluorescence spectroscopy.

B. FLUORESCENCE ANISOTROPY DECAY

Light (electromagnetic waves) and electronic orbitals are coupled by vector quantities called transition moments (respectively, M_a for absorption and M_e for emission). The absorption of light is proportional to the scalar product of the incident electric field and of M_a. Thus, excitation of an isotropic population of fluorescent species by polarized light generally creates a temporary anisotropic population of excited molecules. Molecular motions progressively destroy this anisotropy and affect the polarization of the reemitted fluorescence light. This polarization is studied using pulse fluorometry techniques. The interesting quantity, the fluorescence anisotropy decay following an infinitely short pulse at time zero, is defined by

$$r(t) = [I_V(t) - I_H(t)]/[I_V(t) + 2I_H(t)] \tag{1}$$

where I_V and I_H correspond to fluorescence intensities for analyzer direction parallel and perpendicular to the vertical polarization of the incident beam, respectively. In this expression

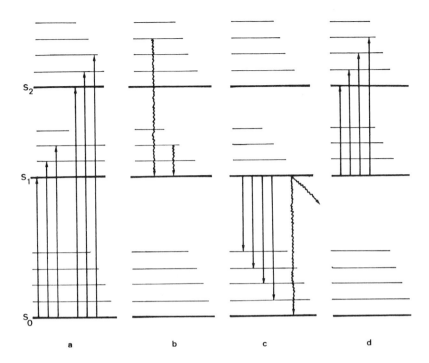

FIGURE 1. Schematic representation of electronic transitions. (a) Absorption, 10^{-15} s, (b) internal conversion, 10^{-12} s, (c) fluorescence, 10^{-10} to 10^{-7} s, and (d) 2° photon absorption, 10^{-15} s.

$(I_V + 2 I_H)$ represents the total fluorescence intensity. r(t) is proportional to the second moment of the OACF of the emission transition moment:

$$r(t) = r_0 M_2(t)/M_2(O) \qquad (2)$$

The fundamental anisotropy r_0 is a time-independent molecular parameter. Thus, the OACF can be quasicontinuously sampled in the time domain directly, provided one can sample r(t) precisely. This property makes FAD a rather unique tool for discussing the different models for the OACF. (In principle, the transient Kerr effect is also able to provide such a sampling, but at the present time this latter technique does not reach the same accuracy).

C. APPARATUS

FAD is a well-known technique[3-5] and the general features of the numerous apparatus available only differ in details. The diagram of the apparatus we used in most of our studies,[6,7] with synchrotron radiation as the exciting source, is displayed in Figure 2. The excitation wavelength is selected by a double holographic grating monochromator, which is vertically polarized. The emission wavelength is selected by a single holographic grating monochromator. The most accurate results are provided by single photon counting:[3,4,6] no more than one fluorescence photon is received by the photomultiplier (PM) for one excitation pulse. On receiving this photon, the cathode of the PM ejects a photoelectron which is amplified in order to produce an electric pulse at the anode. This pulse is sent to the "start" input of a time-to-amplitude converter (TAC) at a time t_1, which is compared with the time t_2 at which the delayed reference pulse is received by the stop input. At its output, the TAC sends a pulse whose amplitude is proportional to t_2-t_1 to an analogic-digital converter and then to the computer, which stores one count in the channel corresponding to the time $t_0 =$

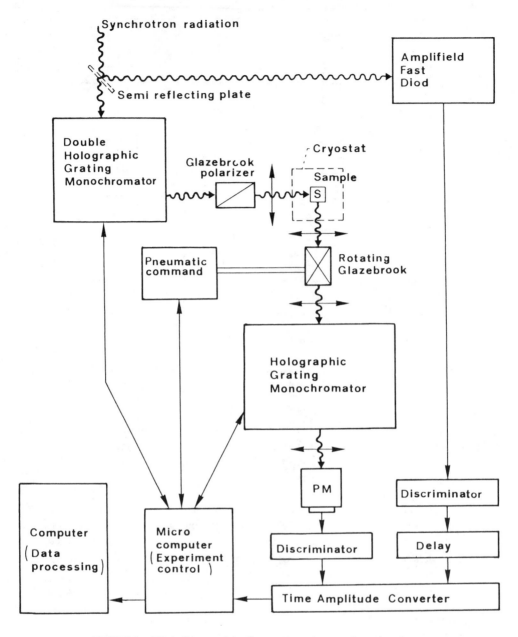

FIGURE 2. Block diagram of the fluorescence anisotropy decay experiment.

$t_2 - t_1$. After many excitation cycles, the number of counts accumulated in one channel is proportional to the probability of emission at a given time after excitation. The errors on these number of counts are random, independent, and they follow the well-known Poisson distribution.[33] The emission decay is recorded alternatively in vertical (parallel) and horizontal (perpendicular) polarizations using a computer-commanded rotating polarizer. At last, the finite width of the excitation pulse is accounted by a nonlinear least-square iterative reconvolution procedure. Thus, a good accuracy on r(t) is achieved if

1. The counting rate is high (or the experimental time is long) to minimize the random relative error proportional to $n_i^{-1/2}$, where n_i is the number of counts in channel i.

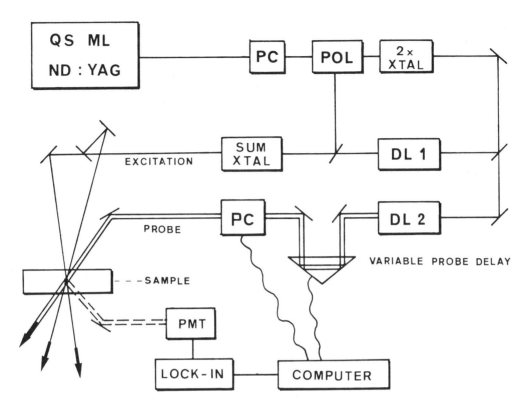

FIGURE 3. Block diagram of the HTPG apparatus used in Reference 8. Q-switched, mode-locked Nd:YAG laser with two synchronously pumped dye lasers: PC = Pockels' cell; POL = polarizer with escape window; DL1, DL2 = cavity dumped dye lasers; and PMT = photomultiplier tube.

2. One remains in the single photoelectron regime.
3. The excitation pulse is as short as possible, as stable in time as possible, and correctly sampled.

The FAD setup is essentially independent of the exciting source chosen, except that the excitation monochromator is not necessary with a laser which is monochromatic.

D. HOLOGRAPHIC TRANSIENT POLARIZATION GRATING

Holographic transient polarization grating (HTPG) is also called picosecond holographic grating in the literature. It has been recently applied to polymer dynamics by Hyde et al.[8] In the HTPG setup, a first mode-lock dye laser (DL1 in Figure 3) is used to create a polarization grating of excited molecules using spatial interference. The contrast of this grating, which is progressively destroyed by molecular reorientation, is probed by the absorption of light issued from a second dye laser (DL2), precisely delayed as regards to DL1, which induces a transition from the first to the second excited electronic state (Figure 1d). Strictly speaking, this method does not involve fluorescence, but several reasons justify that we present it here. The most important is that HTPG gives access to the same OACF than FAD, with picosecond resolution (typically 100 ps). It is another means to obtain the same information. Also, from a molecular point of view, fluorescence and second-photon absorption are both light-mediated transitions from the S_1 level to another. (In many respects, second-photon absorption can be considered as an antifluorescence occurring in a photon bath). Finally, to our knowledge, HTPG is the only laser technique which has been used to study the OACF of polymers.

TABLE 1
Typical Performance of Pulsed UV Sources

Characteristics	Arc lamp	Nondedicated synchrotron (ACO)	Dedicated synchrotron (super-ACO)	Mode-locked cavity dumped laser
Pulse width	1 ns	1 ns	75 — 400 ps	50 ps
Pulse shape fluctuation per hour	$\simeq$ .1 ns	$\leq$ 20 ps	$\leq$ 5 ps	$\leq$ 5 ps
Intensity drift per hour	$\simeq 10^{-2}$	$\simeq 10^{-3}$	$\simeq 10^{-3}$	Negligible
Intensity fluctuation/ 10 s	$\simeq 10^{-2}$	Negligible	Negligible	$10^{-2} \sim 10^{-3}$
Repetition rate	1 — 100 kHz	13 MHz	4 — 400 MHz	0 — 10 MHz
Photon flux (arbitrary units)	1	10^2	$> 10^2\, 10^3$ (magnet) $> 10^3$ (wiggler)	$> 10^4$
Spectral range (nm)	300 — 400 discrete	X-ray-IR continuous	X-ray-IR continuous	250 — 800 (dyes) continuous
Tunability range (nm)	300 — 400	250 — 800	250 — 800	$\leq$50

E. LIGHT SOURCES

Considering the explosive development of laser technology in almost all areas of light spectroscopy (which is well exemplified by this volume!) and the number of picosecond laser single photon apparatus available in the world, it is surprising that none of these have been used to study the OACF of polymers. One of the aims of this chapter is to analyze this situation and to promote progress in this direction.

The presently available light sources for single photon spectroscopy are presented in Table 1 (the values of the characteristics differ from one experiment to the others, and the figures we propose are only orders of magnitude). Laser setups are described in detail elsewhere in this volume. Their major advantages are a very short pulse, a very high intensity, and a narrow bandwidth. Arc lamps can challenge picosecond lasers on no ground except the cost (and to some extent the tunability).

A nondedicated synchrotron (ACO, located at Orsay, France) has been used steadily by our group and others for fluorescence single photon spectroscopy. Its flux is by far not as high as this of lasers, but in the case of dyes with a high quantum yield, we found that the counting rate was limited by the possibilities of the electronics rather than by the light flux. The use of dyes with a lower quantum yield in polymers is generally disappointing anyway, because stray fluorescence is difficult to eliminate. The pulse width of ACO is large compared with that of lasers, but second generation machines dedicated for synchrotron radiation (SR), such as Super-ACO (starting operation this year in Orsay[9]), reach a time resolution comparable with the best microchannel PM and with laser sources. A definite advantage of SR is the continuous tunability throughout the visible UV range (and more). It allows (1) best choice of excitation and emission wavelengths to match absorption peaks of the dye and to avoid these of the matrix, (2) free choice of dyes (3) recording of the excitation pulse at the emission wavelength, and (4) multifrequency measurements (or time-resolved spectroscopy). The wavelength is modified within seconds by tuning the monochromator. In principle, laser setups can span the same range by the use of various primary line dyes and frequency doubler or frequency sum crystals, but all these changes require modifications of the setup and long and delicate realignment and timing. The reasonable choice for laser picosecond anisotropy decay is probably to follow the approach of Hyde et al.[8] for HTPG, that is, to devise a fixed wavelength setup adapted to a well-chosen versatile dye. Finally, it is worth mentioning that small subhertz-intensity fluctuations (typically 1%) in the output power of lasers, which are absolutely benign for fluorescence lifetime measurements, can completely spoil the beautiful potential performance of picosecond lasers of anisotropy if sequential recording of I_V and I_H is performed. Such fluctuations, which are probably associated with thermally driven mechanical drifts, are very difficult to control and they can reduce the base line accuracy to about 10^{-2} (Reference 10) whereas figures of 10^{-3} or 10^{-4} are necessary to take full advantage of a 3- to 4-decade experimental window. This difficulty can be overcome by an accurate intensity monitoring and correction, or by differential synchronous detection.

It is worth quoting that HTPG does not suffer from this limitation, because the anisotropy is proportional to the recorded signal directly and not to a difference of signals. However, this technique suffers from other disadvantages as regards to FAD, such as a higher beam power in the sample which makes temperature and sample stability more difficult to control.

F. DATA ANALYSIS

The anisotropy obtained in actual experiments is deformed by a convolution with the excitation pulse profile, the width of which is finite, and with the electronics response (essentially due to the PM). In precomputer ages, this deconvolution problem was the bugaboo of fluorescence decay and various inexpensive methods, such as moments, modulating functions, Laplace transforms, etc. were predominant.[11] The explosive increase of computing power in the last 10 years has supported a generalized use of least-square iterative reconvolution[11,12] methods which are now available to mini- or even to microcomputers. In this approach, a model of motion is used to generate a reconvoluted decay which is fitted to the data. The Poisson distribution of the noise in single photon mode provides a complete knowledge of the statistics of the data, and criteria such as the weighted residuals, the sum of residuals χ^2, or the autocorrelation of residuals are particularly useful and objective elements for a detailed discussion of models. Varying the excitation frequency[13] or the experimental window[14,15] provides extra physical criteria aimed to discriminate expressions which may fit the data correctly but lead to parameters which behave in a nonphysical way. For instance, if a model is supposed to reflect the true OACF of the chain backbone, the best fit parameters should not vary significantly when this OACF is sampled with a different experimental window. Also, the fundamental anisotropy r_0 cannot exceed 0.4, for theoretical reasons.

The iterative reconvolution (IR) methods require the knowledge of an analytical model for the decay. This limitation has motivated continuing efforts towards a true deconvolution, i.e., obtaining the system response without poostulating any model.[11,16] Important progress has been achieved recently on the basis of the maximum entropy formalism,[17] but the extension to anisotropy is still to be done. This deconvolution approach will probably receive a renewed interest because recent theories often involve steps that cannot be solved on a purely analytical ground. An increasing number of theoretical predictions is available only numerically. Such models are very difficult to account for in an IR method, but they can be directly compared with deconvoluted experimental data.

III. PROBES AND LABELS

A. NEED FOR AN EXTRINSIC LUMINESCENT SITE

Many polymers do not exhibit any luminescence, owing to their chemical structure. Even when the polymer chain contains luminescent groups as does polystyrene or poly(vinylnaphthalenes), the energy-transfer processes occurring between these groups lead to an almost complete depolarization of the emitted light, denying any possibility of studying the chain mobility. Therefore, an extrinsic luminescent molecule is necessary. Several aspects must be considered when choosing such a molecule. For instance, the fundamental anisotropy, r_0, must have a high value in order to yield easily measurable polarizations. In the same way, it is convenient to get a high quantum yield of luminescence that does not vary greatly over a wide range of changes in the properties of the medium (temperature, solvent nature, chemical structure of polymers, etc.). Also, the absorption and emission bands must occur in a spectral range suitable for experiments, typically above 300 nm or more, depending on the absorption of the polymer. Of course, the way the molecule is associated with the polymer is an essential issue. The easiest way is the physical incorporation in a polymer-bulk matrix. This probe technique is useful to study intermolecular correlation and free volume aspects (some examples are provided below), but only labels chemically bonded to the polymer can reflect the nature of chain motions.

B. PROBES

Various probes can be used for fluorescence or phosphorescence polarization studies. Some of the most convenient ones are listed in Table 2 with their optical characteristics. They have in common a rigid skeleton, a reasonably simple geometry, and a well-assigned first transition moment, all features suitable for a simple and unambiguous discussion of dynamics. In order to avoid any energy transfer between probe molecules, their concentration must be sufficiently low, usually less than 1 ppm. This concentration belongs to the range of concentration in which the luminescence intensity is linearly increasing.

C. SIDE-CHAIN LABELING

Clearly, probes cannot provide accurate information on the type of local motions performed by a chain, not to mention the different models proposed for such motions. For this, it is necessary to link the dye covalently to the chain.[18] In this case, we use the term label. A very common label is the anthracene group, for its optical properties are very convenient and many reactive derivatives are available for labeling. A review of the various ways of labeling a polymer chain with anthracene derivatives has been written by Krakoviak.[18] Various types of labeling can be performed: at the chain end; along the main chain or on the side chain, depending on the chemical reaction used; and on the chemical structure of the polymer chain. In each case, the most important aspect is the interdependence between the motion of the transition moment and that of chain backbone segments. It is generally easy to bond the label as a side chain, and reactions have been proposed for polymers with various chemical structures.

With vinyl-type polymers, one way of putting an anthracene group in the side chain is to perform a copolymerization with vinylanthryl monomers. Among them, 1- or 2-vinylanthracene can be introduced into polymers by free-radical or ionic polymerization, but this leads to labeled polymers (Figure 4) in which the transition moment of the anthracene group can undergo rotations independently of chain motions. The best comonomer seems to be 9-10-styrilbenzylanthracene (Figure 4). Indeed, there are no side reactions during copolymerization and the direction of the transition moment is well defined relatively to the chain backbone, independently of rotational motions of the anthracene group around its linkage to the polymer chain.

In the case of polymers bearing chemically reactive functions, such as COOH, OH, NH_2, anthracene derivatives such as 9-anthryl-diazomethane or 9-chloromethylanthracene can be used (see Figure 5) but they always imply a few bonds between the chain skeleton and the anthryl group. One interesting case deals with the reaction of 9-anthrylcarbene with the C-H bond (Figure 5), which allows a labeling of polyethylene and polypropylene chains.

D. MAIN-CHAIN LABELING

When the transition moment of the label lies along the chain backbone, its motion well reflects the orientation autocorrelation function. Such labeling may be obtained by anionic

TABLE 2
Some Typical Fluorescent Probes with Their Absorption and Emission Wavelength Range

Probe	Formula	Absorption range (nm)	Emission range (nm)
Anthracene		300 — 380	380 — 500
Diphenylanthracene (DPA)		310 — 410	390 — 510
Diphenylhexatriene (DPHT)		310 — 390	400 — 580
Diphenyloctatetraene (DPOT)		330 — 410	440 — 660

Note: Double arrows indicate the transition moments.

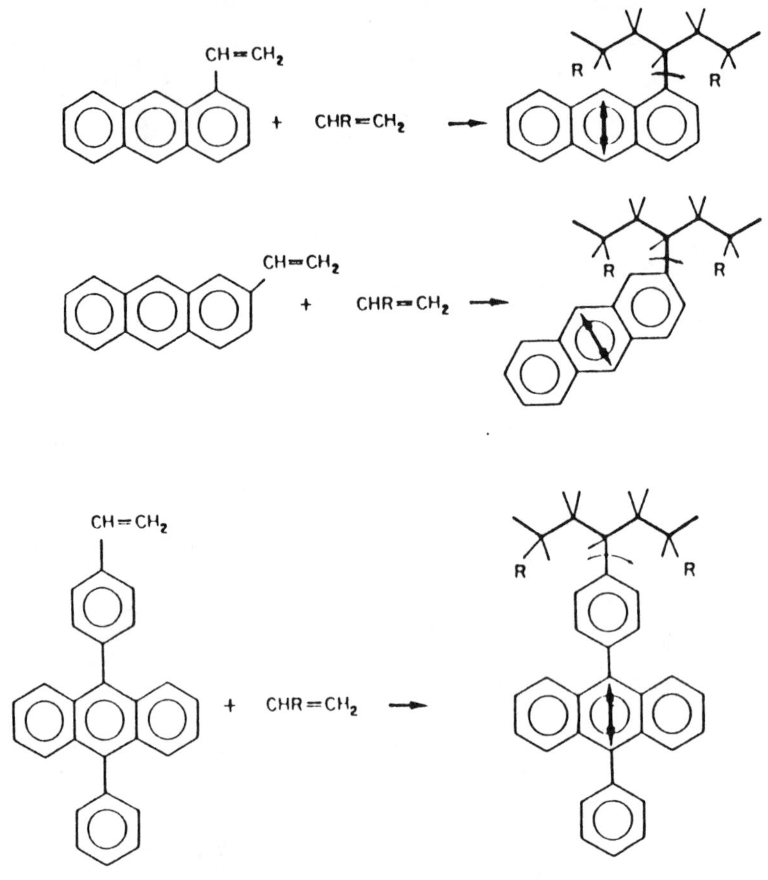

FIGURE 4. Three typical reaction schemes for side-chain labeling using vinylanthryl derivatives.

polymerization when deactivating the living chains with 9,10-(bisbromomethyl) anthracene (Figure 6). Polystyrene, polyisoprene, and other polydienes have been labeled this way.[20] When dealing with condensation polymers, such as polyesters or polyamides, some anthracene derivatives bearing ester, alcohol, or amine groups in position 9,10 can be incorporated (Figure 6) by copolycondensation. It is worth noting that in all these labeling, the transition moment of the anthracene group lies along the local axis of the chain backbone and cannot perform orientation changes independent of the main chain.

E. FLEXIBLE OLIGOMER PROBES

A new type of probe constituted by an anthracene bearing oligomer tails of well-defined size in 9-10 position have been synthetized in our laboratory.[22] In this made to measure approach, each new series is a particular case, and systematic synthesis approaches must be developed. Dialkylanthracene an dialkoxyanthracene probes have been prepared so far. The diakylanthracenes used in FAD experiments[23,24] (see below) were obtained by reaction of an organic-magnesium compound on anthraquinon (Figure 7), followed by hydrolysis and reduction. A series of compounds with tails of 1, 2, 4, 6, 10, 12, 14, and 16 carbon atoms were prepared. We show in the next section that this synthesis work is rewarded by unique information on molecular dynamics.

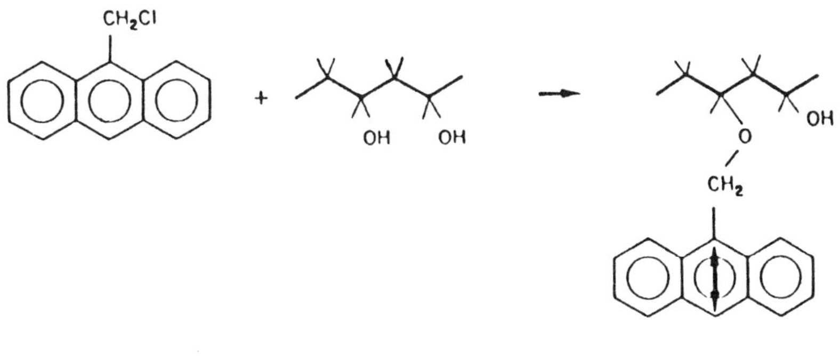

A

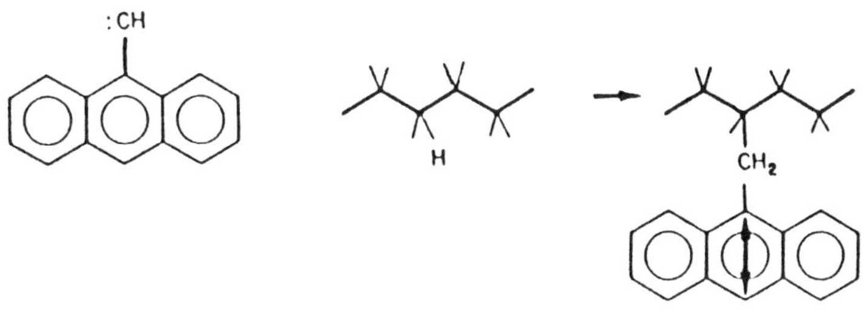

B

FIGURE 5.   Side chain labeling of polymers bearing reactive functions (A) or hydrogen (B).

IV. STUDIES OF LOCAL DYNAMICS

A. CHAIN MOTIONS: THEORETICAL MODELS

The theory of polymer dynamics entered its modern age when Rouse proposed to model the chain by a sequence of beads separated by springs. The random forces exerted by the viscous environment are localized on the beads. In spite of its crudeness, this early model contains the two essential features of polymer dynamics, i.e., the connectivity and the flexibility. It leads to a master equation for the orientation probability of the one-dimension-diffusion type and to a $t^{-1/2}$ long-time dependence of the OACF.[25] This model is valid only on a distance range greater than the statistical unit, i.e., the smallest chain portion large enough to be Gaussian. Thus, more detailed descriptions have been proposed in order to get OACF expressions valid in the whole time and distance range of experiments (summarized in Table 3). A brief historical review of these models have been given in Reference 7. At the present time, the most acknowledged models of motion can be gathered in two families. The first one is based on a 1-D diffusion equation for bond orientation probabilities. The Valeur et al.[26] (VJGM), Bendler and Yaris[27] (BY), Jones and Stockmayer[28] (JS) models, and their extensions[29] belong to this family. These models can be considered as extensions of the Rouse model in which finite size elements or cutoffs are introduced to account for short- and long-time deviations to the ideal Rouse chain. The BY model, which presently seems the most favored model of this family, leads to an OACF:

$$M_2(t) = 1/2(\pi/2)^{1/2}(\tau_2^{-1/2} - \tau_1^{-1/2})\{\text{erfc}[(t/\tau_2)^{1/2}] - \text{erfc}[(t/\tau_1)^{1/2}]\} \qquad (3)$$

This expression presents a satisfactory behavior at short time. However, it seems difficult to correlate the parameters τ_1 and τ_2, which are the inverses of the arbitrary cutoff frequencies, with molecular quantities.

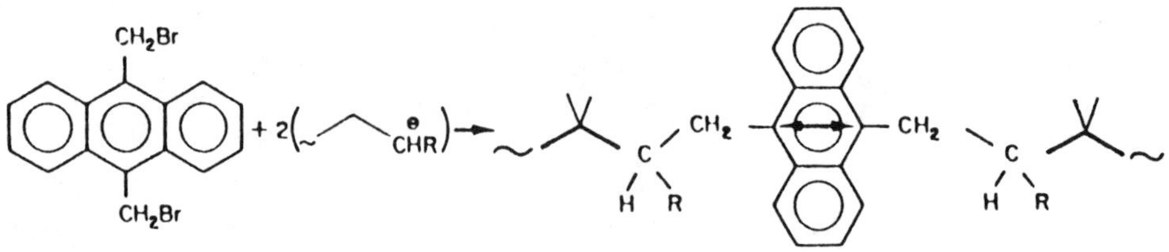

A

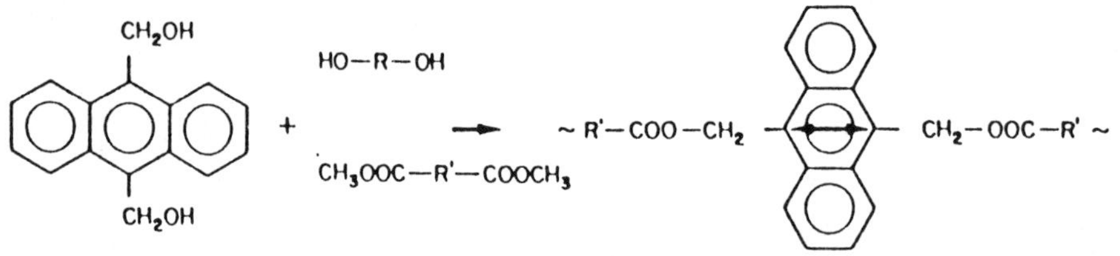

B

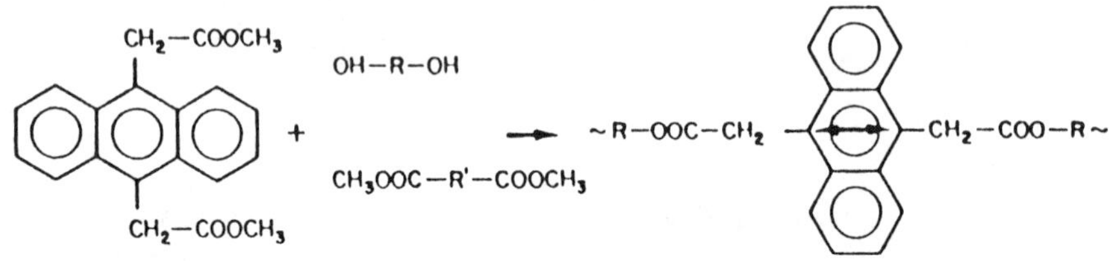

C

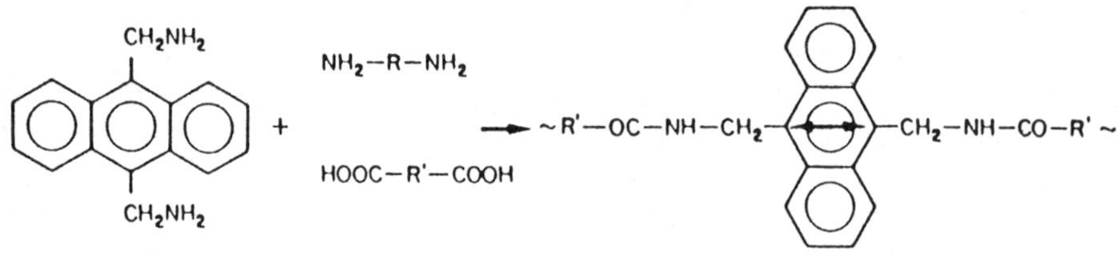

D

FIGURE 6.   Examples of main-chain labeling by anionic polymerization (A) or polycondensation with an anthracene bearing reactive substituents in 9, 10 position (B, C, and D).

The second family of models starts from a more or less idealized description of the conformations on a molecular scale and from exchange rates between these conformations. The most well known is the Hall-Helfand model[30] in which the conformation correlation function (CCF)

$$C_{ii}(t) = \exp(-t/\tau_2)\exp(-t|\tau_1)I_0(t/\tau_1) \qquad (4)$$

stems from the assumption of two types of local conformation jumps (CJ), respectively. correlated CJ with rate $1/\tau_1$, and independent CJ with a rate $1/\tau_2$ (I_0 represents the modified Bessel function of order 0). The concept of correlated CJs arose from detailed molecular

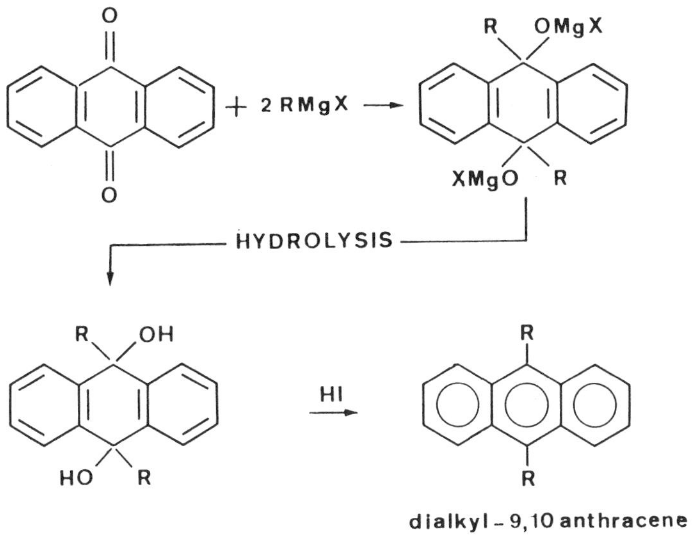

O

O

+ 2 RMgX ⟶

R OMgX

XMgO R

— HYDROLYSIS —

R OH

HO R

HI →

R

R

dialkyl – 9,10 anthracene

FIGURE 7. Reaction scheme for the preparation of 9, 10 diakyl-anthracene probes. (From Yeung, C. K., Ph.D. thesis, Paris, 1982.)

TABLE 3
Analytical Models for the Orientation Autocorrelation of Polymers

Model	Abreviation	Autocorrelation function	Ref.
Williams and Watts	WW	$\exp\left(-\dfrac{t}{\tau_i}\right)\beta$	36
Valeur et al.	VJGM	$\exp\left(-\dfrac{t}{\theta}\right)\exp\left(\dfrac{t}{\rho}\right)\mathrm{erfc}\left(\dfrac{t}{\rho}\right)^{1/2}$	26
Jones and Stockmayer	JS_n	$\displaystyle\sum_{k=1}^{n} a_k \exp\left(-\dfrac{t}{\tau_k}\right)$	28
Bendler and Yaris	BY	$\dfrac{1}{2}\left(\dfrac{\pi}{t}\right)^{1/2}\left[\dfrac{1}{\tau_1^{1/2}}-\dfrac{1}{\tau_1^{1/2}}\right]\left[\mathrm{erfc}\left(\dfrac{t}{\tau_2}\right)^{1/2}-\mathrm{erfc}\left(\dfrac{t}{\tau_1}\right)^{1/2}\right]$	27
Hall and Helfand	HH	$\exp\left(-\dfrac{t}{\tau_2}\right)\exp\left(-\dfrac{t}{\tau_1}\right)I\left(\dfrac{t}{\tau_1}\right)$	30
Lin et al.	LJS	$\exp\left(-\dfrac{t}{\tau_1}\right)\left\{I\left(\dfrac{t}{\tau_1}\right)+I\left(\dfrac{t}{\tau_1}\right)\right\}$	29
Viovy et al.	GDL	$\exp\left(-\dfrac{t}{\tau_2}\right)\exp\left(-\dfrac{t}{\tau_1}\right)\left\{\left(I\dfrac{t}{\tau_1}\right)+I\dfrac{t}{\tau_1}\right\}$	14
Generalization	LJS*	$\exp\left(-\dfrac{t}{\tau_2}\right)\exp\left(-\dfrac{t}{\tau_1}\right)\}I\left(\dfrac{t}{\tau_1}\right)+I\left(\dfrac{t}{\tau_1}\right)\}$	24, 35

dynamics calculations by Weber and Helfand.[31] These authors observed that a single CJ must be associated with bond-angle torsions, which avoid wide-swinging motions of the tails. These torsions tend to favor well defined compensating CJ involving the second neighboring bond (on an sp^3 chain). Therefore, this correlation of conformations is a direct consequence of the connectivity at a molecular level. This is also reflected in the mathematics, since the Pauli matrices appearing in the Hall-Helfand treatment present strong analogies with the Rouse matrix. Hall and Helfand suggested that the CCF they derived should provide a good approximation of the OACF. We later proposed, on a somewhat heuristic basis, that some actual polymers might better obey a generalization of the HH model (called GDL model)

$$M_2(t) = \exp(-t/\tau_2)\exp(-t/\tau_1)[I_0(t/\tau_1) + aI_1(t/\tau_1)] \qquad (5)$$

where I_1 is a modified Bessel function of order 1, and a is assumed to be equal to 1. This expression was put on a sounder molecular basis by Lin et al.[29] Recently, we were able to generalize the HH model in order to account for the presence of a local heterogeneity[32] in the chain, such as a fluorescent label. The corresponding expression for the CCF,

$$C(t) = \exp(-t/t2)\exp(-t/t1) \qquad (6)$$
$$\cdot\{I_0(t/\tau_1) + \alpha(t/\tau_1)[-|1(t/\tau_1) + 2\Sigma(-1)^n|2n+1(t/\tau_1)]\}$$

is presently being compared with experimental decays,[33] but the presence of a fourth parameter α, which reflects the perturbation due to the label, makes Equation 6 more demanding in experimental accuracy than the HH expression.

Finally, Bahar and Erman[34] recently started a theoretical program in which realistic intramolecular potentials are used to quantitatively predict the OACF by means of the rotational isomer model. This approach requires heavy computational work, but it should be rewarded by a direct connection between molecular structure and OACF.

B. DISCUSSION OF MODELS-OF-MOTION USING LABELED CHAINS

The use of FAD for studying the OACF of labeled polymers was pioneered more than 10 years ago by Valeur et al.[26] We exemplify the present power of this approach using recent results on polyisoprene (PI), a favorable polymer for several reasons:

1.   It has a wide technological interest and its mechanical behavior is well studied.
2.   It is possible to label PI with one single anthracene inserted in 1-9 position in the middle of the chain, which is the best topology for studying main-chain motions.
3.   Labeled chains of well-defined microstructures have been prepared.[8,35,41]
4.   Its flexibility and Tg makes this polymer suitable for experiments both in dilute solutions and in bulk, and studies of those two types have been recently performed.[8,24,35]

Hyde et al.[8] have studied labeled PI in solution in hexane and in cyclohexane at room temperature using the HTPG technique. The HH, BY, GDL, and Williams-Watts[36] expressions were fitted to the data (see Table 3). In both solvents, these four models account rather well for the data, but quantitative discrepancies in the values of χ^2 obtained are in favor of the BY and HH models (Figure 8). This is in agreement with previous studies on polystyrene in solution, for which the HH, BY, and GDL models closely accounted for the data, with a slight preference going to one model or the other depending on the solvent or on the temperature (see Table 4).

Polyisoprene main-chain dynamics in bulk was studied by our group. In this case, the values of χ^2 are 0.99 (GDL), 1.01 (BY), and 1.02 (HH). The agreement is very good for all models (Figure 9) and the significant discrepancy between the GDL model and data apparently observed by Hyde et al. in the case of solutions is not recovered. Since our sample and theirs do not have the same microstructure, we also studied our labeled chains, diluted in decalin (a study in cyclohexane or hexane was impossible because of the different experimental windows of the Hyde et al. apparatus and ours). In this case, the HH and BY lead to better values of χ^2 than the GDL model (1.00, 1.02, and 1.05, respectively). We conclude that the inversion in the order of best fit models is essentially due to the change of environment. In any case, it must be emphasized that these differences are at the limit of experimental accuracy, and that the three models HH, GDL, and BY can be considered as reflecting the OACF of polymers very well in a wide variety of environments, temperatures, and correlation times. We tend to favor the HH or GDL models over the BY one

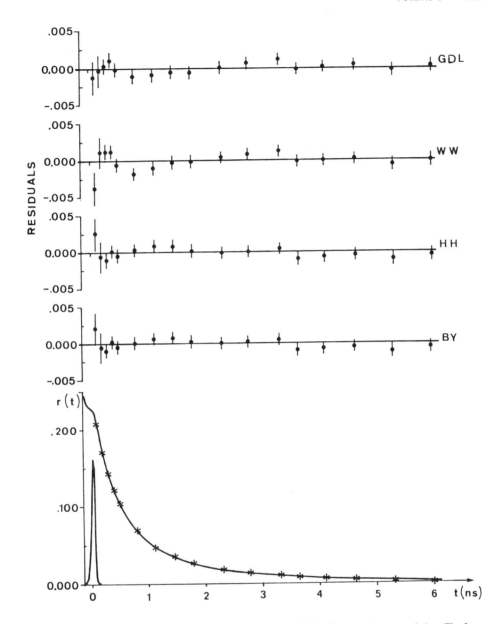

FIGURE 8.   Time-dependent anisotropy for anthracene labeled PI in dilute cyclohexane solution. The four models fitted lead to best fit reconvoluted curves (full lines) indistinguishable by the eye, but the residuals (upper curves) show slight differences. The reduced-least squares are 0.62 (HH), 0.66 (BY), 0.68 (GDL), and 1.23 (WW), respectively.

because they open a more direct route to molecular interpretations, but BY is as good on a purely experimental ground.

Recent FAD studies also emphasized the very important role of the parameter τ_2/τ_1. For a given model this dimensionless factor reflects the shape of the OACF regardless of the absolute values of the correlation times. Indeed, our studies have revealed that this parameter is essentially independent of temperature. On the other hand, it seems to depend on the polymer chain and on the environment (e.g., see the results for polystyrene in various solvents[14] or the comparison between bulk and dilute PI presented above). This behavior sheds some doubts on Hall and Helfand's original interpretation which attributed τ_1 and τ_2

TABLE 4
Summary of Experiments, Best Fit Models, and Values of τ_2/τ_1 for Various Polymers

Polymer	Solvent	Temp(°C)	Best fit model	τ_2/τ_1	Ref.
Polystyrene	Ethyl acetate	25	GDL (BY,HH)	2.6 (HH)	14
				4 (GDL)	
	Glyceryl tripropanoate	25	BY (HH,GDL)	$>10^2$(HH)	14
				11 (GDL)	
	Toluene-PS mixtures	25	GDL (HH,BY)	6 — 15 (GDL)	37
	Styrene-PS reaction bath	90	GDL	15 — 30 (GDL)	38
Polybutadiene	Bulk	− 33/80	BY (GDL,HH)	20 — 30 (HH)	15
				30 — 40 (GDL)	
Polyisoprene 39% *cis*, 36% *trans*, 25% 1-2	Cyclohexane	21	BY (HH)	3.6 (HH)	8
				7.8 (GDL)	
	Hexane	21	BY (HH)	3.6 (HH)	8
				8.9 (GDL)	
Polyisoprene 92% *cis*, 5% *trans*, 3% 1-2	Decalin	25	HH (BY)	4 (HH)	24
				9.9 (GDL)	
	Bulk	21/91	GDL (HH,BY)	6 — 8 (HH)	24
				8 — 13 (GDL)	
C16 (polyethylene)	Polybutadiene	− 50/50	GDL	30 (GDL)	23
PMMA	Methylcellosolve	26	GDL (HH)	7.5 (HH)	39
				9 (GDL)	

Note: Models which closely approach the results of the best one are indicated within parentheses in the Best fit model column.

with the rates of different conformational changes; in such a case, one would expect no effect of the environment, since the basis of the HH model is purely intramolecular (apart from a structureless external brownian activation). Moreover a dependence of τ_2/τ_1 upon temperature is expected in the HH model, since different CJs are likely to have different activation energies. The independence of τ_2/τ_1 upon temperature indicates that one single activation process drives the dynamics in the temperature range available to FAD (typically 50 to 150° above Tg). The temperature dependence of this process is reflected by that of τ_1. Figure 10 provides a plot of τ_1 (GDL) as a function of $10^3/(T-T_\infty)$ for PI, which is linear within experimental error, in agreement with the WLF relationship:

$$\log \tau_1/\tau_1(\tau_g) = - C_2^g + C_1^g C_2^g/(T - T\infty) \tag{7}$$

the slope, 874, is close to 900, the value is derived from viscoelastic data. A similar agreement has been obtained for other polymers. This confirms that the viscous slowing down responsible for the glass transition is directly related with the decreasing rate of elementary conformational jumps of chain backbones.

C. USE OF FLEXIBLE PROBES

The measured OACF of labeled polymers is fully consistent with an interpretation in terms of segmental dynamics, but more general models lead to similar correlation functions and no direct information about the scale of the motions actually observed by FAD is given by this approach. Flexible fluorescent probes such as those described in Section III.E are unique tools for this. We performed FAD measurements on probes of the C_n series (di-alkylanthracene) with n equals 1, 6, 10, 12, 14, and 16, dissolved in a polybutadiene matrix.[23] The FAD curves at 29.8°C for C_6 to C_{14} probes are displayed in Figure 11. The data show quite clearly that the orientation decorrelation rate decreases when the size of the alkyl

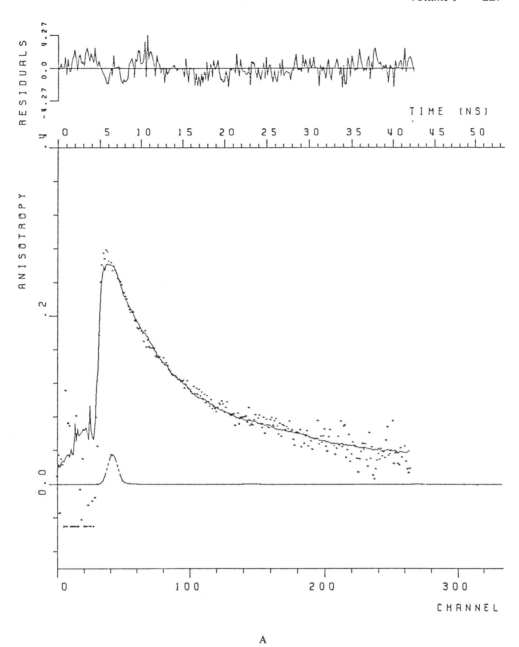

A

FIGURE 9. Reconvoluted best fit for HH (A), GDL (B), and by (C) models, compared with the experimental FAD of anthracene-labeled PI in bulk PI.

substituents is increased. During the time of the experiment, the anisotropy decreases to zero in the case of C_6, but it remains significantly higher than zero for C_{14}. Furthermore, the shape of the OACF also varies with the length of the tails: the longer the alkyl tail, the more pronounced is the nonexponential character.

Since the alkyl substituents are similar to very short polyethylene chains, it seemed interesting to check if their motions could be described by models characteristic of polymer dynamics. Among those, the Jones-Stockmayer (JS) expression[28] is of particular interest since it describes a finite chain with an odd (and generally moderate) number of bonds

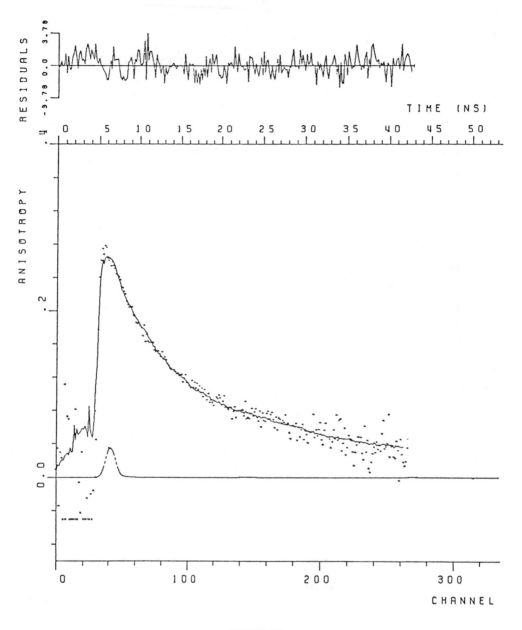

FIGURE 9B.

performing 3-bond motions in a tetrahedral lattice. The damping of the orientational memory along the polymer chain is replaced by a finite diffusion involving U kinetic units on each side of the central bond. The resulting expression for the time dependence of the anisotropy is

$$r(t) = r_o \sum_{i=1}^{u} a_i \exp(- t/\tau_i) \tag{8}$$

where the a_i coefficients are given in Reference 28. The dependence of the number of kinetic units yielding the best fit with the number n of CH_2 groups in the probe tails is shown in Figure 12. The slope indicates that a kinetic unit in the JS model corresponds approximately to three carbon atoms of an aliphatic chain. The intercept provides an evaluation of the scale of the perturbation due to the anthracene group, i.e., about six carbon atoms on each side.

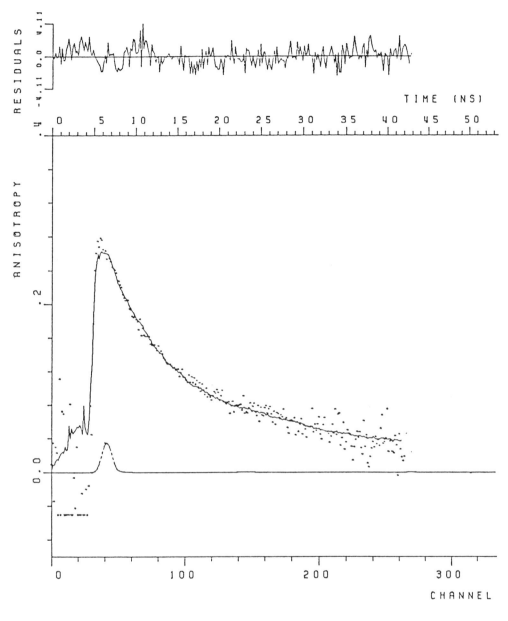

FIGURE 9C.

Besides this, it is interesting to examine how the HH or GDL expressions, used for polymer chains in solution or in bulk, can fit the FAD curves of the various Cn probes. It appears that both functions yield a very good fit for C_{14} and C_{16} and that, for this latter probe, the characteristic times τ_1 and τ_2 are essentially the same as those of labeled polybutadiene. This gives an estimate of the size of the sequence whose motion is reflected through the FAD of the anthracene label: about 30 bonds. For shorter tails the quality of the fit rapidly decreases because of a higher contribution of the overall motions compared to the internal (polymer-like) motions. For such probes, the dynamics should be intermediate between that of a polymer and that of a rigid anisotropic molecule, and no tractable model is available. The important point, however, is that probes as short as 30 carbon-carbon bond show definite polymer-like local dynamics, and that increasing the tails further does not

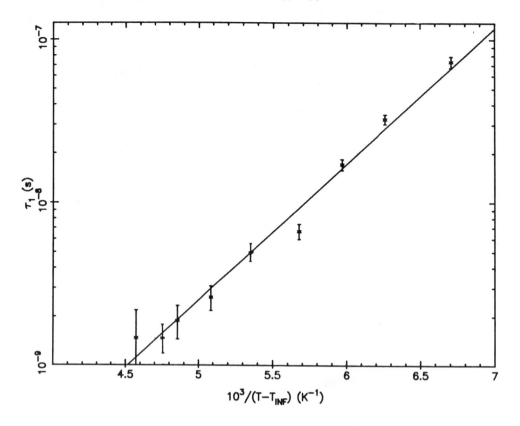

FIGURE 10.   Log (τ_1) vs. $10^3/(T-T_\infty)$ for labeled PI (GDL model).

lead to a measurable increase in conformational memory. Also, it is worth quoting that the temperature evolution of the correlation times of the probes C_6 to C_{16} is the same as that of labeled polybutadiene in the same matrix (i.e., it follows the WLF behavior). These properties definitely answer the question, how local are the motions observed in FAD? and they also open the route to new studies. In particular, the short length of these probes allows a very good compatibility with many matrices. Such molecules have all the practical convenience of extrinsic probes, but they retain the specific aspects of polymer segmental dynamics. Therefore, they are unique tools for discussing the intermolecular contributions to polymer dynamics.

D. RIGID PROBES IN A POLYMER MELT OR ELASTOMER

The extrinsic-probe approach has been extensively used in ESR and in continuous-excitation fluorescence. The advantages of FAD are a very good accuracy associated with the reliable statistics of the data and a continuous sampling of the OACF, which gives access to the nature of the motions of the probe (e.g., FAD can unambiguously tell apart a fast limited motion and a slow unlimited one). Various polymer melts such as polybutadiene,[40] PI,[41] polyisobutylene,[42] and polypropylene oxide[43] have been studied by this technique. We exemplify this approach using a recent study of polypropylene oxide (PPO) compounds of molecular weights 425 and 4000 and the corresponding monomer (propanediol 1,2) (PD) as probed by diphenylhexatriene (DPH).[43] This probe (Figure 4) essentially behaves like a rigid rod with its transition moment along the long axis, as shown by a monoexponential anisotropy decay in various molecular solvents. This behavior is recovered in PD and PPO, in the temperature range 250 to 350 K, indicating a purely viscous-like behavior on the scale of DPH. The dependence upon the temperature shows no significant difference between

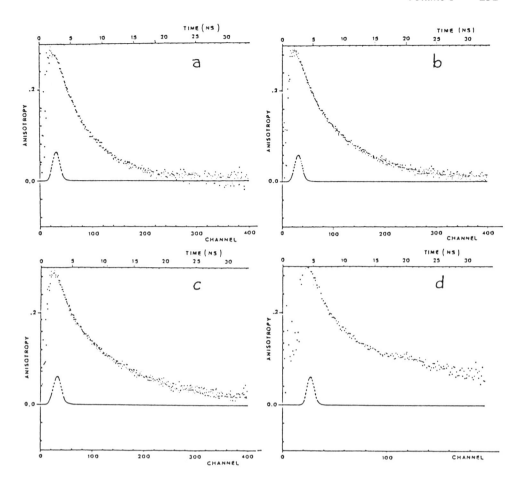

FIGURE 11. Experimental fluorescence anisotropy decay for different C_n probes in polybutadiene. a/n = 6; b/ n = 8; c/n = 10; and d/n = 14.

the two polymers PPG425 and PPG4000. Comparison with photon correlation and Fabry Perot interferometry studies of the same samples show a very good agreement with the WLF equation on 8 decades as revealed by the classical plot of log τ as a function of $10^3/(T-T_\infty)$ (Figure 13). The parameters $C_1C_2 = 417 \pm 10$ and $T_\infty = 171 \pm 3$, obtained on an entirely molecular basis, agree with the viscoelastic parameters within experimental error. This comparison confirms the local origin of the Tg slowing down and provides an evaluation of the scale of the motions, typically around 15 Å. In spite of a similar Tg, the monomer has very different WLF parameters, corresponding to a much smoother slowing down than for PPO.

V. CONCLUSIONS

In this chapter, we tried to give an idea of the interest of luminescence anisotropy decay studies in polymers. Using a few examples, we have shown that high-flux sources, with a pulse width no more than 1 ns, are able to tell apart theoretical models which differ in a rather subtle way and to provide unambiguous information on the orientational autocorrelation function.

The limitations of the method are also worth discussing. The first one is the need for an extrinsic fluorescent site. In the case of rigid probes, the scale of the motions explored can be tuned by changing the size and/or the geometry of the probe, but no direct information

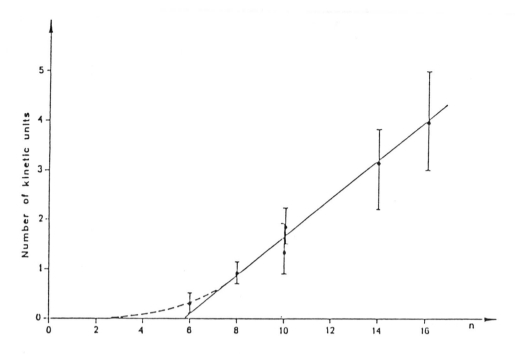

FIGURE 12.   Number of kinetic units leading to the best fit for each C_n probe vs. the number n of carbon atoms in the alkyl tail.

on chain dynamics is provided. By covalently bonding a fluorescent label to the chain, main-chain dynamics are observed. The question then arises of the perturbation due to the label. The first, qualitative, answer is given by comparing the dynamics of the free dye with that of the labeled chains in the same environment. In the case of anthracene inserted in various polymer chains, 9,10 position, the motions are at least ten times slower than those of the free probe in the same matrix, which indicates that the label essentially reflects chain motions. Experiments using oligomers of well-defined size provide a quantitative evaluation of the range of the perturbation (typically a few carbon atoms). Further progress is expected from recent theoretical calculations, which explicitly takes into account the label in polymer dynamic theory. Combining this approach with the unique continuous sampling of the OACF should make the comparison of FAD with other nonperturbative spectroscopic methods, such as NMR, particularly fruitful.[47]

The disadvantage of labeling can be a strong advantage as soon as one wants to differentiate the intramolecular and the intermolecular contributions to the dynamics. Since the experiment is sensitive to the labeled chain only, the dynamics of a given chain in various environments can be studied. This has been applied to study molecular weight and concentration effects in solution,[37] dynamics in the course of polymerization,[38] and quality-of-solvent effects.[48] Studies on intermolecular effects in bulk polymers are also under progress.

Another important limitation of FAD is its relatively narrow experimental window (typically 3 decades for the synchrotron source we used, including deconvolution). It has been seen that a more detailed connection between the molecular structure and the OACF requires that additional parameters be fitted to the data, and that functions be more critically differentiated. For instance, we have some indications that the ratio τ_2/τ_1 is an important parameter for molecular dynamics; the measure of this parameter would greatly benefit from an improvement in the experimental window. Since the long time limit of this window is essentially given by the fluorescence lifetime, the improvement must come from a reduction in the pulse duration (this is confirmed by the experiments of Hyde et al.[8] although these

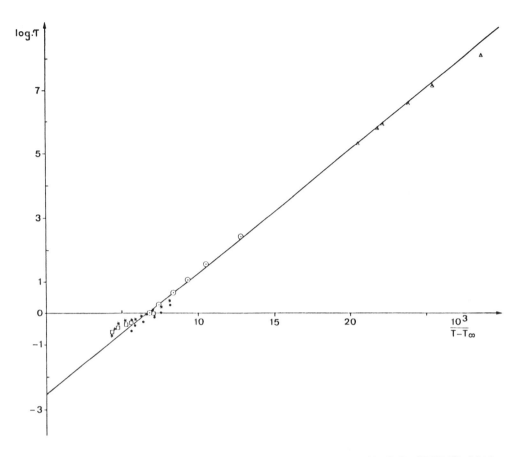

FIGURE 13.   Logarithm of correlation time vs. $10^3/(T\text{-}T_\infty)$ for polypropylene oxide: FAD of DPH ($\bigcirc$).[43] Light scattering from references[42] (*, polarized, $\square$ depolarized), [45]($\bullet$), and [46]($\triangle$).

authors did not take full advantage of their improved time window because of low viscosity solvents). Dedicated synchrotron machines of the new generation will probably become very powerful tools for the study of local motions in polymers because they associate a short pulse with a perfect tunability. However, no such machine is presently available for FAD and their availability to the polymer community will be limited to a few weeks per year in the best of cases. This is by far insufficient for a steady progress of molecular understanding. On the contrary, picosecond lasers are present in numerous laboratories and their time resolution is outstanding. The results obtained by Hyde et al.[8] using HTPG demonstrates that the relative lack of tunability of laser systems is not an overwhelming disadvantage and that the development of a nontunable system well adapted to a versatile label such as anthracene is a perfectly valid strategy. In this context, it is surprising that, to our knowledge, no laser FAD apparatus optimized for polarization has been applied to polymer dynamics yet. We hope that this chapter will help to promote this approach.

ACKNOWLEDGMENTS

An important part of the experimental material displayed in this chapter was obtained thanks to a continuous support of the Laboratoire pour l'Utilisation du Rayonnement Electromagnétique (LURE) and of the CNRS Centre Interrégional de Calcul Electronique (CIRCE). We also acknowledge friendly discussions with M. Ediger.

REFERENCES

1. **Becker, R. S.**, *Theory and Interpretation of Fluorescence and Phosphorescence,* John Wiley & Sons, New York, 1969; **Soleillet, P.**, *Ann. Phys. N.Y.*, 12, 23, 1929.
2. **Parker, C. A.**, *Photoluminescence of Solutions,* Elsevier, Amsterdam, 1968.
3. **Winnik, M. A., Ed.**, *Photophysical and Photochemical Tools in Polymer Science,* D. Reidel, Dordrecht, Netherlands, 1986.
4. **Phillips, D., Ed.**, *Polymer Photophysics,* University Press, Cambridge, U.K., 1985.
5. **Wahl, P.**, Decay of fluorescence anisotropy, in *Concepts in Biochemical Fluorescence,* Chen, R. F. and Edelhoch, H., Eds., Marcel Dekker, New York, 1975.
6. **Brochon, J. C.**, Protein structure and dynamics by polarized pulse fluorometry, in *Protein Dynamics and Energy Transduction,* Shin'ichi Ishiwata, Ed., Taniguchi Foundation, Japan, 1980.
7. **Viovy, J. L. and Monnerie, L.**, Fluorescence anisotropy technique using synchrotron radiation as a powerful method for study in the orientation correlation functions of polymer chains, *Adv. Polym. Sci.,* 67, 99, 1985.
8. **Hyde, P. D., Waldow, D. A., Ediger, M. D., Kitano, T., and Ito, K.**, Local segmental dynamics of polyisoprene in dilute solution: picosecond holographic grating experiments, *Macromolecules,* 19, 2533, 1986.
9. Service de Documentation du CEA, LURE Annu. Rep. 1985 to 1987, Saclay, France, 1987.
10. **Merola, F.**, private communication.
11. **Bouchy, M., Ed.**, Deconvolution and Reconvolution of Analytical Signals, ENSIC-INPL, Nancy, France, 1982; **Cundall, R. B. and Dale, R. E.**, Eds., Time-resolved Fluorescence Spectroscopy in Biochemistry and Biology, Vol. 69, (NATO-ASI Ser. A), Plenum Press, New York, 1983.
12. **Wahl, P.**, Analysis of fluorescence anisotropy decays by a least square method, *Biophys. Chem.,* 10, 91, 1979.
13. **Barkle, M. D., Kowalczyk, A. A., and Brand, L.**, Fluorescence decay studies of anisotropic rotations of small molecules, *J. Chem. Phys.,* 75, 3581, 1981.
14. **Viovy, J. L., Monnerie, L. and Brochon, J. C.**, Fluorescence polarization decay study of polymer dynamics: a critical discussion of models using synchrotron data, *Macromolecules,* 16, 1845, 1983.
15. **Viovy, J. L., Monnerie, L., and Merola, F.**, Fluorescence anisotropy decay studies of local polymer dynamics in the melt. I. Labeled polybutadiene, *Macromolecules,* 18, 1130, 1985.
16. **André, J. C., Bouchy, M., Viovy, J. L., Vincent, L. M., and Valeur, B.**, Use of regularization operators together with Lagrange multipliers in numerical deconvolution of fluorescence decay curves, *Comp. Chem* 65, 1982.
17. **Livesey, A. K. and Brochon, J. C.**, Analyzing the distribution of decay constants in pulse-fluorometry using the maximum entropy method, submitted.
18. **Monnerie, L.**, Dynamic depolarization of luminescence as cited in **Phillips, D., Ed.**, *Polymer Photophysics,* University Press, Cambridge, U.K., 1985, 279.
19. **Anufrieva, E. V. and Gotlib, Y. Y.**, Investigation of polymers in solution by polarized luminescence, *Adv. Polym. Sci.,* 40, 1, 1981.
20. **Valeur, B. and Monnerie, L.**, Dynamics of macromolecular chains. III. Time-dependent fluorescence polarization studies of polystyrene in solution, *J. Polym. Sci. Polym. Phys. Ed.,* 14, 11, 1976.
21. **Kasparyan-Tardiveau, N., Valeur, B., Monnerie, L., and Mita, I.**, Comparison between end-bond and internal-bond mobility of polystyrene in dilute solution: an investigation by time-dependent fluorescence polarization, *Polymer,* 24, 205, 1983.
22. **Yeung, C. K.**, Ph.D. thesis, Université Pierre et Marie Curie, Paris, 1982.
23. **Viovy, J. L., Frank, C. W., and Monnerie, L.**, Fluorescence anisotropy decay studies of local polymer dynamics in the melt. II. Labeled model compounds of variable chain length, *Macromolecules,* 18, 1606, 1985.
24. **Veissier, V.**, Thesis, Paris, 1987.
25. **Rouse, P. E.**, A theory of the linear viscoelastic properties of dilute solutions of coiling polymers, *J. Chem. Phys.,* 21, 1272, 1953.
26. **Valeur, B., Jarry, J. P., Gény, F., and Monnerie, L.**, Dynamics of macromolecular chains. I. Theory of motions on a tetrahedral lattice, *J. Polym. Sci., Polym. Phys. Ed.,* 13, 667, 1975. Dynamics of macromolecular chains. II. Orientation relaxation generated by elementary three-bond motions and notion of an independent kinetic segment, *Polym. Sci., Polym. Phys. Ed.,* 13, 675, 1975.
27. **Bendler, J. T. and Yaris, R.**, A solvable model of polmer main-chain dynamics with applications to spin relaxation, *Macromolecules,* 11, 650,1978.
28. **Jones, A. A. and Stockmayer, W. H.**, Models for spin relaxation in dilute solutions of randomly coiled polymers, *J. Polym. Sci., Polym. Phys. Ed.,* 15, 847, 1975.
29. **Lin, Y. Y., Jones, A. A., and Stockmayer, W. H.**, Comparison of two models for local chain motions, *J. Polym. Sci. Polym. Phys. Ed.,* 22, 2195, 1984.

30. **Hall, C. K. and Helfand, E.**, Conformational state relaxation in polymers: time-correlation functions, *J. Chem. Phys.*, 77, 3275, 1982.
31. **Weber, T. A. and Helfand, E.**, Time-correlation functions from computer simulations of polymers, *J. Phys. Chem.*, 87, 2881, 1983.
32. **Veissier, V. and Viovy, J. L.**, Effect of an inhomogeneity on local chain dynamics: conformational autocorrelation function, *Macromolecules*, 21, 855, 1988.
33. **Ediger, M.**, private communication.
34. **Bahar, Y. and Erman, B.**, Investigation of local motions in polymers by the dynamic rotational isomeric state model, *Macromolecules*, 20, 1368, 1987.
35. **Veissier, V., Viovy, J. L., and Monnerie, L.**, Local dynamics of cis-polyisoprene in dilute solution and in the melt: a fluorescence anisotropy decay study, *Polymer*, in press.
36. **Williams, G. and Watts, D. C.**, Non-symmetrical dielectric relaxation behavior arising from a simple empirical decay function, *Trans. Faraday Soc.*, 66, 80, 1970.
37. **Viovy, J. L. and Monnerie, L.**, A study of local chain dynamics in concentrated polystyrene solutions using fluorescence anisotropy decay, *Polymer*, 27, 181, 1986.
38. **Viovy, J. L. and Monnerie, L.**, Segmental dynamics during the thermal polymerization of styrene: a kinetic fluorescence anisotropy decay study, *Polymer*, 28, 1547, 1987.
39. **Sasaki, T., Yamamoto, M., and Nishijima, Y.**, Dynamics of anthracene-labelled PMMA chains studied by the fluorescence depolarization methods, *Makromol. Chem. Rapid Comm.*, 7, 345, 1986.
40. **Queslel, J. P., Jarry, J. P., and Monnerie, L.**, Stationary fluorescence depolarization study of mobility of rigid probes in bulk elastomers: motion of dimethylanthracene and three trans-diphenylpolyenes inserted in polyisoprene, polybutadiene and random butadiene-styrene copolymers, *Polymer*, 27, 1228, 1986.
41. **Jarry, J. P. and Monnerie, L.**, Fluorescence depolarization study of the glass-rubber relaxation in a polyisoprene, *Macromolecules*, 12, 927, 1979.
42. **Fofana, M.**, Ph.D. thesis, Université Pierre et Marie Curie, Paris, 1988.
43. **Fofana, M., Viovy, J. L., Veissier, V., Monnerie, L. and Johari, G. P.**, Studies of the mobility of probes in polypropylene oxide. I. Fluorescence anisotropy decay, *Polymer*, in press.
44. **Huang, Y. Y. and Wang, C. H.**, Brillouin, Rayleigh and depolarized Rayleigh scattering studies of polypropylene glycol, *J. Chem. Phys.*, 62, 120, 1975.
45. **Yano, S., Rahalkär, R. R., Hunter, S. P., Wang, C. W. and Boyd, R. H.**, Studies of molecular relaxation of polypropylene oxide solutions by dielectric relaxation and Brillouin scattering, *J. Polym. Sci., Polym. Phys. Ed.*, 14, 1377, 1976.
46. **Wang, C. W., Fytas, G., Lilye, D., and Dorfmüller, T.**, Laser light beating spectroscopic studies of dynamics in bulk polymers: polypropylene glycol, *Macromolecules*, 14, 1363, 1981.
47. **Monnerie, L., Viovy, J. L., Dejean de la Batie, R., and Lauprêtre, F.**, Spectroscopic investigation of local dynamics in polymer melts, *Polym. Prep.*, 27, 297, 1986.
48. **Rioka, J., Gysel, H., Schneider, J., Nyffenegger, R., and Binkert, T.**, Mobility of fluorescent polyacrylamide derivatives in water-acetone mixtures, *Macromolecules*, 20, 1407, 1987.

Chapter 8

LASER STUDIES OF ENERGY TRANSFER IN POLYMERS CONTAINING AROMATIC CHROMOPHORES

Donald B. O'Connor and Gary W. Scott

TABLE OF CONTENTS

I.   Introduction...238

II.   Polystyrene Spectra and Structure...238
       A.   Polystyrene Spectra..238
       B.   Polystyrene Excimers and Polystyrene Structure.........................239

III.   Theory of Electronic Energy Transfer...241
       A.   Mechanisms of Electronic Energy Transfer..............................242
             1.   Radiative Transfer 242
             2.   Nonradiative Energy Transfer...................................242
                    a.   Dipole-Dipole Energy Transfer..........................242
                    b.   Electron Exchange Energy Transfer 244
       B.   Kinetic Models of Electronic Energy Transfer..........................244
             1.   Stern-Volmer Kinetics 244
             2.   Perrin Kinetic Model...245
             3.   Förster Kinetics 245
             4.   Other Kinetic Models 246
       C.   Electronic Energy Migration...246

IV.   Energy Transfer Studies in Polystyrenes 247
       A.   Pure Polystyrene 247
       B.   Polystyrene Copolymers Containing Only Intrinsic Traps 248
       C.   Copolymers of Polystyrene Containing Extrinsic Traps..................248

Acknowledgments 255

References...255

I. INTRODUCTION

Study of electronic energy transfer in polymers is of great importance in understanding and describing the photophysical and photochemical properties of macromolecular structures. In naturally occurring, biological photosynthetic systems, absorbed light energy may be transferred extremely efficiently from absorption sites to reaction centers, units which convert the light energy to chemical energy.[1] Polymer chemists have long hoped to develop synthetic polymers which would possess this photon harvesting capability and to apply these techniques to the promotion of photochemical reactions. Nonradiative electronic energy transfer is also of great importance in the mechanisms of photostabilization of synthetic polymers. Polymers resistant to photodegradiation are of great utility in many practical areas. For example, polymeric materials are used extensively in the solar energy industry for solar collectors, optical elements, and protective containers.[3] Knowledge gained from the investigation of the nature of electronic energy transfer will promote better designs of polymeric materials for the particular environmental conditions either on earth or in deep space.

This review will describe current understanding of the nature of electronic energy transfer in aromatic polymer systems, with particular emphasis on the polystyrenes, both in solution and as solid films over a wide temperature range. Our present understanding of energy transfer in these systems has been elucidated by transient and photostationary fluorescence experiments. Special emphasis will be placed on the role of laser spectroscopy in the study of these materials. The tenets of the electronic energy transfer theories used in the interpretation of this data will be briefly reviewed. In order to relate these theories to the particular polystyrene systems, we will begin by briefly reviewing the relationship of polystyrene emission spectra and kinetics to the composition and temperature of the polymer system.

II. POLYSTYRENE SPECTRA AND STRUCTURE

Polystyrene, shown in Figure 1, is a prototype of macromolecules which contain pendant aromatic chromophores attached to the polymer backbone. The interactions between aromatic chromophores in such polymers introduce a wide range of energy transfer processes that can be studied by conventional and laser spectroscopic methods. The modification of the macromolecular structure of polystyrene by its copolymerization with other polymeric materials of both lower and higher excited electronic state energies provides an additional approach to the study of energy transfer in these polymers. Since polystyrene and such copolymers of polystyrene are the prototypical synthetic macromolecular system containing aromatic chromophores for which electronic energy transfer is most widely studied, the present review will concentrate on this particular macromolecular system.

A. POLYSTYRENE SPECTRA

The absorption spectrum of polystyrene (Figure 2) very closely resembles that of toluene with an absorption maximum near 260 nm. On the other hand, the room temperature emission spectrum of polystyrene in solution consists of two bands (Figure 3). The normal Stokes shifted emission band, $\lambda_{max} \sim 280$ nm, is attributed to fluorescence from excited state monomers, which are single phenyl groups bearing localized excitation. The highly Stokes-shifted emission band, $\lambda_{max} \sim 330$ nm, is properly assigned to fluorescence from excimers.[4,5] excited state species each of which consist of two interacting phenyl units. In room temperature films the excimer to monomer steady-state fluorescence intensity ratio is greater than it is in solution (Figure 3). The temperature dependence of the steady-state polystyrene film emission spectra is shown in Figure 4. The monomer emission intensity increases significantly at low temperature while the excimer emission decreases slightly with decreasing temperature.[6] The lowest energy band whose intensity increases with temperature in Figure 4 is due to phosphorescence of the monomer species.

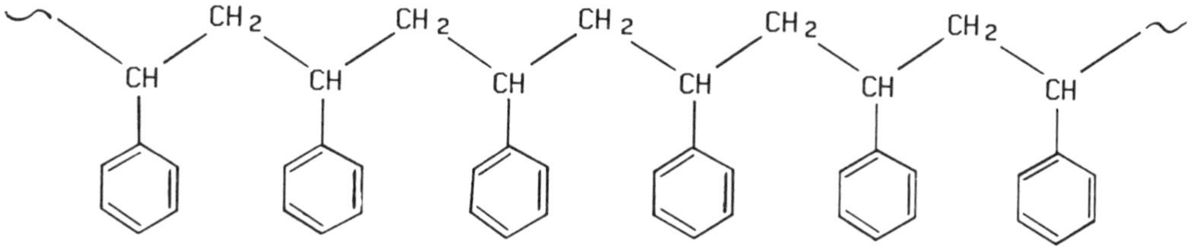

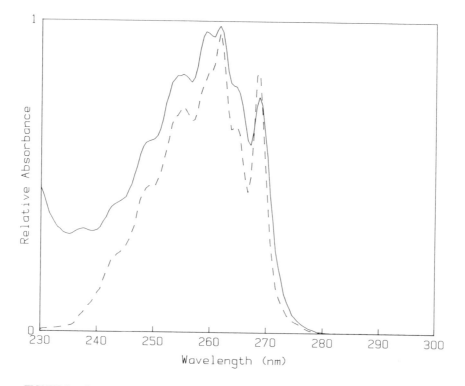

FIGURE 1.   Polystyrene structure. (No particular tacticity or conformation is implied.)

FIGURE 2.   Room temperature absorption spectra of polystyrene (——) and toluene (---) in cyclohexane solution.

The room temperature monomer fluorescence lifetime is approximately 1 ns in solution[7,8] or in a solid film.[6] The room temperature excimer fluorescence lifetime is about 13 ns in solution.[7,8] The excimer fluorescence lifetime in a room temperature film is somewhat longer, $\tau_E \sim 21$ ns, and shows little dependence on temperature.[6] The monomer lifetime in solid polystyrene films increases to greater than 20 ns at 20 K.[6]

B. POLYSTYRENE EXCIMERS AND POLYSTYRENE STRUCTURE

In general, excimers are formed as result of the interaction of an electronically excited, planar aromatic molecule such as pyrene, benzene, or naphthalene with an identical molecule in its ground state. Experimental evidence from studies of crystals of aromatic molecules[9] and paracyclophanes[10] suggests that the two molecules form a sandwich structure in which the molecular planes are parallel to each other and separated by approximately 3.3 Å or less. The excimer emission is characterized by two distinct features. First, the excimer emission is red shifted from the corresponding monomer emission[11] by ~ 6000 cm^{-1}, and second, the excimer emission is broad and structureless under any conditions due to the unbound, dissociative nature of the excimer ground state.

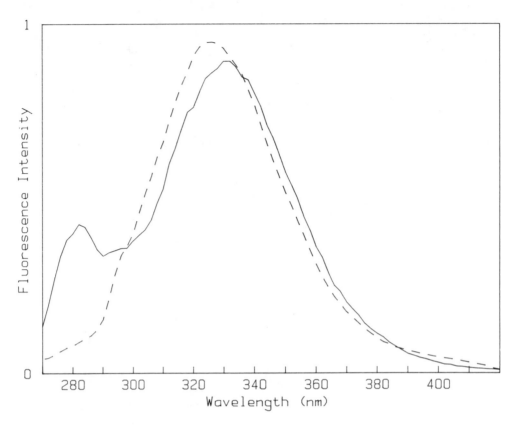

FIGURE 3.   Room temperature emission spectra of polystyrene in cyclohexane solution (———) and neat polystyrene film (---). (λ_{exc} = 260 nm).

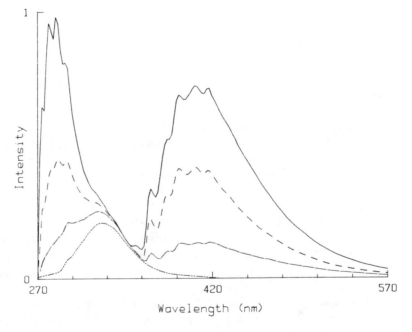

FIGURE 4.   Emission spectra of polystyrene film (λ_{exc} = 260 nm) at 15 K (———), 58 K (---), 93 K (- · -), and 296 K (- · · -).

In polystyrene the excimer may, in principle, be formed by any two adjacent phenyl units provided they can adopt a geometry conducive to excimer formation, i.e., parallel molecular planes separated by 3 to 4 Å. Studies[12] of diphenyl and triphenyl alkanes in solution reveal that intramolecular excimers are formed only if the two interacting phenyl units are connected by a C-C sigma bonding chain containing three intervening carbon atoms between the two phenyls. This result is known as the Hirayama n = 3 rule.[12] In polystyrene, although each adjacent pair of pendant phenyl units in the polystyrene chain is separated by three carbon atoms (see Figure 1), only certain, very specific chain conformations will permit neighboring phenyl units to meet the geometrical requirements of excimer formation. The probability of these conformations occurring in a particular polystyrene sample necessarily depends on factors such as tacticity, solvent, molecular weight, and temperature of that sample.

A study[13] in which the ratio of excimer to monomer steady-state fluorescence intensities at 77 K in atactic and isotactic polystyrene was measured found that this ratio was significantly greater in the regularly oriented isotactic polymer. Isotactic polystyrene may form a crystal consisting of 3-1 helices.[14] and the excimer to monomer fluorescence ratio increases with the extent of crystallinity. Thus, it is thought that purely crystalline isotactic polystyrene contains the highest fraction of conformations favorable to excimer formation while amorphous atactic polystyrene contains many conformations in which phenyl units cannot form excimers.

Conformational energy calculations[15,16] of meso and racemic dyads of polystyrene chain sequences have been performed as a function of skeletal bond rotations, and these calculations give the fraction of conformations that may be conducive to excimer formation at thermodynamic equilibrium. In solid polystyrene films, the intensity of excimer emission shows a dependence upon the temperature at which the film was cast as skeletal bond rotations are largely inhibited due to steric effects in the solid state. Hence, the conformational equilibrium is established at the temperature of casting. Since the ratio of excimer to monomer emission intensity increases with increasing casting temperature,[17,18] it was concluded that the conformations suitable for excimer formation are not the thermodynamically favored, lowest energy ones.

Several solutions studies[19,20] have indicated that the ratio of excimer to monomer emission intensity is proportional to the molecular weight of polystyrene. Itagaki et al.[19] studied the emission of styrene oligomers in cyclohexane solution for which the number of monomer units varied from 2 to 60. They found a dramatic increase in the excimer to monomer emission ratio from two to eight monomer units and a leveling off of this ratio above ten monomer units. Ishii et al.[20] examined polystyrene in dichloroethane solution and found a similar trend, although in their study this ratio did not level off until the number average molecular weight of the polymer, $\overline{M}_n$, was approximately 10,000.

In pure polystyrene films there is no apparent dependence of the excimer to monomer emission intensity ratio vs. molecular weight over the range $1950 \leq \overline{M}_n \leq 50,000$.[18] Apparently this ratio has not been investigated in films of shorter styrene oligomers as it was in solution.

III. THEORY OF ELECTRONIC ENERGY TRANSFER

The electronic energy transfer that we will discuss can be generally depicted

$$D^* + A \rightarrow D + A^* \tag{1}$$

in which D and A represent a donor molecule and an acceptor molecule, respectively, and the asterisk indicates electronic state excitation. As a result of the interaction of an acceptor molecule in its ground electronic state with a donor molecule possessing electronic excitation energy, the acceptor molecule can become electronically excited while the donor simultaneously returns to its ground electronic state. Note that this process excludes energy transfer in which the electronic energy of a donor is transferred to an acceptor in the form of vibrational and rotational energy as may occur in the nonradiative decay of the excited state of a solute molecule in the presence of a solvent bath.

In the energy transfer studies in polystyrene systems that are discussed in this chapter, the phenyl monomer unit may actually act as both a donor and an acceptor in the case of energy migration along or between polymer chains. Another case which has been studied involves the excited monomer unit as a donor and another different chromophore, either copolymerized into the polystyrene structure or simply separately present in solution, acting as the energy acceptor. The excimer may also in some systems act as the donor, but it may not act as the acceptor since, by definition, the excimer is only an excited-state species (see Section II.B).

A. MECHANISMS OF ELECTRONIC ENERGY TRANSFER
1. Radiative Transfer

The process of radiative transfer can be express by Equation 2

$$D^* \rightarrow D + h\nu, \qquad \text{and}$$

$$h\nu + A \rightarrow A^* \tag{2}$$

This two step mechanism involves the absorption of a photon by the acceptor which had been emitted from an excited-state donor. Clearly this process can provide energy transfer over a long range. It can be shown[21] that the probability of radiative energy transfer, P, is given by

$$P \simeq 1/q_D \int_0^\infty F_D(\bar{\nu})[1 - 10^{-\epsilon_A(\bar{\nu})[A]x}]d\bar{\nu} \tag{3}$$

in which [A] is the concentration of acceptor molecules, q_D is the quantum efficiency of fluorescence of the donor, F_D is the normalized emission spectrum of the donor (i.e., $\int F_D (\bar{\nu})d\bar{\nu} = 1$), ϵ_A is the absorption extinction coefficient of the acceptor, and x is the sample thickness.

Although this mechanism is often referred to as "trivial" electronic energy transfer, it may be the dominant mechanism of energy transfer in dilute solutions.[22] Its effect on the manifestations of energy transfer in polymers should always be considered in interpreting the data, even in the case of more concentrated systems.

2. Nonradiative Energy Transfer
a. Dipole-Dipole Energy Transfer

This mechanism of energy transfer derives its name from the nature of the interaction between the donor and acceptor molecules. First detailed by Förster,[23,24] the mechanism assumes a Coulombic (electrostatic) interaction between two isolated molecules. Förster proposed that, for an allowed transition, the dipole-dipole interaction energy would dominate all other interactions such as dipole-quadropole. The classical expression of the interaction energy E of two point dipoles is given by

$$E = \frac{\kappa \mu_D \mu_A}{n^2 R^3} \tag{4}$$

in which μ_D and μ_A are the magnitudes in Debyes of the respective donor and acceptor transition dipoles, R is the D-A separation distance, n is the refractive index of the medium, and κ is the orientation factor for μ_D and μ_A. Förster associated μ_D and μ_A with the measured quantities obtained from the oscillator strengths, and thus μ_D and μ_A may be calculated from the pure radiative rate constants of the donor and acceptor. He was then able to derive the following expression for the rate constant of dipole-dipole energy transfer

$$k_{DA} = \frac{9000\kappa^2 \ln 10 \phi_D}{128\pi^5 n^4 N \tau_D R^6} \int_0^\infty F_D(\bar{\nu}) \epsilon_A(\bar{\nu}) \frac{d\bar{\nu}}{\bar{\nu}^4} \qquad (5)$$

in which ϕ_D is the quantum yield of emission of the donor in the absence of the acceptor, n is the index of refraction of the solvent, N is Avagadro's number, τ_D is the excited state lifetime of the donor in the absence of the acceptor, R is the donor-acceptor separation, $F_D(\bar{\nu})$ is the emission spectrum of the donor again normalized such that $\int F_D(\bar{\nu}) d\bar{\nu} = 1$, and $\epsilon_A(\bar{\nu})$ is the molar extinction coefficient of the acceptor as a function of ν. The orientation factor κ^2 is usually taken to be $2/3$ such as would be expected for a random distribution of the transition dipole orientations of the donor and acceptor molecules. However, in a polymer such as polystyrene in which orientational averaging of the donor and acceptor may not occur during the lifetime of the donor for steric reasons, the value to be used for κ^2 must be carefully considered.[25]

If we define an average donor-acceptor distance $R = R_0$ at which the rate of energy transfer is equal to the fluorescence decay rate of the donor in the absence of the acceptor, i.e., $k_{ET} = \tau_D^{-1}$, then it follows that

$$R_0^6 = \frac{9000\kappa^2 \ln 10 \phi_D}{128\pi^5 n^4 N} \int_0^\infty F_D(\bar{\nu}) \epsilon_A(\bar{\nu}) \frac{d\bar{\nu}}{\bar{\nu}^4} \qquad (6)$$

This parameter R_0 is called the Förster critical radius and is indicative of the distance over which dipole-dipole energy transfer may be significant. Förster critical radii calculated using Equation 6 for donor-acceptor pairs selected from over 200 compounds are tabulated by Berlman.[26]

Combining Equations 5 and 6 we obtain

$$k_{DA} = \tau_D^{-1}(R_0/R)^6 \qquad (7)$$

which indicates the R dependence of the energy transfer rate. This relationship was confirmed for a system in which the donor and acceptor separation was varied from 12 to 46 Å by attaching a donor to one end and an acceptor to the other end of oligomers of varying lengths.[27]

A closer scrutiny of Equation 6 indicates that transfer processes such as

$$D^*(S_1) + A(S_0) \rightarrow D(S_0) + A^*(T_1) \qquad (8)$$

in which the specific electronic states of the donor and acceptor are indicated in parenthesis for each, are always inefficient, regardless of the spin states of the donor, because the acceptor transition is forbidden, i.e., $\epsilon_A(\nu)$ is small. However, a process in which the *donor* undergoes a change in spin state may be efficient. For example, Bennett et al.[28] studied the energy transfer from the phenanthrene $-d_{10}$ triplet state to rhodamine B and determined a Förster critical radius of $R_0 = 45$ Å.

The values of Förster critical radii calculated using Equation 6 are called theoretical values even though several of the parameters in this expression are experimentally determined. For systems in which the donor emission overlaps significantly with strongly allowed acceptor absorption and the emission quantum yield of the donor is near unity, Equation 6 predicts R_0 values on the order of 5 to 50 Å.[26]

1. Electron Exchange Energy Transfer

Electronic energy transfer may also occur between the donor and acceptor molecules by the exchange of electrons while the electronic clouds of the two molecules overlap. Being identical particles, electrons in this area of overlap cannot be exclusively assigned to either the donor or acceptor so an excited electron initially on the donor before collision may exchange with one initially on the acceptor and end up on the acceptor after collision. To preserve the neutrality of the molecules, an unexcited electron on the acceptor is transferred to the donor. Dexter[29] derived an equation quantifying the rate of energy transfer for such a process.

$$k_{DA} = JKe^{-2R/L} \tag{9}$$

in which K is a constant, R is the donor-acceptor separation, L is the effective, average Bohr radius, and J is the spectral overlap integral in which the donor emission spectrum F_D $(\bar{\nu})$ and the acceptor absorption spectrum $\epsilon_A (\bar{\nu})$ are both normalized to unity.

Since the oscillator strength of transitions for neither the donor nor the acceptor bear on the rate of exchange energy transfer, the triplet-triplet energy transfer process,

$$D^*(T_1) + A(S_0) \rightarrow D(S_0) + A^*(T_1) \tag{10}$$

is allowed. Since this mechanism requires significant overlap of the electron clouds, the sharp exponential decrease of the transfer rate with increasing donor-acceptor separation is understandable.

It is not possible, as it is in the case of dipole-dipole energy transfer (see Equation 5), to predict the energy transfer rate of the exchange mechanism from experimental quantities. Assumptions regarding the shape of the electronic clouds must be made to determine the average Bohr radius L.

B. KINETIC MODELS OF ELECTRONIC ENERGY TRANSFER
1. Stern-Volmer Kinetics

Stern-Volmer kinetics are applicable when there is a nonzero probability of collision between the donor and acceptor molecules during the lifetime of the donor. The quenching (or energy transfer) process is bimolecular and in competition with the unimolecular decay rate of the donor. When these conditions occur the Stern-Volmer expression[30] holds,

$$\phi_0/\phi = 1 + k_{ET}\tau_D[A] \tag{11}$$

in which ϕ_0 and τ_D are the donor emission quantum yield and lifetime, respectively, in the absence of the acceptor and ϕ is the donor emission quantum yield at the acceptor concentration [A]. A plot of ϕ_0/ϕ vs. [A] will yield a straight line of slope k_{ET}. Thus, if the donor lifetime is known the energy transfer rate may be determined.

For some systems, the rate constant k_{ET} may approach the diffusion controlled rate constant, k_{diff}. This suggests that the rate determining step in the energy transfer is the mutual diffusion of the donor and the acceptor to each other and not the actual energy transfer mechanism such as electron exchange. An estimate of the diffusion rate constant can be obtained from the Debye equation,

$$k_{diff} = 8RT/(3000\eta) \tag{12}$$

in which R is the gas constant, T, the temperature, and η, the viscosity of the solvent.

2. Perrin Kinetic Model

When the donor and acceptor are fixed and cannot diffuse relative to each other during the excited-state lifetime of the donor, such as in glassy matrices, then the Stern-Volmer equation is inappropriate. In such cases the Perrin expression,[31]

$$\ln(\phi_0/\phi) = VN[A] \tag{13}$$

in which ϕ_0, ϕ, and [A] are as described in the Stern-Volmer expression, N is Avogadro's number, and V is the volume of the quenching sphere of the donor, is often applied. If an acceptor is within this quenching sphere, the donor is quenched with 100% efficiency. If the acceptor lies beyond this quenching sphere, the donor is not quenched. A plot of ln (ϕ_0/ϕ) vs. [A] will yield the quenching sphere volume V.

3. Förster Kinetics

For systems which undergo dipole-dipole energy transfer, as discussed earlier (Section III.A.2.a), the Förster critical radius can be determined by either of two methods. One technique requires the measurement of steady state emission intensities of the donor in the presence of varying amounts of acceptor. The equations[27,32,33] which relate these quantities are

$$f = \pi^{1/2}\gamma\exp(\gamma^2)[1 - \mathrm{erf}(\gamma)]$$
$$f = (\phi_0 - \phi)/\phi_0$$
$$\gamma = C/C_0$$
$$C_0 = (3000)/(2\pi^{3/2}NR_0^3) \tag{14}$$

in which ϕ_0 and ϕ are the emission quantum yields of the donor in the absence and presence of an acceptor of concentration C, γ is termed the quenching efficiency, and N is again Avogadro's number.

The other technique for determining R_0 involves analysis of the emission decay of the donor as a function of acceptor concentration. The expression[27,34] describing this relationship is

$$I(t) = I(0)e^{-t/\tau_D}\exp[-2\gamma(t/\tau_D)^{1/2}] \tag{15}$$

in which τ_D is the donor lifetime in the absence of acceptor and the quenching efficiency γ is the same as defined above. Fitting this expression to fluorescence decays of the donor will yield the Förster critical concentration, C_0, and hence the Förster critical radius, R_0.

The value of R_0 determined by either the steady-state emission spectra (Equation 14) or the fluorescence decay kinetics (Equation 15 is termed the experimental Förster critical radius. A deviation between this value and the theoretical critical radius calculated using Equation 6 is indicative of the presence of either an additional energy transfer mechanism or perhaps significant diffusion of the donor and acceptor during the excited-state lifetime of the donor. In many polymer systems the determination of a experimental critical radius exceeding the theoretical value has been purported to be the result of an energy migration process which is discussed in the next section.

4. Other Kinetic Models

The applicability of Förster dipole-dipole kinetics (see Section III.B.3) as opposed to Stern-Volmer kinetics (see Section III.B.1) is best determined by analyzing the magnitude of

$$Q = R_0^2/(\tau_D D) \qquad (16)$$

in which R_0 is the calculated Förster critical radius, τ_D is the donor fluorescence lifetime in the absence of quencher, and D is the relative diffusion coefficient of the donor and acceptor. For systems possessing very small values of Q (i.e., $Q < 10^{-3}$), such as might occur in low viscosity solutions, Stern-Volmer kinetics are applicable and the donor fluorescence will decay exponentially at a rate given by $1/\tau_D + k_{ET}$ [A]. In cases where the donor and acceptor remain relatively fixed during the donor lifetime, Q will be large (i.e., $Q > 10^2$) and the case of Förster kinetics is appropriate.

Energy transfer kinetics in many systems of moderate Q values are not adequately modeled by either of these two limiting cases. Several theoretical models[35-38] have been proposed to explain the emission kinetics in which both a significant diffusion-controlled transfer as well as dipole-dipole energy transfer occurs during the lifetime of the donor. An analytical expression for the emission decay kinetics in such systems has been proposed by Yokota and Tanimota:[37]

$$I(t) = I(0)e^{-t/\tau_D}exp[-2B\gamma(t/\tau_D)^{1/2}] \qquad (17)$$

This expression is very similar to the Förster kinetic equation (see Section III.B.3) and the parameters τ_D and γ are as defined earlier. The effect of diffusion in this formulation is seen in the expression for the parameter B:

$$B = \left(\frac{1 + 10.87x + 15.5x^2}{1 + 8.743x}\right)^{3/4} \qquad (18)$$

in which $x = D(\tau_D/R_0)^{1/3} t^{2/3}$ and D is again the relative donor-acceptor diffusion constant. When diffusion is insignificant during the lifetime of the donor, x approaches zero, B approaches one, and the expression of Yokota and Tanimota reduces to the Förster kinetic Equation 15.

C. ELECTRONIC ENERGY MIGRATION

Electronic energy migration consists of a series of energy transfer processes by which electronic excitation, initially localized on a particular chromophore of the polymer, is transferred in steps to another chromophore which may or may not be connected to the same polymer backbone. During each of the energy transfer steps, the excitation is localized on a single chromophore for some short, but finite, period of time. This situation is in contrast with one in which energy is delocalized over many chromophores, such as occurs in organic crystals, and is modeled by exciton theory.[39] In polystyrene, for example, these energy transfer stesp may be made between neighboring monomer units on the same polymer chain, between nonneighboring monomer units on the same chain, or between monomer units of different polymer chains. The mechanism of energy transfer in each step of electronic migration may be dipole-dipole, high order multipole-multipole, or electron exchange. Nevertheless, such energy migration in polymers is sometimes referred to in the literature as exciton transfer.

Single step energy transfer between nonneighboring units of polystyrene in dilute solutions of a good solvent is unlikely as a result of the separation distance (see Sections

III.A.2.a. and b). In a poor solvent or a solid film in which the polymer chain may coil, the distance between nonneighboring units may be relatively small and the likelihood of a single step transfer between nonneighboring units is much higher. Energy transfer steps may occur between neighboring phenyl units by either the dipole-dipole mechanism or the electron exchange mechanism, the latter providing there is some overlap of the electronic orbitals. The validity of using the simple dipole-dipole mechanism for describing transfer between neighboring phenyl units is questionable for the following reason: the simple dipole-dipole transfer theory assumes that the donor and acceptor are point dipoles, and this becomes an increasingly poor assumption as the donor-acceptor separation decreases.

In a disordered system of donors and acceptors in which both long- and short-range single step energy transfer as well as multistep energy migration may occur, the kinetics of the entire energy transfer process are expected to be complex and, thus, very difficult to model. Burshstein[40] has recently reviewed the kinetic models of energy transfer in liquid or solid solutions of randomly distributed donors and acceptors of concentrations ρ and c, respectively. Briefly, the results of this modeling are if a low concentration of donors are surrounded by a large concentration of acceptors, then the quenching is static, i.e., the excitations are quenched at the sites of original absorption if $c/\rho >> 1$. Otherwise, the quenching of the excitation is accelerated by donor-donor energy migration steps. When the excitation can make several energy migration steps during its lifetime, the excitation may then reach a donor site where the quenching rate is higher than it was at the original site. The degree of this self-quenching increases with increasing donor concentration ρ.

Fredrickson et al.[41,42] have developed a model of long-range energy migration based on dipole-dipole energy transfer between neighboring chromophores in a polymer such as polystyrene. The model may be useful in interpreting the transient fluorescence experiments and steady-state emission spectra of either the donors or acceptors on polymer chains. Interpretation of transient and steady-state fluorescence experiments of polymers containing small concentrations of a donor and a trap utilizing this model may permit the determination of microscopic features of the polymer such as the characteristic ratio, which is a measure of the statistical chain flexibility.[43]

IV. ENERGY TRANSFER STUDIES IN POLYSTYRENES

A. PURE POLYSTYRENE

If a molecule is excited with polarized exciting light and its emission is subsequently observed through a polarizer, the degree of polarization p, given by

$$p = \frac{I_{\parallel} - I_{\perp}}{I_{\parallel} + I_{\perp}} \qquad (19)$$

is indicative of the extent of rotational reorientation in the absence of energy transfer, provided the emission and absorption dipoles are parallel. In this expression $I_{\parallel}$ and $I_{\perp}$ are the emission intensities polarized parallel and perpendicular to the excitation beam, respectively. Gupta et al.[8] recorded the values of the polarization of excimer emission in films under steady-state illumination as well as during the 2-ns time frame immediately following pulsed excitation of polystyrene. The emission was totally depolarized in both cases, suggesting rapid energy migration of the excitation prior to finding an excimer forming site. The energy migration, believed to be a multistep hopping process, results in a random orientation of the excitation polarization, providing the energy transfer process does not require parallel alignment of the transition dipoles of the donor and acceptor in each transfer step. (On the other hand, MacCallum and Rudkin[44] also examined the fluorescence polarization of the excimer emission in thin films of polystyrene and obtained a polarization value of 0.77, in obvious contradiction to the results of Gupta.) It should be noted that Gupta et al.[8] found the polarization to increase with the extent of photodegradation of the polystyrene film.

The molecular weight dependence of the excimer to monomer emission intensity ratio is often cited as evidence of energy migration in polystyrenes, Itagaki et al.[19] asserted that the onset of the plateau of this ratio at seven or eight monomer units for styrene oligomers in solution indicates that the energy migration length is seven or eight units. In this model, a phenyl unit ten units away from the initial. excitation cannot contribute to the excimer emission yield because the excitation cannot migrate that far. Assuming a simple one-dimensional random walk model of the hopping process, the authors were able to calculate a rate constant for excitation hopping between nearest neighbors of $k_{mig} \sim 3 \times 10^{10} s^{-1}$, ($\tau_{hop} \simeq 30$ ps).

The excimer to monomer intensity ratio of pure polystyrene in miscible blends with poly(vinylmethyl ether) also depends upon the polystyrene molecular weight.[45] The authors of that study were able to effectively model this molecular weight dependence based on a one-dimensional random walk model.[46] For blends of polystyrene and poly(vinylmethyl ether), a lattice model based on a three-dimensional random walk was also developed to model the fluorescence intensity ratio as a function of molecular weight and composition of blend.[47] For pure polystyrene the authors estimated that the excitation only makes approximately two hops before either forming an excimer or being deactivated.

B. POLYSTYRENE COPOLYMERS CONTAINING ONLY INTRINSIC TRAPS

Phillips et al.[48] examined the emission kinetics of a series of polystyrene polymers and polystyrene-butadiene block copolymers in dichloromethane solutions in which the number of phenyl chromophores was varied. Kinetic analysis revealed an excimer formation rate dependent upon the number of phenyl units in a block sequence up to approximately 25 units at which point the rate leveled off with increasing length. This is essentially the same as a molecular weight effect and is therefore similar to the results of Ishii et al.[20] and Itagaki et al.[19] However, using the reasoning of Itagaki one would estimate an energy migration rate constant $k_{mig} \sim 3 \times 10^{11} s^{-1}$, ($\tau_{hop} \simeq 3$ ps), an order of magnitude larger than that of Itagaki et al.[19]

Polarization studies[49,50] of the polystyrene monomer emission of polystyrene-methyl methacrylate copolymers in methyltetrahydrofuran glasses at 77 K were conducted to determine the degree of depolarization which was found to increase with an increasing phenyl fraction. This is understandable if energy migration occurs and each step of the migration process results in a change in the orientation of the emission dipole. Reid and Soutar[50] found that p^{-1} varied linearly with the mean length of the polystyrene sequence in the copolymer.

Sienicki and Bojarksi[51] studied the polystyrene monomer and excimer fluorescence quantum yields in copolymers of styrene with methyl methacrylate, ethyl acrylate, and acrylonitrile in dichloroethane solutions. Based on a model assuming spectroscopically active and inactive monomers, they found the number of excimer forming sites to be proportional to the fraction of linkages between phenyls. This result confirms that excimers are indeed formed between neighboring phenyl units, at least in dilute solutions.

C. COPOLYMERS OF POLYSTYRENE CONTAINING EXTRINSIC TRAPS

Some very compelling evidence of energy migration in polystyrene was presented by Phillips et al.,[52] who studied a copolymer of polystyrene containing 0.01 mol % 2,5-diphenyloxazole (PPO) chromophores. PPO is expected to be an energy acceptor and a localized energy trap in such a system as a result of the significant overlap of its absorption spectrum, $\lambda_{max} \sim 370$ nm, with polystyrene emission (Figure 3). The steady-state fluorescence spectrum of the copolymer in methylene chloride at room temperature resulting from excitation at 257 nm exhibited polystyrene monomer, excimer, and the PPO emission. The

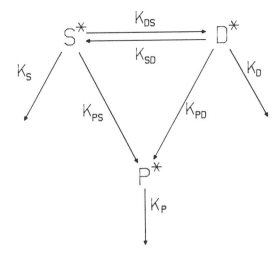

FIGURE 5. Energy transfer kinetic scheme for copolymer of polystyrene and 2,5-diphenyl-oxazole (PPO).

authors fit multiple exponential decays to the fluorescence decays in three regions near 290, 325, and 420 nm, respectively. In the monomer region, $\sim$ 290 nm, the best fit gave biexponential fluorescence decay lifetimes of 0.68 and 12.15 ns. These values are similar to but slightly shorter than the lifetimes recorded under identical experimental conditions in the absence of any PPO.[53] The presence of the longer lifetime is believed to be a result of excimer dissociation to form excited-state monomer. The decay profile near 420 nm was best fit by a triexponential decay function with lifetimes of 0.70, 1.65, and 13.16 ns, values which are characteristic of the monomer, PPO, and excimer emissions, respectively. In the excimer region, $\lambda \sim$ 325 nm, the emission data also yielded three lifetimes of 0.68, 1.65, and 14.7 ns. This kinetic data is consistent with the kinetic scheme of Figure 5 where S and D refer to the polystyrene monomer and excimer and P represents PPO. The exact solution of the kinetics implied by this general mechanism yielded expressions consistent with the multiexponential decays described above. However, the data from the copolymer alone does not permit determination of the relative magnitude of monomer to PPO energy transfer rate (k_{PS}) vs. the excimer to PPO energy transfer rate (k_{PD}). The authors were able to generate an estimate of this ratio by comparing the kinetics of pure polystyrene under identical conditions[53] with the kinetics of the copolymer. The calculated estimate was $k_{PS}/k_{PD} \sim 30$.

The authors also analyzed gated spectra of the copolymer emission and found the following: during the laser excitation pulse duration ($\Delta t \sim$ 7ns) the monomer and PPO emissions are stronger than the excimer. After the excitation pulse is over, the excimer emission becomes the dominant feature and both monomer and PPO emission are weak. That the PPO emission is stronger in the presence of strong monomer emission and weak in the presence of strong excimer emission suggests that the monomer to PPO energy transfer is much more significant than the excimer to PPO energy transfer.

The authors reason that if energy migration were not existent in polystyrene, then the excimer population would have to be relatively higher in polystyrene as indicated by the large relative emission intensity of the excimer. It follows that energy transfer from the excimer to PPO would then have to be more significant relative to the monomer-PPO transfer due to the longer relative lifetime of the excimer and far greater overlap of the excimer emission with PPO absorption than the monomer emission with PPO absorption (see Equation 5). However, their data indicate quite the opposite. Thus, they conclude energy migration occurs in polystyrene in dilute dichloromethane solutions.

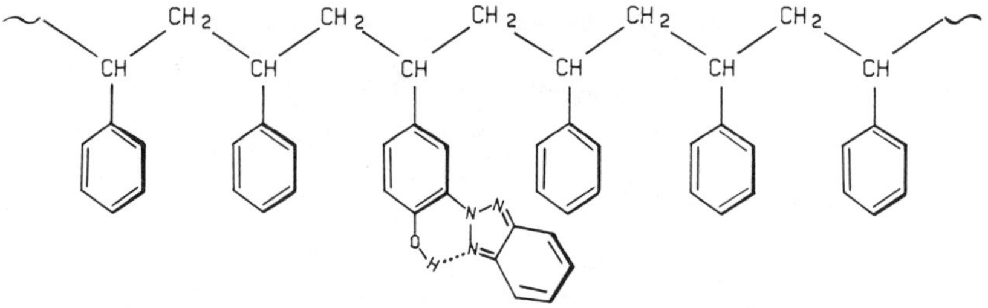

FIGURE 6.   Structure of styrene-co-VHPB copolymer.

A polystyrene copolymer incorporating a molecule known as a photostabilizer has been synthesized[54] and its photophysics investigated[6,55,56] to examine the role of energy migration in polystyrene. The copolymers were prepared by thermal copolymerization of styrene with varying amounts of 2-(2'-hydroxy-5'-vinylphenyl)-2H-benzotriazole (VHPB). The structure of the copolymer is shown in Figure 6.

Photostabilizers such as the chromophore of VHPB have long been incorporated into polymers to inhibit photochemical degradation of the polymer resulting from exposure to UV light.[2] This protection is due in part to selective absorption of UV light by the photostabilizer, thereby preventing its absorption by photochemically active portions of the polymer. An ideal photostabilizer would enhance the photophysical deactivation pathways of a polymer, thereby decreasing the extent of unwanted photochemical degradation processes such as photooxidation. Molecules such as the chromophore of VHPB accomplish this by functioning as energy traps and rapidly degrading UV energy absorbed by other polymer chromophores before photochemical damage can occur.

In order to understand the photophysics of these copolymers it is necessary to review the photophysics of the chromophore of this photostabilizer. The room temperature absorption spectrum of 2-(2'-hydroxy-5'-methylphenyl)-2H-benzotriazole (MeHPB) in methylcyclohexane (Figure 7) exhibits a broad band in the region 280 to 360 nm. The emission of MeHPB in polar, hydrogen-bonding solvents[57-62] consists of a moderately Stokes-shifted blue fluorescence ($\lambda_{max} \sim 410$ nm) while in nonpolar, hydrocarbon matrices at low temperature[58,63-67] a highly Stokes-shifted red fluorescence ($\lambda_{max} \sim 600$ nm) is observed. The red fluorescence has been attributed to emission from an excited-state, proton-transferred tautomer of MeHPB and the blue fluorescence to emission from an excited species which does not undergo proton transfer. The relationship between the fluorescence spectrum of MeHPB and this process is depicted in Figure 8.

The room temperature emission spectrum of a copolymer film of styrene with 3.4 mol% VHPB (styrene-co-3.4% VHPB) excited at 355 nm (Figure 9) shows a broad, highly Stokes-shifted red fluorescence, $\lambda_{max} \sim 630$ nm. This band is attributed to the excited-state, proton-transferred tautomer of the VHPB chromophore. The lifetime of this emission is short, $\tau_{fl} = 28 \pm 3$ps, and its rise time is less than 10 ps.[55] This very short rise time corresponds to the excited-state intramolecular proton transfer rate. The combination of this very rapid excited-state proton-transfer process and the rapid excited state decay of this tautomer explains the utility of the VHPB chromophore as a photostabilizer. The rapid proton transfer process efficiently degrades the UV excitation and the rapid return of the VHPB chromophore to its ground state makes it rapidly available to repeat the process.

In order to elucidate the energy transfer mechanisms and efficiencies in these copolymers, the steady-state emission and fluorescence kinetics of a series of 13 different copolymers of polystyrene incorporating VHPB with concentrations ranging from 0 to 4.78 mol% VHPB were investigated.[56] The molecular weight in these copolymers far exceeded the value at which a molecular weight dependence of polystyrene excimer emission intensity is observed (see Section II.B).

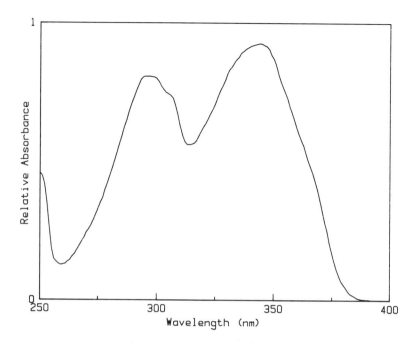

FIGURE 7.   Room temperature absorption spectrum of MeHPB in methylcyclohexane solution (λ_{exc} = 355 nm).

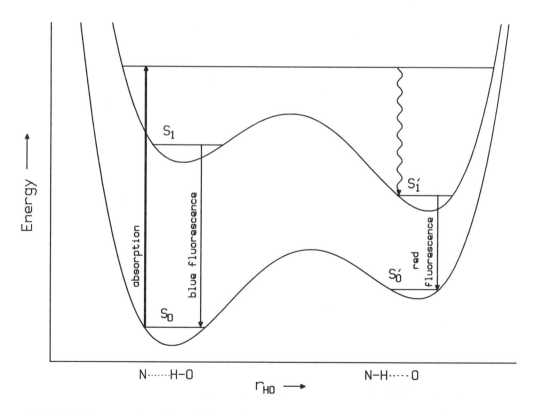

FIGURE 8.   Representation of the excited-state intramolecular proton-transfer mechanism of MeHPB in methylcyclohexane (λ_{exc} = 355 nm).

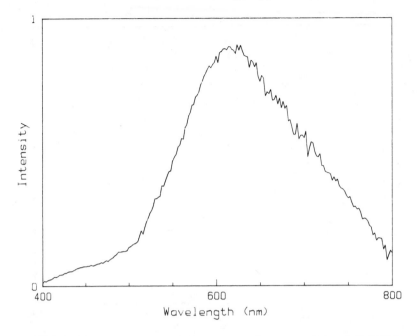

FIGURE 9.   Room temperature emission spectrum of styrene-co-(3.4 mol % VHPB) copolymer film excited at 355 nm.

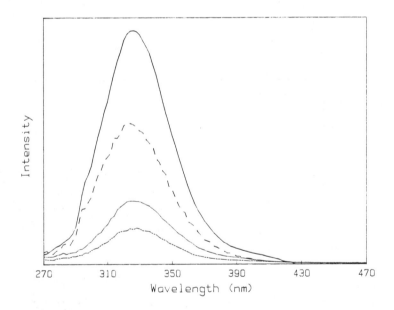

FIGURE 10.   Room temperature emission spectra (λ_{exc} = 260 nm) of styrene-co-VHPB copolymer films; 0 mol % VHPB (——), 0.014 mol % VHPB (---), 0.0527 mol % VHPB (- · -), and 0.154 mol % VHPB (- · · -). Spectra are corrected for reabsorption (trivial energy transfer) and competitive absorption by VHPB.

The room temperature, steady-state emission spectra of several of the styrene-co-VHPB copolymer films excited at 260 nm are shown in Figure 10. The spectra show a marked decrease in the fluorescence intensity of the polystyrene excimer emission ($\lambda_{max} \sim 320$ nm) with increasing mole percent VHPB. This decrease is due in part to increasing competitive absorption of 260-nm photons by VHPB (screening) and trivial energy transfer (see Section

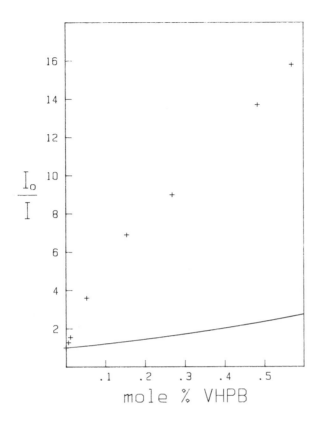

FIGURE 11.   Dependence of the ratio of the excimer emission at room temperature in a pure polystyrene film, I_0, to the excimer emission in styrene-co-VHPB copolymers, I, upon the amount of VHPB incorporated into the polystyrene film. Samples were excited at 260 nm and the emission intensity was monitored at 320 nm. The lower curve indicates the contribution of dipole-dipole energy transfer from the excimer to VHPB.

III.A.1) from the polystyrene excimer and monomer to the VHPB chromophore. The effect of screening is minimal in all samples since 260 nm corresponds to a minimum in the VHPB chromophore absorption and a maximum in polystyrene absorption (Figures 2 and 7). Correction factors for screening and trivial energy transfer were calculated, and the spectra in Figure 10 reflect the quenching efficiency of the polystyrene excimer and monomer emissions by various nonradiative energy transfer mechanisms involving VHPB. The dependence of the ratio of excimer emission intensity in pure polystyrene, I_0, to the excimer emission intensity in a copolymer, I, vs. the mole percent of VHPB in the copolymer is depicted in Figure 11.

The excimer fluorescence decay profiles ($300 < \lambda < 350$ nm) of the copolymer films excited with 266-nm laser pulses were analyzed. The excimer emission decay profiles for copolymers containing from 0 up to 0.0527 mol % VHPB were all well fit by a single exponential decay yielding a lifetime of 21.5 ± 0.5 ns, in good agreement with previous measurements[8] in pure polystyrene (section II.A). The excimer fluorescence decay profiles for films of higher mole percent VHPB could not be adequately fit by a single exponential decay. They were fit instead using the Förster kinetic equation (Equation 15). The quenching efficiency λ as a function of mole percent VHPB was plotted and the experimental Förster critical radius of 16 Å was obtained from the slope. A calculation of the theoretical Förster critical radius using Equation 6 gave $R_0 = 17$ Å. The degree of excimer fluorescence quenching resulting from long-range, dipole-dipole energy transfer from the excimer to the

VHPB chromophores can be calculated from Equation 14. The resulting dependence of I_0/I on the mole percent VHPB is shown in the lower curve of Figure 11. Examination of Figure 11 clearly indicates that Förster energy transfer is not a particularly significant mechanism of quenching of the excimer emission in these copolymer films. The fact that the excimer lifetime in the styrene-co-(0.0527 mol % VHPB) is the same as that in pure polystyrene even though $I_0/I \sim 4$ indicates that the dominant mechanism of quenching involves inhibiting the formation of excimers, i.e., quenching the precursor of the excimer.[56]

This data indicates that energy migration is extensive in polystyrene films by the following reasoning: let us assume that energy migration does not occur. Then the excitation energy is localized at the sites of absorption and excimers can be formed only at those sites. Effective quenching of the excimer emission can only occur by long-range, dipole-dipole energy transfer from either the excimer *or* the monomer to the VHPB chromophore. (Electron exchange energy transfer (Section III.A.2.6) is expected to be very inefficient in systems with a low concentration of acceptors and low mobility.) The Förster critical radius for monomer to VHPB transfer was calculated using Equation 6 to be 9.6 Å. Using this value and Equation 14 we would predict the quenching efficiency of the monomer to be negligible in the films investigated. The Förster quenching of the excimer by the VHPB chromophore is not significant either as seen in Figure 11. Since our data indicate very efficient quenching of the excimer emission, the assumption that energy migration does not occur in polystyrene films must be incorrect.

An estimate of the excimer concentration in polystyrene films was also obtained from an analysis of the data obtained on styrene-co-VHPB copolymers. When the ratio $I_0/I = 2$ (0.025 mol % or 2.5×10^{-3} *M* VHPB), the probability of energy transfer from the migrating monomer to the VHPB chromophore or to an excimer site must be equal. Förster quenching of the excimer at this concentration is negligible (see Figure 11). If quenching or excimer formation involves migration followed by a one-step Förster transfer to either an VHPB chromophore or an excimer site, respectively, then it follows that

$$C_E = C_{HPB}(R_0^{HPB}/R_0^E)^3 \tag{20}$$

in which C_E and C_{HPB} are the concentrations of excimer-forming sites and VHPB chromophores and $R_0^{HPB} = 9.6$ Å is the monomer-VHPB chromophore critical radius. R_0^E is the critical radius for polystyrene monomer-monomer transfer which is reported[68] to be 6 Å. Plugging the values into Equation 20 yielded a concentration of excimer-forming sites of $C_E \simeq 1 \times 10^{-2}$ *M* or approximately 1 excimer-forming site per 950 monomer units.[56]

ACKNOWLEDGMENTS

Some of the research by the authors reviewed in the present work was partially supported by the Committee on Research of the University of California, Riverside and by the Jet Propulsion Laboratory, California Institute of Technology under contract with the National Aeronautics and Space Administration.

REFERENCES

1. **Clayton, R. K.**, *Photosynthesis: Physical Mechanisms and Chemical Patterns,* Cambridge, University Press, Cambridge, 1980.
2. **Ranby, B. and Rabek, J. F.**, *Photodegradation, Photooxidation, and Photostabilization of Polymers: Principles and Applications,* John Wiley & Sons, New York, 1975.
3. **Gebelin, C. G., Williams, D. K., and Deanin, R. D., Eds.**, Polymers in Solar Energy Utilization, ACS Symp. Ser. 220, American Chemical Society, Washington, D.C., 1983.
4. **Basile, L. J.**, Effect of styrene monomer on the fluorescence properties of polystyrene, *J. Chem. Phys.,* 36, 2204, 1962.
5. **Yanari, S. S., Bovey, F. A., and Lumry, R.**, Fluorescence of styrene homopolymers and copolymers, *Nature,* 200, 242, 1963.
6. **Coutler, D. R., Gupta, A., Miskowski, V. M. and Scott, G. W.**, Excited state singlet energy transport in polystyrene, in *Photophysics of Polymer Systems,* Hoyle, C. and Tochelson, J., Eds., (ACS Symp. Ser., Vol. 358), 1987, chap. 22.
7. **Ghiggino, K. P., Roberts, A. J., and Phillips, D.**, Energy relaxation in synthetic polymer solutions, *J. Photochem.,* 9, 301, 1978.
8. **Gupta, M. D., Gupta, A., Horwitz, J., and Kliger, D.**, Time-resolved fluorescence and emission depolarization studies of polystyrene: photochemical processes in polymeric systems. IX, *Macromolecules,* 15, 1372, 1982.
9. **Stevens, B.**, Effects of molecular orientation on fluorescence emission and energy transfer in crystalline aromatic hydrocarbons, *Spectrochim. Acta,* 18, 439, 1962.
10. **Vala, M. T., Haebig, J., and Rice, J. A.**, Experimental study of luminescence and excitation trapping in vinyl polymers, paracyclophanes and related compounds, *J. Chem. Phys.,* 43, 886, 1965.
11. **Birks, J. B.**, *Photophysics of Aromatic Molecules,* Wiley-Interscience, London, 1970, 354.
12. **Hirayama, F.**, Intramolecular excimer formation I. Diphenyl and triphenyl alkanes, *J. Chem. Phys.,* 42, 3163, 1965.
13. **David, C., Putmen-de-Lavarielle, N., and Geuskens, G.**, Luminescence studies in polymers. IV. Effect of orientation, tacticity and crystallinity on polystyrene and polyvinylcarbazole fluorescence, *Eur. Polym. J.,* 10, 617, 1974.
14. **Natta, G., Corradini, P., and Bassi, I. W.**, The crystalline structure of several isotactic polymers of α-olefins, *Rend. Accad. Naz. Linceri. VIII,* 19, 404, 1955.
15. **Yoon, D. Y., Sundararajan, P. R., and Flory, P. J.**, Conformational characteristics of polystyrene, *Macromolecules,* 8, 776, 1975.
16. **Stegen, G. E. and Boyd, R. H.**, Conformational properties of polystyrene, *Polym. Prep.,* 19, 595, 1971.
17. **Fox, R. B., Price, T. R., Cozzens, R. F., and McDonald, J. R.**, Photophysical processes in polymers. IV, Excimer formation in vinylaromatic polymers and compounds, *J. Chem. Phys.,* 57, 534, 1972.
18. **Franck, C. W. and Harrah, L. A.**, Excimer formation in vinyl polymers. II. Rigid solutions of poly (2-vinylnaphthalene) and polystyrene, *J. Chem. Phys.,* 61, 1526, 1974.
19. **Itagaki, H., Horie, K., Mita, I., Washio, H., Tagawa, S., and Tabota, Y.**, Intramolecular excimer formation of oligostyrenes from dimer to tridecamer: the measurements of rate constants for excimer formation, singlet energy migration, and relaxation of internal rotation, *J. Chem. Phys.,* 79, 3996, 1983.
20. **Ishii, T., Handa, T., and Matsunaga, S.**, Effect of molecular weight on excimer formation of polystyrenes in solution, *Macromolecules,* 11, 40, 1978.
21. **Birks, J. B.**, *Photophysics of Aromatic Molecules,* Wiley-Interscience, London, 1970, 521.
22. **Basile, L. J. and Weinreb, A.**, Transfer of excitation energy in solid solutions of anthracene-polystyrene and 9,10-diphenylanthracene-polystyrene, *J. Chem. Phys.,* 33, 1028, 1960.
23. **Förster, Th.**, Zwischenmolekulare energiewanderung und fluoreszenz, *Ann. Physik.,* 2, 55, 1948.
24. **Föster, Th.**, Experimentelle und theoretische untersuchung des zwischennomolekularen übergangs von electronenanregsenerige, *Z. Naturforsch,* 49, 321, 1949.
25. **Dale, R. E. and Eisinger, J.**, Intramolecular energy transfer and molecular conformation, *Proc. Natl. Acad. Sci. U.S.A.,* 73, 271, 1976.
26. **Berlman, I. B.**, *Energy Transfer Parameters of Aromatic Compounds,* Academic Press, New York, 1973.
27. **Stryer, L. and Haugland, R. P.**, Energy transfer: a spectroscopic ruler, *Proc. Natl. Acad. Sci. U.S.A.,* 58, 719, 1967.
28. **Bennett, R. G., Schwenker, R. P., and Kellogg, R. E.**, Radiationless intermolecular energy transfer. II. Triplet-singlet transfer, *J. Chem., Phys.,* 41, 3040, 1964.
29. **Dexter, D. L.**, A theory of sensitized luminescence in solids, *J. Chem. Phys.,* 21, 836, 1953.
30. **Stern, O. and Volmer, M.**, The extinction period of fluorescence, *Phys. Z.,* 20, 183, 1919.
31. **Perrin, F.**, Law governing the dimunition of fluorescent power as a function of concentration, *C. R. Acad. Sci. Ser. C.,* 178, 1978, 1924.

32. **Förster, Th.,** 10th Spiers memorial lecture: transfer mechanisms of electronic excitations, *Discuss. Faraday Soc.,* 27, 7, 1959.

33. **Eisenthal, K. B. and Siegals, S.,** Influence of resonance transfer on luminescence decay, *J. Chem. Phys.,* 41, 652, 1964.

34. **Bennett, R. G.,** Radiationless intermolecular energy transfer. I. Singlet-singlet transfer, *J. Chem. Phys.,* 41, 3037, 1964.

35. **Voltz, R., Laustrait, G., and Coche, A.,** Transfert d'energie dans les liquides aromatiques, *J. Chim. Phys., Phys. Chim. Biol.,* 63, 1253, 1966.

36. **Klein, U. K. A., Frey, R., Hauser, M., and Gössele, U.,** Theoretical and experimental investigations of combined diffusion and long range energy transfer, *Chem. Phys. Lett.,* 41, 139, 1976.

37. **Yokota, M. and Tanimoto, O.,** Effects of diffusion on energy transfer by resonance, *J. Phys. Soc. Jpn.,* 22, 779, 1967.

38. **Guarino, A.,** Electronic energy transfer: a coherent formalism for pulsed and steady state irradiation, *J. Photochem.,* 11, 243, 1979.

39. **Davydov, A. S.,** *Theory of Molecular Excitons,* Plenum Press, New York, 1971.

40. **Burshstein, A. I.,** Energy transfer kinetics in disordered systems, *J. Lumin.,* 34, 167, 1985.

41. **Fredrickson, G. H., Andersen, H. C., and Frank, C. W.,** Electronic excited-state transport and trapping on polymer chains, *Macromolecules,* 17, 54, 1984.

42. **Fredrickson, G. H., Andersen, H. C., and Frank, C. W.,** Electronic excited-state transport on isolated polymer chains, *Macromolecules,* 16, 1456, 1983.

43. **Fredrickson, G. H., Andersen, H. C., and Frank, C. W.,** Electronic excited state transport and trapping as a probe of intramolecular polymer structure, *J. Chem. Phys.,* 79, 3572, 1983.

44. **MacCallum, J. R. and Rudkin, L.,** Nonradiative energy transfer in polyenes, *Nature,* 266, 338, 1967.

45. **Gelles, R. and Frank, C. W.,** Energy migration in the aromatic vinyl polymers. II. Miscible blends of polystyrene with poly (vinyl methyl ether), *Macromolecules,* 15, 741, 1982.

46. **Fitzgibbon, P. D. and Frank, C. W.,** Energy migration in the aromatic vinyl polymers, I. A one-dimensional random walk model, *Macromolecules,* 15, 783, 1982.

47. **Gelles, R. and Frank, C. W.,** Energy migration in the aromatic vinyl polymers. III. Three-dimensional migration in polystyrene/poly(vinyl methyl ether), *Macromolecules,* 15, 747, 1982.

48. **Phillips, D., Roberts, A. J., Rumbles, G., and Soutar, I.,** Transient decay studies of photophysical processes in aromatic polymers. VII. Studies of molecular weight dependence of intramolecular excimer formation in polystyrene and styrene-butadiene block copolymers, *Macromolecules,* 16, 1597, 1983.

49. **David, C., Baeyens-Volant, D., and Geuskens, G.,** Polarization of the fluorescence and phosphorescence of copolymers in glass solution, *Eur. Polym. J.,* 12, 71, 1976.

50. **Reid, R. F. and Soutar, I.,** Intramolecular excimer formation in macromolecules. I. Energy migration and excimer formation in copolymers exhibiting nearest-neighbor excimer interactions, *J. Polym. Sci., Polym. Phys. Ed.,* 16, 231, 1978.

51. **Sienicki, K. and Bojarski, C.,** Studies of the electronic energy migration and trapping in styrene copolymers, *Macromolecules,* 18, 2714, 1985.

52. **Phillips, D., Robert, A. J., and Soutar, I.,** Intramolecular energy transfer, migration, and trapping in polystyrene, *Macromolecules,* 16, 1593, 1983.

53. **Soutar, I., Roberts, A. J., Phillips, D., and Rumbles, G.,** Transient decay studies of photophysical processes in aromatic polymers. VI. Intramolecular excimer formation and dissociation in polystyrene and styrene-methyl methacrylate copolymers, *J. Polym. Sci., Polym. Phys. Ed.,* 20, 1759, 1982.

54. **Yoshida, S., Lillya, C. P., and Vogl, O.,** Functional polymers. XIII. Synthesis and polymerization of 2(2-hydroxy-5-methylphenyl)-5-vinyl-2H-benzotriazole, *J. Polym. Sci., Polym. Chem. Ed.,* 20, 2215, 1982.

55. **O'Connor, D. B., Scott, G. W., Coulter, D. R., Gupta, A., Webb, S. P., Yeh, S. W., and Clark, J. H.,** Direct observation of the excited-state proton transfer and decay kinetics of internally hydrogen-bonded photostabilizers in copolymer films, *Chem. Phys. Lett.,* 121, 417, 1985.

56. **Coulter, D. R., Gupta, A., Yavrouian, A., Scott, G. W., O'Connor, D. B., Vogl, O., and Li, S.-C.,** Electronic energy transfer and quenching in copolymers of styrene and 2-(2'-hydroxy-5' vinylphenyl)-2H-benzotriazole: photochemical processes in polymeric systems. X, *Macromolecules,* 19, 1227, 1986.

57. **Gupta, A., Scott, G. W., and Kliger, D.,** *Mechanisms of Photodegradation of Ultraviolet Stabilizers and Stabilized Polymers,* (Am. Chem. Soc., Symp. Ser., Vol. 151), American Chemical Society, Washington D.C., 1980, chap. 3.

58. **Huston, A. L. and Scott, G. W.,** Picosecond kinetics of excited state decay processes in internally hydrogen-bonded polymer photostabilizers, *Proc. Soc. Photo-Opt. Instrum. Eng.,* 322, 215, 1982.

59. **Huston, A. L., Scott, G. W., and Gupta, A.,** Mechanism and kinetics of excited-state relaxation in internally hydrogen-bonded molecules: 2-(2'-hydroxy-5' methylphenyl)-benzotriazole in solution, *J. Chem. Phys.,* 76, 4978, 1982.

60. **Huston, A. L., Merritt, C. D., Scott, G. W., and Gupta, A.,** Excited-state absorption spectra and decay mechanisms in organic photostabilizers, in *Picosecond Phenomena,* Vol. 2, Shank, C. V., Hochstrasser, R. M., and Kaiser, W., Eds., Springer-Verlag, Berlin, 1980, 232.

61. **Huston, A. L. and Scott, G. W.,** Spectroscopic and kinetic investigations of internally hydrogen-bonded (hydroxyphenyl)benzotriazoles, *J. Phys. Chem.,* 91, 1408, 1987.

62. **Lee, M. Yardley, J. T., and Hochstrasser, R. M.,** Dependence of intramolecular proton transfer on solvent friction, *J. Phys. Chem.,* 91, 4621, 1987.

63. **Otterstedt, J. A.,** Photostability and molecular structure, *J. Chem. Phys.,* 53, 5716, 1973.

64. **Werner, T.,** Triplet deactivation in benzotriazole-type ultraviolet stabilizers, *J. Phys. Chem.* 83, 320, 1979.

65. **Woessner, G., Goeller, G., Kollat, P., Stezowski, J. J., Hauser, M., Klein, U. K. A., and Kramer, H. E. A.,** Photophysical and photochemical deactivation processes of ultraviolet stabilizers of the (2-hydroxyphenyl)benzotriazole class, *J. Phys. Chem.* 88, 5544, 1984.

66. **Huston, A. L.,** Mechanism and Kinetics of Excited-State Relaxation in Internally Hydrogen-Bonded Aromatic Molecules, Ph.D. dissertation, University of California, Riverside, 1983.

67. **Flom, S. R. and Barbara, P. F.,** The photodynamics of 2-(2'-hydroxy-5' -methylphenyl)-benzotriazole in low temperature organic glasses, *Chem. Phys. Lett.,* 94, 488, 1983.

68. **Polacki, Z.,** Some remarks on donor properties of polystyrene, *J. Photochem.,* 28, 135, 1985.

Chapter 9

FLASH PHOTOLYSIS ELECTRON SPIN RESONANCE AND ELECTRON POLARIZATION

K. A. McLauchlan

TABLE OF CONTENTS

I.   Introduction..260

II.   Experimental Methods...261

III.   Flash-Photolysis ESR and Magnetic Resonance...............................268
        A.   Time Integration of Transient Signals.............................274

IV.   Chemically Induced Dynamic Electron Polarization (CIDEP)...................276
        A.   An Overview...276
        B.   The Triplet Mechanism...278
        C.   The Radical Pair Mechanism..282
                1.   The Patterns of Geminate RPM Polarization.................287
                2.   Polarization in F Pairs...................................288
                3. ST_{-1} RPM Polarization................................290
                4.   Polarization When Motion is Restricted....................292
        D.   Secondary Polarization..292
        E.   Measurement of Polarization Ratios................................293

V.   Calculation of CIDEP Spectra..293

VI.   CIDEP in Practice — The Photolysis of Maleimide Solutions.................293

VII.   The Effects of Exchange in Spin-Polarized Radicals........................296

VIII.   Concluding Remarks..300

References..301

I. INTRODUCTION

Flash photolysis methods have long proved valuable to the study of free radicals in solution, including those involved in radical-initiated polymerizations. Yet in their common form, using UV optical detection, they suffer from the severe limitation of the low resolution associated with UV spectra in solution. Although they can be very fast they can identify only a chromophore that a radical contains, and not the actual radical itself. What is needed is a high-resolution detection method which can be applied to the highly transient species created by flash photolysis.

The established technique with the highest resolution for the direct observation of stable free radicals is electron spin resonance (ESR). This depends upon the excitation of transitions between the states of the unpaired electron, made nondegenerate by the application of a magnetic field, B. Each radical possesses its own characteristic ESR spectrum with its own g-value and its own hyperfine structure due to through-bond coupling (with coupling constant a) to nearby nuclei. In the fields normally used (about 3300 gauss) the transitions largely satisfy the equation

$$\nu = g\mu_B B + \sum_i a_i m_i \tag{1}$$

where ν is their frequency, μ_B the Bohr magneton, and m_i the nuclear magnetic quantum number of nucleus i. In such fields $\nu \sim 9.6$ GHz, in the microwave region, and resonance is almost always approached by keeping this constant and changing the field of the spectrometer.

For a nucleus of spin I there exist $(2I + 1)$ values of m_i ranging to positive and negative values about zero. The simple (first-order) spectrum of an electron coupling to n-equivalent spin-$^1/_2$ nuclei consequently consists of $(n + 1)$ lines half of which originate in radicals with positive, and half in negative, values of m_i; this is significant later. The relative intensities of the lines are in the ratios of the coefficients in the expansion of $(1 + x)^n$, a result simply of degeneracies. The absolute intensities of radicals at thermal equilibrium with their surroundings are determined by the Maxwell-Boltzmann distribution. These simple rules usually allow direct radical identification, although the transitions become further split by second-order effects when the second term in Equation 1 becomes appreciable compared with the first. This is observed commonly in alkyl radicals, whereas in phosphorus-centered ones, for example, even the g-value can no longer be measured directly from the spectrum without computation.

Hyperfine coupling in these latter types of radicals causes the spectrum to extend for up to 1 to 200 G for organic radicals and over 1000 G for P-centered ones. This implies that the magnetic field of the spectrometer must be changed by a significant fraction of its overall value to display it. This is accomplished by the use of electromagnets, and varying the current through their coils, or by using separate sweep coils. In either case the inductive nature of the electrical coils prevents the current through them being changed very quickly and makes the observation of highly transient radicals impossible. In consequence past ESR studies have been limited to stable radicals or those created continuously to yield steady-state concentrations, implying radical lifetimes of 1 ms or greater. Truly transient species have been studied indirectly by trapping them to form stable radicals, but this raises difficulties in identification of the primary radical and in possible selective scavenging. As a consequence, the application of ESR spectroscopy to the direct observation of the radicals produced by flash photolysis has depended upon the development of new methods (described below). These can be used to yield complete spectra for radical identification at times to within 20 ns of radical creation. Whereas this is slower than the fastest optical methods it is an optimal period for observing free radicals rather than their precursors, although the nature of these may also be deduced, as described below.

The advent of the new methods produced a surprise in that radicals observed within a few microseconds of their creation almost invariably exhibit ESR spectra of unusual appearance. Although the line positions are exactly as expected, i.e., the g- and a-values are unchanged, the intensities deviate from those expected for thermally equilibrated radicals. Whole spectra appear in enhanced absorption (A) or in emission (E), while in others some of the hyperfine lines appear in each phase. Spectra of the second type had in fact been observed previously from steady-state studies on H· atoms.[1] It is apparent that the hyperfine states of the radicals acquire nonequilibrium populations, a phenomenon known as chemically induced dynamic electron polarization (CIDEP).[2,3] Its importance is twofold. First, the increase in absolute intensities of the lines as compared with those from similar concentrations of thermally equilibrated radicals facilies radical detection at low concentration. Second, the phases of the polarized signals allow the multiplicity of the radical precursor to be deduced directly from the spectrum with no necessity for further experimentation. CIDEP causes ESR spectra to have unusual appearances and their analysis to yield molecular information depends upon an understanding of the origins of CIDEP and their effects. This constitutes a major part of this paper.

Polarization theory allows the intensities of ESR lines to be calculated while the g- and a-values yield their positions. To calculate spectra for comparison with experimental ones requires one further step, the calculation of the line shape. This is obtained conveniently from an analysis based upon the Bloch equations. It is shown below that when continuous wave (CW) methods are used to display the spectra of polarized radicals the lines do not show the Lorentzian line shapes associated with nonsaturated equilibrated ones, and they depend upon several molecular and experimental parameters. The line shapes change continuously in time after the creation of polarized radicals in a way without analogy in conventional spectra.[4] Some very remarkable shapes occur when initially emissive lines change their phase in time. All of these variations must be understood to analyze observed spectra without error.

Further complications in the line shapes and time dependencies of polarized lines arise if the radicals undergo electron, proton, or site exchange processes. The application of magnetic resonance techniques to their study is well established for normal radicals and it is shown here that they may be extended to study highly transient species. New phenomena occur in the time dependence of radicals undergoing slow electron exchange whilst polarization transfer effects allow detection of electron exchange processes at very slow rates, slower than with previous CW ESR methods.

II. EXPERIMENTAL METHODS

Several problems, of both fundamental and technical nature, arise in obtaining the ESR spectrum of a highly transient free radical. The first is that the radical may be identified only if its hyperfine structure can be resolved, i.e., if the line width is exceeded by the splittings due to hyperfine coupling. The controlling factor here is the Uncertainty Principle which suggests that the line width should be of the order of 1 G (the typical size of small couplings) within about 10 ns of radical creation, although in a CW experiment it decreases for about 1 μs thereafter (see below). The lower limit sets the design criterion for the response time of the spectrometer and immediately implies that the radicals should be produced with as short a photolysis flash as possible. Currently this is the 10- to 20-ns duration output from an excimer laser, which can be considered instantaneous on the microsecond time scale without great error.

The next problem is to ensure that the detection system of the spectrometer can respond to a signal which varies on this time scale. This is not the case for a conventional spectrometer which utilizes field modulation at 100 KHz and phase-sensitive detection of the resulting signal, where the narrow frequency band limits the response time to about 40 μs. As a consequence, transient ESR spectrometers do not use field modulation but rather rely upon direct detection of the signal in a broad-band system with a short response time. The signals are not displayed consequently in the derivative form familiar in normal ESR studies and any changes of phase in the signal have fundamental, and not methodological, origins. Broad banding a spectrometer in this way decreases the signal-to-noise ratio (S/N) of the spectrum as compared to one operating with field modulation, but this is largely recovered by use of pulse-correlated detection methods and signal-averaging techniques. The radicals are created inside a resonant microwave cavity (usually operating in the TE102 mode) and the microwave signals are rectified with a detector crystal whose output is applied to a wide-band preamplifier and amplifier before being input to either a digital (transient recorder) or analogue (boxcar averager) fast recording device. All of these units have their own characteristic frequency response, band width, and response time.

The characteristics of a resonant circuit element are usually summarized in terms of a quality factor (Q-factor) defined, for the cavity, for example, as

$$Q = 2\pi \text{ (energy stored in the cavity)/(energy dissipated per cycle of radiation)}$$

A cavity with a high Q-factor has a narrow, high-frequency response leading to good S/N but, through the narrow band width, poor response time. In conventional ESR spectrometers a TE102 cavity is normally used with Q ~ 4500 and slow response but it happens that it can be detuned to a value of Q ~ 800 when the response time is approximately 10 ns, the target figure. In a well-designed transient system it is the cavity which limits the response time of the detection system, all subsequent units having larger band widths. The concomitant limitation in S/N necessitates use of low noise components, particularly at the start of the detection chain. It is common to use band-pass filters to limit the band widths whenever the full time resolution of the technique is not required. The recording device is triggered from the laser pulse circuitry so that it receives input for only a brief period, typically a few microseconds, following the photolysis flash, thereby limiting the noise input. In practice it is usually used in a signal-averaging mode to improve the S/N further (see below).

Having ensured that the radicals can be created quickly enough and that the spectrometer can respond to the resulting signals on the time scale required, there remains the problem as to how to sweep the magnetic field to display the spectrum on the same time scale. The answer is not to try to but rather to discover new ways of displaying it. An attractive possibility, not yet really practical for transient radicals, is to keep the field constant and apply a powerful short microwave pulse to excite the entire spectrum, with subsequent Fourier transformation of the resulting free-induction decay. This is technically a much more difficult experiment that the equivalent one in nuclear magnetic resonance (NMR) spectroscopy due to the resonant frequency being in the microwave region and to the free-induction decay being very short lived. In consequence the methods which have been developed are based upon the idea that the radicals are produced not once, in a single flash, for their spectrum to be recorded but repetitively in a number of flashes that occur at different values of the magnetic field.[5,6] Following each flash the signal is sampled either essentially continuously, as in the two-dimensional technique[7] or for some specific period, as in the time-integration spectroscopy (TIS)[8] and boxcar methods.[9] The spectrum at any specific time following the flash is reconstructed form the data obtained from several thousand separate creations of the radical. These techniques place two further stringent requirements on the apparatus besides the obvious necessity to flow the solution through the irradiation region to minimize depletion of the photoactive material.

First, it is imperative that the intensity of the photolysis flash remains essentially constant throughout the period required to record the spectrum, which at present varies from about 20 min to 2 h depending upon the technique used and the field resolution required. A well-designed and well-maintained excimer laser has a 2% intensity variation between successive pulses but this easily degrades to 10% in practice. The effects of these random pulse-to-pulse variations can be eliminated largely by signal averaging the effects of several pulses at the same magnetic field position, and the lasers are run at typically 20 Hz. This figure is determined, in techniques which involve digital data recording, by the need to transfer the data from the primary transient recorder data store into a backing computer, often with additional data manipulation. Signal averaging of this form cannot however compensate for the characteristic fall off in the intensity of an excimer laser with time. For this reason in the latest experiments a laser has been used with continuous output monitoring with feedback so as to maintain the long-term average output constant.

Second, the magnetic field of the spectrometer must be extremely stable. This is accomplished usually by using a Hall-effect probe to monitor and control it via feedback to the power supply, although when the highest field resolution is required in the spectrum a separate probe is used to operate a field/frequency lock. The change in the field needed to display the spectrum can be performed either continuously, at so slow a rate that the field essentially does not change during data acquisition following each pulse (although this is often compromised in practice), or in discrete steps, taking advantage of the digital field sweep available on Bruker spectrometers. It so happens that all the experiments which have involved the use of boxcar detectors, either for direct-detection or spin-echo applications, have been performed with continuous sweep, while two-dimensional and TIS studies have used discrete steps. The difference is not fundamental in that the former techniques could be performed in the discrete mode. This does have the advantage that the data is acquired at a fixed value of the field and then the experiment is open to full theoretical interpretation.

If the magnetic field is such that when a photolysis flash occurs the radical produced has a hyperfine transition on resonance, a signal is detected which decays in time. It decays first due to relaxation, for as mentioned above, the radical usually is spin polarized, and second due to reaction. At the normal radical concentrations involved ($10^{-5} M$ in the Oxford experiments) the former is much faster than the latter and completely dominates the early decay behavior. In the two-dimensional transient ESR experiment[7] this decay curve is sampled at 100 MHz and stored in the memory of a backing computer which then cancels the store contents of the transient recorder. The next flash produces another decay curve which is added in the memory to the first, a process continued for typically 128 flashes for signal averaging. At this stage the averaged curve is output, together with information on the field position and a scaling factor to allow comparison with subsequent data, to a second computer where the data is stored at a chosen location on a Winchester disc. The magnetic field of the spectrometer is then incremented under software control and the whole cycle is repeated. This continues for at least 512 field positions. With each decay curve containing 2048 channels of 16-bit data this corresponds to about 2 Mb of information stored on the disc. As presently configured this takes about 2 h of running time but this is due to inefficient use of the equipment and is being reduced to 10 min. The store contains a matrix of intensity, time, and field position which allows the decay curves to be output side-by-side in order of the field values at which they were obtained.

A two-dimensional plot of this type is shown in Figure 1. It gives a particularly clear illustration of the different phases of the signals due to the CIDEP phenomenon and of the time dependence of each line. From the surface the conventional spectrum at any given time after radical creation can be obtained by taking a cross-section perpendicular to the time axis. This method is unique among the transient ESR ones in preserving all the information from the radical system from the time the radical is created until it disappears as a result of reaction, and the spectrum at any time can be recovered directly from the data base, off-line from the spectrometer. At a given temperature and concentration there is no need to repeat the experiment to investigate the spectrum variation in time, as there is with TIS and

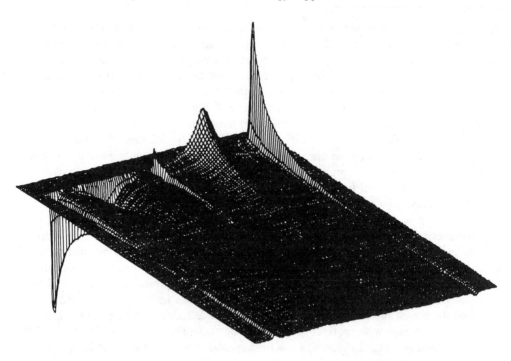

FIGURE 1.   A two-dimensional ESR spectrum of cyclohexyl-l-ol radicals created by photolysis of cyclohexanone in cyclohexanol solution. The surface is composed of 512 individual decay curves stacked side-by-side in the order of the field values at which they were obtained. The lines exhibit alternate line broadening due to modulation of the β-ring couplings by ring-flipping between degenerate conformers. The spectrum exhibits CIDEP due to the ST_0 RPM mechanism, in E/A phase. (From McLauchlan, K. A. and Stevens, D. G., *J. Chem. Soc. Faraday Trans. 1*, 83, 29, 1987. With permission.)

boxcar methods. This has several advantages: less material is needed for investigation, fewer laser pulses are needed overall (and laser fills and components are expensive), and the spectrometer becomes available for further experimentation. The conventional spectrum extracted as described above is however essentially point sampled and prone to the effects of noise variations which limit the period for which a satisfactory spectrum can be extracted as the signal falls in time. This can be overcome by sampling each decay curve not at one point of time but over a range of times and adding the typically 100 samples in a 1-μs period together. In this way a satisfactory spectrum can be obtained long after the point-sampled signals have decayed into the noise level, in favorable cases up to 200 μs after the photolysis flash (Figure 2). This principle is the basis of the TIS method too and, in analogue from, of boxcar methods, in both of which this summation/averaging is performed on-line.

The surface shown in Figure 1 is in fact the result of some data manipulation besides the technique described above. In the first place each decay curve has been drawn in a way which shows only pulse-correlated information. The experiment is performed in a fashion which discriminates against any stable radical present or against a steady-state concentration of radicals set up at the 20-Hz photolysis rate. In this technique, and in the TIS one below, this is accomplished by recording some baseline signal before the flash occurs and subtracting its level from that of the subsequent decay curve. This is valid except where very long periods are used for recording the curve. A similar result is achieved by using dual boxcars[10] to measure the two relevant parts of the curve and subtracting the output of one from the other. The raw output from the transient recorder actually shows a decay curve of the type discussed above superimposed upon a much more slowly changing background signal which is observed even when the magnetic field of the spectrometer is set to a position where there is no ESR signal. Its origin lies in the fact that each laser pulse dumps typically 50 mJ of

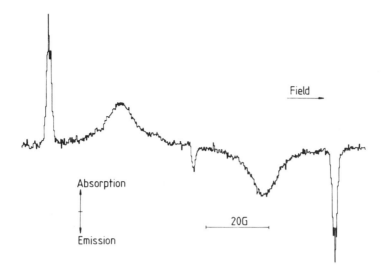

FIGURE 2. A time integration spectroscopy (TIS) spectrum of the cyclohexyl-1-ol radical obtained with sampling and summation between 160.0 and 175.2 μs after radical creation, after the point-sampled spectrum has disappeared into the noise level. I exhibits A/E polarization whose origin is not fully understood and may be spurious (see text). (From McLauchlan, K. A. and Stevens, D. G., *J. Chem. Soc. Faraday Trans. 1, 83,* 29, 1987. With permission.)

energy into a resonant cavity which normally operates at microjoules of microwave energy. This causes the sample and the cavity to warm and unbalances the microwave system. There is a further short-lived contribution to the off-resonance signal which originates in the biphotonic release of photoelectrons as the laser beam hits the cavity surfaces. The overall off-resonance signal is completely reproducible at constant laser-pulse energy and it is subtracted from all the decay curves obtained to display the spectrum.

Earlier methods of transient ESR spectroscopy based upon TIS and boxcar techniques are more widely used than two-dimensional techniques which require more sophisticated equipment and software. They yield the ESR spectrum at some particular time after the flash and are sampling methods in two senses: not only is the signal sampled at a particular field value but also the decay curve is itself sampled over a chosen period after the flash. In time-integration spectroscopy the decay curve is, as above, stored in a transient recorder and transferred between photolysis flashes into a dedicated microcomputer. Here, under software control, a period is chosen over which the signal is summed digitally before this integrand is output to a store location corresponding to a given magnetic field value. The decay curve is then destroyed before the magnetic field is changed and the process repeated. A plot of integrand vs. field position constitutes the ESR spectrum. This method requires little computer memory and is fast in operation but has the disadvantage, mentioned above, that the entire field sweep with its associated thousands of laser pulses must be repeated in order to obtain the spectrum at a different time after the photolysis flash. Time integration has two purposes. One is to improve the S/N of the spectrum, as indicated above, but the other is more subtle. It will be shown below that under a wide range of operating conditions the ESR signal from a transient radical oscillates in time (the Torrey oscillations[11]). The original principle in designing the TIS method was to sum the signal over the period of these oscillations so as to tend to average them out. This is extremely important in practice since the onset of oscillations in the time-domain signal leads to spurious side bands in the ESR spectrum. The occurrence of these oscillations, and their frequency, depends *inter alia* upon the offset from the resonance position and the microwave field strength at the sample. In obtaining Figure 1 this was set at too low a level to cause them but it should be realized that they are a normal feature of point-sampled spectra and can lead to spectral misinterpretation.

In boxcar methods the digital sampling described above is replaced by an analogue method.[9,10] The signal is input into a boxcar averager with which the part of the decay curve required is selected and this part is applied to a capacitor as part of a sample and hold circuit. The charge stored by this depends upon the amplitude of the signal and the sample period and, in an experiment involving repetitive laser flashes, it reaches an asymptotic value in a time related to the short time constant of the input circuitry. The voltage across the capacitor is continuously read out through a larger time constant as the field is swept and this time constant is the origin of the improved S/N of such a device. As opposed to the TIS experiment where sampling is at 100 MHz here it is continuous and the absolute S/N may be greater. Furthermore, the boxcar experiment is extremely simple to set up: besides being broad banded a conventional ESR spectrometer requires only a photolysis source and a boxcar averager to yield spectra of transient radicals. There are however two disadvantages. First, as presently performed, the field is swept continuously so that while the spectrum is being obtained the offset of the spectrometer from resonance changes continuously. This implies that the frequency of the Torrey oscillations changes and while these are removed as in TIS the line shape becomes more complex, making detailed spectral analysis more difficult. This could be overcome by using possibly impractically slow sweeps or a stepwise sweep, where the full theory established for the TIS experiment could be used. This aspect of the boxcar experiment requires careful attention in those applications, such as exchange studies (see below), where line-shape analysis must be performed. The second disadvantage of using the boxcar is that at no point is a complete decay curve recorded, only a part of it being sampled and stored. It cannot therefore be used in two-dimensional work.

The three related techniques above all involve use of CW microwave radiation but there is a further method, spin-echo spectroscopy,[12] in which the microwaves are pulsed. In this experiment two pulses are used. An initial strong pulse of as short a duration as possible (currently about 30 ns) is applied to rotate the magnetization created by the photolysis flash in the direction of the field into the perpendicular measurement direction (a 90° pulse). It is followed at a time interval τ (typically of 0.2- to 1.0-μs duration) by a second longer pulse which inverts the magnetization completely in this direction (a 180° pulse) and at a time 2τ a spin echo is observed. This is input into a boxcar averager and the spectrum is obtained by outputting the signal from this, measured from the echoes produced at successive radical creations, as the field is swept continuously. This method allows observation on the time scale of the first pulse and is a little faster than the other techniques in which the microwave power level is less and it takes longer to rotate the magnetization so as to obtain a component in the measurement direction. This is one advantage of the spin-echo technique, the other being that a delay is introduced between the time that the radicals are produced and the time at which they are observed. This obviates difficulties from short-term noise components associated with the photolysis flash.

However, there are disadvantages to this method, some of which are of a fundamental nature.[7] First, the ESR lines of small radicals in solution are usually homogeneously broadened and the echo amplitude is consequently small and may be overlapped by the free-induction decay of the magnetization. This can be overcome at the expense of loss of resolution by deliberate degradation of the magnetic field homogeneity of the spectrometer. Second, the echo amplitude falls in time after the first pulse exponentially with a time constant T_2, the spin-spin relaxation time of the system, which in normal liquids is often 1 μs or less. Third, the echo itself essentially consists of two free-induction decays back-to-back and has a width related to T_2. In practice this has led to boxcar sampling periods of only a few hundred nanoseconds. All of these factors limit the S/N attainable and spin-echo measurements have provided useful only in the first few microseconds of the existence of

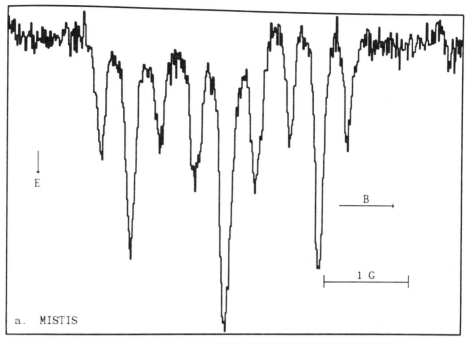

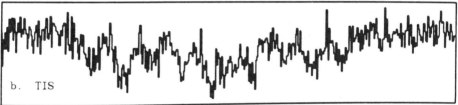

FIGURE 3. The MISTIS and TIS spectra of the radical trianion from benzene-2,3,5,6-tetra-carboxylic acid, both observed with time integration between 34.0 to 35.0 μs after the flash. In the former, the microwaves were applied 31.5 μs after the flash while in the latter they were applied continuously. The considerable increase in the S/N of the MISTIS spectrum as compared with the TIS one results from the short spin-spin relaxation time, T_2, in the system.

the radical. This limits the application of the method as compared with CW methods although there is no doubt that the spin-echo technique has important use at early times and in the accurate determination of the spin-lattice relaxation time T_1 and T_2. As opposed to the CW techniques it also requires the building or purchase of a special pulse spectrometer.

One further transient ESR method should be mentioned, microwave switched time-integration spectroscopy (MISTIS), which is a mixture of CW and pulse methods.[7] The experiment consists simply in interposing a delay between the time that a radical is created in the photolysis flash and the microwaves are switched on to observe it. As opposed to the spin-echo method only the low levels of microwave field strength associated with normal ESR spectrometers are switched, and no precautions are needed to safeguard the detector crystal from its effect. What this experiment accomplishes is a considerable improvement in S/N of the spectrum obtained somewhat after radical creation as compared with the conventional CW methods (Figure 3). It works because in these experiments the signal magnitude depends to a large extent on the phase coherence in the transverse magnetization set up by the microwave field, a coherence which diminishes with the rate constant T_2^{-1}. This implies that the signal observed is the stronger the closer in time the microwaves are applied to the observation time required. It is effective in initially spin-polarized systems rather than those in which polarization occurs during the lifetime of the radical.

III. FLASH-PHOTOLYSIS ESR AND MAGNETIC RESONANCE

As mentioned in Section I the ESR specra of transient radicals observed using the new time-resolved methods almost invariably exhibit CIDEP. The polarization serves to accentuate a feature of magnetic resonance in that the signal intensity is not necessarily a direct measure of radical concentration. In this way it differs from optical detection methods used under normal, unsaturated, conditions. It is consequently necessary to consider the experiment in the magnetic resonance context and to derive expressions for the variation of the signal height at a given magnetic field value in time. This leads to an understanding of the Torrey oscillations and to the possibility of extracting relaxation times from the observations. From calculation of these decay curves as a function of offset from resonance the line shapes pertaining to both point-sampled and TIS spectra can be computed in turn. Our approach will be to take the spin polarization as phenomenological at this stage and we shall consider only that initial polarization which is created before the radicals are first observed, P_I. It is sufficient here to realize it may be typically 10 to 100 P_{eq} in practice, where P_{eq} is the polarization at thermal equilibrium. These quantities are defined in terms of the generalized polarization

$$P = \{(n_\beta - n_\alpha)/(n_\beta + n_\alpha)\} \qquad (2)$$

where n_β and n_α are the populations of the lower ($m_s = -\frac{1}{2}$) and upper ($m_s = +\frac{1}{2}$) electronic states, respectively. P may depend upon the hyperfine state of the radical, that is, each hyperfine component in the spectrum may have its own characteristic value. This point, the origins and types of spin polarization, and their effects on observed spectra are discussed below.

Flash photolysis ESR is a unique magnetic resonance experiment in that the magnetization is created essentially instantaneously in the photolysis flash, and it is to be expected that the observations will depend upon the pulse response of the system. There is a close analogy with a pulsed NMR or ESR experiment in which the response of a sample at thermal equilibrium is observed when a sudden resonance field is applied. Either situation is analyzed conveniently in terms of the classic Bloch equations. It should be realized that when the radicals are created in the flash-photolysis experiment their spins are aligned in the direction of the magnetic field of the spectrometer, the conventional z-direction, to yield a zero-time bulk magnetization $M_z(0)$. However, in any magnetic resonance experiment a component of the magnetization has to be obtained in a direction perpendicular to this for observation, and this is accomplished by applying a fluctuating magnetic field at right angles to the applied field (Figure 4). This fluctuating field causes $M_z(t)$ to deviate from its value at time t and creates components $M_x(t)$ and $M_y(t)$ in the perpendicular plane. The Bloch equations are simply the coupled differential equations which relate the three components. They are most conveniently written in a coordinate system which rotates at the angular frequency of the applied radiation:

$$\underline{M}_x(t) = -T_2^{-1} \cdot \underline{M}_x(t) + \Delta\omega \cdot \underline{M}_y(t) \qquad (3)$$

$$\underline{M}_y(t) = -\Delta\omega \cdot \underline{M}_x(t) - T_2^{-1} \cdot \underline{M}_y(t) + \omega_1 \cdot \underline{M}_z(t) \qquad (4)$$

$$\underline{M}_z(t) = -\omega_1 \cdot \underline{M}_y(t) - T_1^{-1} \cdot \underline{M}_z(t) + n \cdot P_{eq} \qquad (5)$$

or in matrix form,[11,14,15]

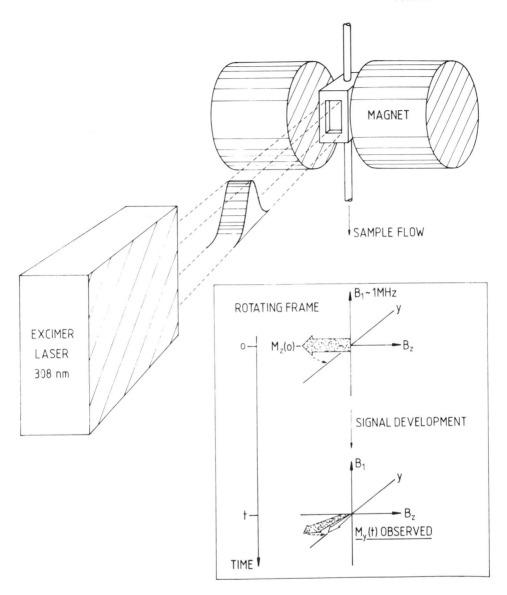

FIGURE 4. A schematic diagram of the flash-photolysis experiment showing that the radicals are created with their spins aligned to the applied field of the spectrometer to yield a finite value of the bulk magnetization at this time in its direction ($M_z(0)$). This has to be rotated by applying a resonant microwave field, B_1, perpendicular to the main magnetic field so as to yield a component, at a later time, $M_y(t)$ in the measurement direction. This process is shown in a frame rotating at the resonance frequency in the figure. The observed signal is proportional to this component.

$$\underline{M}(t) = L\underline{M}(t) + T_1^{-1} \cdot \underline{M}_{eq} \tag{6}$$

where

$$L = \begin{bmatrix} -T_2^{-1} & \Delta\omega & 0 \\ -\Delta\omega & -T_2^{-1} & \omega_1 \\ 0 & -\Delta\omega & -T_1^{-1} \end{bmatrix} \tag{7}$$

$$\underline{M}(t) = \begin{bmatrix} \underline{M}_x(t) \\ \underline{M}_y(t) \\ \underline{M}_z(t) \end{bmatrix} \quad \text{and} \quad \underline{M}_{eq} = n \cdot P_{eq} \begin{bmatrix} 0 \\ 0 \\ 1 \end{bmatrix} \tag{8}$$

Here T_1 and T_2 are the spin-lattice and spin-spin relaxation times, $\Delta\omega$ is the offset of the applied radiation frequency from resonance in angular frequency units, ω_1 is the incident microwave power in angular frequency units, and n is the concentration of free radicals in the system. It happens that at normal radical concentrations in the flash-photolysis experiment (10^{-6} to 10^{-4} M) the early-time behavior of the magnetization is dominated by relaxation effects which occur on the microsecond time scale. In consequence n can often be taken as constant throughout the observation period. When this is not so the effect of radical decay by chemical reaction can be incorporated by adding a further term to Equation 6:[13,15]

$$- \frac{\dot{n}(t)}{n(t)} \underline{M}(t) \tag{9}$$

This comes from the assumption that, although the concentration of the radicals is varying continuously, $M_z(t)$ always relaxes towards the instantaneous equilibrium value, $n(t) \cdot P_{eq}$.

It remains to introduce the effects of the radicals being formed with initial spin polarization. since this occurs rapidly within the time scale of the experiment it can be introduced simply as an initial boundary condition in the magnetization:

$$\underline{M}(o) = n(o) \cdot P_I \begin{bmatrix} 0 \\ 0 \\ 1 \end{bmatrix} \tag{10}$$

where n(0) is the concentration of radicals created in the photolysis flash. It will be seen below that not all polarization is of this initial variety but that it can be generated throughout the lifetime of reactive radicals in the system. This needs a further magnetization-generation term to be added to Equation 6.

The matrix equation, including the radical-decay term, is a first-order linear ordinary differential equation with constant coefficients and has the general solution[16]

$$\underline{M}(t) = \exp\left[\int_0^t L'(t)dt\right]\left\{\underline{M}(o) + \int_0^t \exp\left[-\int_0^t L'(t)dt\right]Q(t)dt\right\} \tag{11}$$

with $\quad Q(t) = T_I^{-1} \cdot \underline{M}_{eq} \quad$ and $\quad L'(t) = L - \frac{n(t)}{n(t)} \cdot \mathbf{1}$

where $\mathbf{1}$ is the unit matrix. This can be evaluated using either Laplace transform[15,17] of eigenvalue[15] techniques. The equation can be solved analytically only in some limiting cases but these are useful in illustrating the type of behavior which may be expected.

The signal observed is proportional to the y-component of the magnetization which when relaxation is much faster than reaction can be expressed as

$$\underline{M}_y(t) = [P_I \cdot g_y(t) + P_{eq}T_I^{-1} \cdot G_y(t)]n(t) \tag{12}$$

where
$$G_y(t) = \int_0^t g_y(t)dt \tag{13}$$

and $g_y(t)$ is a function of T_1, T_2, ω_1 and $\Delta\omega$:

$$g_y(t) = \frac{\omega_1 \exp(at)}{b} \sin(bt) \tag{14}$$

giving
$$G_y(t) = \frac{\omega_1}{T_1(a^2 + b^2)} \left[\left[\frac{a}{b} \sin(bt) - \cos(bt) \right] \exp(at) + 1 \right] \tag{15}$$

The parameters a and b have differing values in the various limiting cases:

1. On resonance

$$a = -\tfrac{1}{2}(T_1 + T_2), \qquad b = [\omega_1^2 - \tfrac{1}{4}(T_2^{-1} - T_1^{-1})^2]^{1/2}$$

2. On resonance, and $T_1 = T_2$

$$a = -T_1^{-1}, \qquad b = \omega_1$$

3. Off resonance and $T_1 = T_2$

$$a = -T_1^{-1}, \qquad b = (\omega_1^2 + \Delta\omega^2)^{1/2}$$

The latter two cases show that at high enough microwave powers the observed decay curve will show oscillations in the time domain, exactly analogous to the Torrey wiggles, and the oscillations will be at their lowest frequency exactly on resonance. Here the microwave field strength can be measured directly from the oscillation frequency. It is notable too that even under these limiting conditions the signal does not decay with a single exponential time constant, but is in general biexponential.[17] Considerable care must be taken therefore in attempting to extract the spin-lattice relaxation time from the decay curve although it is instinctively tempting to do so.

When $T_1 \gg T_2$ another limiting-case analysis shows that the magnetization does decay with a single exponential time constant, T_1^{eff}, given by

$$\frac{1}{T_1^{eff}} = \frac{1}{T_1} + \frac{\omega_1^2 T_2}{1 + \Delta\omega^2 T_2^2} \tag{16}$$

Even here, and on-resonance, observed values of T_1^{eff} must be extrapolated to zero microwave power to extract the true value of T_1. In practice, CW techniques used on transient polarized radicals yield T_1 values accurate to $\pm 10\%$ at best. More accurate values can be obtained using spin-echo methods or the microwave-switched time-integration (MISTI) technique.[19]

The MISTI technique is closely related to a similar one used in the solid state.[20] It consists in interposing a delay between the time the radicals are created and the time that the microwaves are turned on to observe them. The resulting decay curves are obtained as a function of the delay time and are integrated between chosen limits to improve the S/N ratio. The variation of this integrand with delay time occurs in a single-exponential fashion with the time constant T_1. This is independent of the actual form of the decay curve, and the spectrometer can be operated under the conditions of optimum S/N.[19]

Although the analytical solutions to the Bloch equations are informative in their comparative simplicity, more general ones can also be obtained semianalytically and with fast numerical techniques these are the ones used in practice. Under the condition of slow radical decay it may be shown that[15,18,21]

$$g_y(t) = \sum_{i=1}^{3} A_{iy}\exp(\lambda_i t) \tag{17}$$

where

$$A_{1y} = \omega_1(\lambda_1 + T_2^{-1})/(\lambda_1 - \lambda_2)(\lambda_1 - \lambda_3) \tag{18}$$

and similarly for A_{2y} and A_{3y} by cyclic permutation.

$$\lambda_1 = \alpha_+ + \alpha_- - \Theta \tag{19}$$

$$\lambda_2 = \beta\alpha_+ + \beta^*\alpha_- - \Theta \tag{20}$$

$$\lambda_3 = \beta^*\alpha_+ + \beta\alpha_- - \Theta \tag{21}$$

$$\alpha_\pm = \left[\frac{-q \pm (q^2 + (4/27)p^3)^{1/2}}{2} \right]^{1/3} \tag{22}$$

$$\beta = -\frac{1}{2} + \frac{\sqrt{3}}{2}i \tag{23}$$

$$\Theta = \frac{1}{3}(T_1^{-1} + 2T_2^{-1}) \tag{24}$$

$$p = -\frac{\delta^2}{3} + \omega_1^2 + \Delta\omega^2 \tag{25}$$

$$q = -\frac{2}{27}\delta^3 + \frac{1}{3}\omega_1^2\delta - \frac{2}{3}\Delta\omega^2\delta \tag{26}$$

and

$$\delta = T_2^{-1} - T_1^{-1} \tag{27}$$

Although these equations appear formidable it should be noted that they are functions of only four variables, T_1, T_2, $\Delta\omega$, and ω_1. Values of these are consequently needed to compute decay curves (or spectra) for comparison with experimental ones. The corollary is that the observations are potential sources of the parameters, as discussed briefly above. It turns out that at low values of ω_1 and with observations in the first few microseconds after radical creation the line width is sensitive only to T_2 to a great degree, and accurate values of it can be obtained from line fitting.

It is instructive to examine the implications of Equation 17 on the observed ESR line by plotting the dependence of $M_y(t)$ as a function of offset (i.e., magnetic field in practice) and time. Such plots yield directly, within a scaling factor, the information obtained experimentally with the two-dimensional technique; in observations on single lines it has been our experience that they reproduce completely the experimental behavior. Figure 5a and b

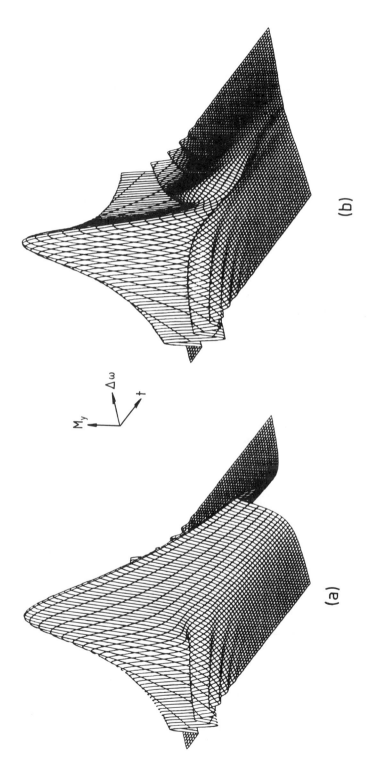

FIGURE 5. Computed two-dimensional field-time surfaces for single lines from radicals with $T_1 = 5.0$ μs, $T_2 = 0.8$ μs, $(P_t/P_{eq}) = +10$, and time and field spans of 1.0 G and 20 μs, respectively, (a) in a low microwave field, $\omega_1 = 1.0$ rad MHz and (b) in a high microwave field, $\omega_1 = 1.0$ rad MHz. The figures show that the line sharpens as time evolves and that oscillations are predicted in the behavior of the magnetization, leading to spurious side bands on the spectral lines. These oscillations occur even at exact resonance in (b). They are removed by the TIS technique. This predicted behavior for individual lines has been verified in detail in (unpublished) experiments in Oxford.

show two such plots calculated for values of T_1 and T_2 typical for a free radical in solution (note that these are not equal) and for different values of the microwave field strength both easily obtainable with a standard spectrometer. It is clear that in these general cases the effects of the oscillations mentioned above are still evident and that their frequencies and amplitudes vary with the offset and the size of ω_1. Only if the latter is large are they observed on resonance (Figure 5b). A cross-section perpendicular to the time axis yields the line shape at a particular time and it is seen that the effects of oscillations in the time domain is to cause side bands to appear about the main peak.[22] In a spectrum these may be misinterpreted as splittings due to hyperfine couplings. It must be realized that their occurrence is inherent to the CW technique used and as such they are potentially observable in both boxcar and transient-recorder experiments. They have been eliminated in the actual two-dimensional spectrum illustrated in Figure 1 by operating at too low a value of ω_1 to cause them. However, this value depends upon the magnitudes of T_1 and T_2 and differs from system to system. In general too, low signal strengths may make it necessary to operate at microwave field strengths which optimize S/N and may cause oscillations, and they should be regarded as a normal part of point-sampled spectra (i.e., those obtained by sampling the signal at one specific time after the photolysis flash). The object of time-integration spectroscopy (TIS) is to eliminate them to a large extent by time integration of the decay curve over their period of oscillation. This works, provided that the period of oscillation is not too long, at the expense of losing some of the time precision of the measurement. However, this is usually unimportant since the integration period can be chosen so as to be short as compared with the variation in the magnetization being sampled.

TIS was invented specifically to accomplish this task, although it gives the added advantage of increasing the S/N of the spectra. Boxcar methods were developed empirically but accomplish the same result in two ways mentioned in the previous section. First, they too sample the signal over a period rather than instantaneously. Second, as presently performed, the field is swept continuously as the sample is taken, also continuously. In consequence the offset from resonance changes while the measurement is taken and so does the frequency of the oscillations. This helps to remove their effects by destructive interference but at the expense of making line shape analysis more complicated. The equations derived here apply to boxcar experiments only if the sweep rate is sufficiently slow for these effects to be neglected.

An interesting feature of Figure 5 is that it can be seen that the width of the resonance line varies in time, being typically 3 to 4 G wide in the first few tens of nanoseconds after the photolysis pulse and sharpening thereafter with a time constant dominated by T_2; this is typically complete in about 1 μs. This is of fundamental origin to the way the CW experiment is performed. As stated above the magnetization is created in the z-direction and rotated into the y-one for detection. During this period it processes about the resultant of the two fields. The line broadening at early times is no more than that typical of a spin-echo experiment and still allows radical identification of a wide variety of species.

A. TIME-INTEGRATION OF TRANSIENT SIGNALS[23]

In Figure 6 are shown for comparison the shapes of single lines taken first as point samples at specific times and second with time integration over a chosen period. It is seen that the side bands in the former are removed by the integration but the effect is to produce a composite line shape consisting of a comparatively sharp feature upon a broader one; their relative magnitudes depend upon the field strength. In practice the existence of the broad component has not proved a problem in resolving hyperfine structure.

The time-integrated signal is obtained simply from Equations 10 and 15 but in practice in the experiment the signal is digitally summed rather than integrated. The observed signal is proportional to

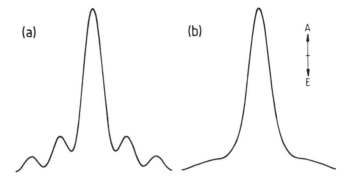

FIGURE 6.   The effect of TIS on the appearance of a line using data taken from Figure 5(b); (a) shows the sidebands evident in a point-sampled spectrum 2.0 µs after radical creation and (b) shows the effect of sampling and summing the signal from 1.0 to 3.0 µs after the flash. The oscillations are removed although a broad component results in the line shape.

$$I_y(t_1, t_2) = \int_{t_1}^{t_2} M_y(t)dt = \lim_{\substack{\Delta t \to 0 \\ n \to \infty}} \Delta t \cdot \sum_{i=0}^{n} M_y(t_i + i\Delta t) \tag{28}$$

where t_1 and t_2 are the integration limits and $\Delta t = (t_2 - t_1)/n$, where n is the number of samples; Δt is usually 10 ns. The integration and the sum are equivalent only if Δt is small and n large, and in general theoretical equations for magnetization variation in time should be similarly summed for comparison with the experiment. This is especially important if rapidly changing signals are observed (as in exchange studies) when the integration period is restricted.

Reference back to Equation 12 shows that the two terms can be integrated separately so that the observed signal may still be thought of as a linear combination of independent subspectra due to the polarized and unpolarized terms in P_I and P_{eq}, respectively. These terms have different line widths and different time dependencies so that their relative contributions vary in time, as does the resultant line shape.[4,23] There is no analogous behavior in normal unpolarized radicals. This causes no great variation in the appearance of spectral lines under most conditions but some extreme effects are observed when the integration period includes the time when an initially emissive signal relaxes through zero to absorption. At low microwave field strengths the two terms are then in opposite phase but are similar in magnitude: a composite line shape is observed when has a broad absorptive outer component but a sharp emissive inner one (Figure 7a). If the field is increased, low-frequency oscillations occur in the observed magnetization which are incompletely averaged by the TIS technique. At earlier times than when the above form is observed a different composite form is obtained which is the qualitative inverse of the former in that the central part is now in absorption and the outer part in emission (Figure 7b). It is stressed that these line shapes are inevitable consequences of the method and the creation of emissively polarized radicals. No such qualitative features affect absorptively polarized ones. The precise line shapes depend upon the relative contributions of the polarized and equilibrated terms and are consequently sensitive functions of the polarization ratio (P_I/P_{eq}); they provide the most accurate method for its determination[24] (see below). Examples of spectra in which all the lines show these effects are given in Figure 8.

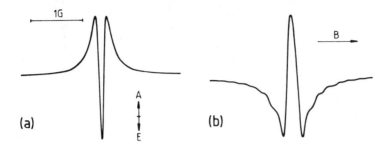

FIGURE 7.   The line shapes predicted near the time when an initially emissive signal relaxes through zero (a) at low microwave field strength (0.05 rad MHz) (b) at a higher value (1.75 rad MHz) where low-frequency oscillations are not removed by the summation process. The two line shapes are inverses of one another but both show a composite form comprised of absorptive and emissive components. The spectral parameters used were $T_1 = 5.0$ μs, $T_2 = 0.6$ μs, $(P_I/P_{eq}) = -6.5$, and an integration period of 10 to 11 μs. The line shapes are sensitive functions of the polarization ratio.

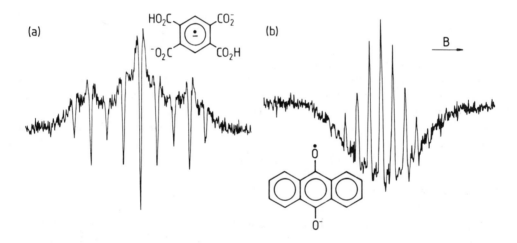

FIGURE 8.   Room temperature TIS spectra of (a) the radical trianion of benzene-2,3,5,6-tetra-carboxylic acid and (b) anthrasemiquinone which show every hyperfine line in their spectra exhibiting the line shape behavior at phase crossover predicted in Figures 7a and b, respectively. The former was obtained on photolysis of a 0.1 *M* solution in a 3:1 cyanomethane/triethylamine mixture and was obtained using an $\omega_1 = 0.1$ rad MHz and integration between 24.0 to 49.0 μs post flash. The latter resulted from photolysis of 0.2 *M* anthraquinone in a 4:1 propan-2-ol/triethylamine mixture at $\omega_1 = 0.5$ rad MHz and integration between 25.0 to 30.0 μs postflash. The precise patterns observed depend upon the values of these quantities, T_1, T_2, and (P_I/P_{eq}). At earlier or later times the spectra appear normal.

IV. CHEMICALLY INDUCED DYNAMIC ELECTRON POLARIZATION (CIDEP)

A. AN OVERVIEW

Almost invariably when radicals are created by flash photolysis in solution and observed in the first few microseconds thereafter they are discovered to be spin polarized; CIDEP is a general and normal phenomenon. The exceptions are those radicals with orbitally degenerate ground states, such as ȮH, in which spin polarization may be removed, or indeed not even created, as a result of extremely fast relaxation processes.[25] Many different spectral behaviors are observed. The spectrum may appear absolutely normal in the relative intensities of the hyperfine components but with non-Boltzmann absolute intensities and be either in absorption (A) or emission (E). In another extreme half of the hyperfine lines may be in absorption and the other half, with equal overall intensity, in emission. Most often spectra exhibit some lines in absorption and some in emission but with an excess of one phase over the other. Sometimes spectra are observed with all their lines in a single phase but with their relative hyperfine intensities distorted. It is not possible to rationalize all these phenomena in terms of a single polarization-producing mechanism but they are explicable by combinations of just two main mechanisms which are independent in their action.[26]

The triplet mechanism (TM) may occur if the radicals are produced by reaction of a molecule in a triplet state and yields spectra entirely in a single phase (see Figure 17), with no distortions to the relative intensities of the lines.[27,28] The phase depends upon the specific molecule involved. The radical pair mechanism (RPM) owes its existence to the fact that radicals are created, or react, in pairs.[29-31] In its most common (ST_0) form (see below) it causes half the spectral intensity to be in absorption and half in emission; it may be either E/A (Figure 13) or A/E in nature, where the first symbol designates the phase of the low-field half of the spectrum, depending upon the spin multiplicity of the radical precursor. The polarization, P, is hyperfine dependent and changes sign about the midpoint of the spectrum if the radicals are produced as an identical pair. If not, both radicals exhibit polarizations in the same sense, but one now has more emissive and the other more absorptive character. These are designated, for example, E*/A and E/A*, where the asterisk denotes a preponderance of one phase.

A typical photochemical route to the production of radicals is via the reaction of a triplet state:

$$^1M \xrightarrow{h\nu} {}^1M* \xrightarrow{ISC} {}^3M* \tag{29}$$

where ISC stands for ''intersystem crossing'', followed by

$$^3M* + S \xrightarrow{k} R_1{\cdot} + R_2{\cdot} \tag{30}$$

If the TM operates, and this depends upon conditions discussed below, the radicals are created with single-phase spin polarization. However, they are produced as a pair and so RPM effects are also expected. Both polarization mechanisms occur within 10 ns of radical creation in organic systems and so both affect the ESR spectrum when first observed using the fastest transient methods. This is why most spectra show their combined effects. Even in a symmetric radical pair the spectrum may exhibit more intensity in one phase than the other (Figure 16). There is a real difference as compared with the pure RPM situation described above with different radicals since here the overall polarized intensity is nonzero. It is often described as the spectrum showing (E/A + E) characteristics to distinguish it from an E*/A pattern in an individual radical, but this description is meaningless in terms of the level populations which control the intensities. Similarly the common description of the RPM causing multiplet-effect polarization and the TM net-effect polarization must be treated with caution.

In detail the RPM occurs due to the fact that spin alignment is conserved in the radical creation step (Equation 30) and, in this case, a triplet geminate spin-correlated radical pair is formed. Such correlated pairs are also produced by the random encounter of freely diffusing radicals (F-pairs) created in separate geminate events. By simple statistics these radicals have equal probabilities of encountering with either singlet S or triplet (T_0) spin correlations (see below) but the singlet pairs are usually more likely to react, leaving the T_0 ones to evolve qualitatively as they would if created as geminate triplet pairs. Whereas the polarization created essentially instantaneously in the geminate step relaxes from that time within, typically, 10 μs, F-pair polarization is generated for as long as reactive radicals persist in the solution. It would be expected therefore that polarizations observed long after radical creation originate from them.

The development of spin polarization depends upon the initial excess of one radical pair state over the other, their subsequent evolution in time to acquire a mixed singlet and triplet character, and the influence of the distance-dependent electron exchange interaction $J(r)$ between the electrons on the radicals. In viscous solutions and in micelles early-time spectra observed by fast transient techniques are of the radical pair itself rather than of the separated radicals, and they show additional lines due to splitting by the $J(r)$ interaction[32,33] (see below). Each component of the split lines is polarized in an opposite phase to yield spectra in which, for typical values of $J(r)$, each line appears to exhibit, for example, E/A polarization. Normal spectra are seen later in time when the radicals have separated and $J(r)$ has gone to zero.[34]

Although the ST_0 form of the RPM is the most common, under various circumstances ST_{-1} polarization also occurs.[35-37] This happens in radicals with particularly large hyperfine coupling constants, radicals at low temperatures and in viscous solutions, and in micelles. Its characteristics are quite different in that it is a genuine generator of spin polarization and yields spectra in net absorption or emission depending upon the spin multiplicity of the precursor. Its action can be distinguished from that of the TM in that it has a hyperfine-dependent component which distorts the relative intensities of the lines in the spectrum.

The origins of all these types of CIDEP are discussed below, as is their influence on the intensities of the observed spectra. Often more than one mechanism contributes to the populations of the levels, and contributions from spin-equilibrated radicals also occur. It is consequently essential to predict the precise patterns which may arise from them besides understanding their (important) qualitative implications. In theory the relative contributions from the different mechanisms are calculable but in practice spectra have to be synthesized by empirical addition of the patterns from the various mechanisms. These resultant intensities must then be convoluted with the line shape functions discussed in the previous section to obtain theoretical spectra for comparison with experimental ones.

Finally, spin polarization persists in radicals until it is removed by relaxation in about 10 μs after the polarization process. During this period the primary radical has time to encounter another molecule and form a secondary radical. If this does happen conservation of spin alignment in this step yields a spin-polarized secondary radical. Spin polarization can in fact be used as a nonintrusive label to follow a chemical reaction.[38]

B. THE TRIPLET MECHANISM[3,39,40]

TM polarization originates in a spin disequilibrium established in the triplet molecule itself before it reacts to create spin-polarized radicals. This is possible because even in the absence of an external magnetic field the degenerate energy levels of the triplet are split by the zero-field interaction, which is usually the dipolar interaction between the electrons. The ISC (Equation 29) obeys molecular selection rules and often occurs preferentially into one or other of the triplet sublevels. Subsequent coupling of these levels to the external field of the ESR spectrometer then yields a triplet state which is spin polarized in the laboratory frame of reference. If this molecule reacts before spin-lattice relaxation within it has removed the spin disequilibrium, two spin-polarized radicals are formed and they are polarized in the same sense (both E or both A) and to the same extent. Since relaxation within the triplet occurs on the nanosecond time scale,[41] only the fastest chemical reactions are able to compete with it, and TM polarization is observed only in electron and proton-transfer reactions and bond-scission ones. The overall production of TM polarization is summarized in Figure 9 where the molecular-frame states are labeled T_x, T_y, and T_z. This simple outline requires expansion at each stage of the process.

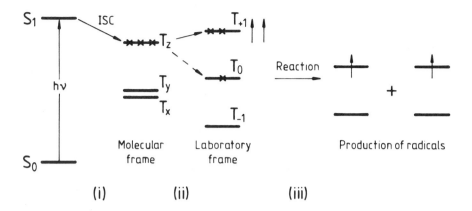

FIGURE 9.   The three stages in the origin of triplet mechanism CIDEP (1) An excited singlet state formed on light absorption by the ground state molecule undergoes a selective ISC to one zero-field-split level (here taken arbitrarily as T_z) of a triplet state of the molecule, (2) the polarization created in the molecule appears in the laboratory Zeeman states, and (3) fast reaction of the laboratory-polarized triplet with a suitable molecule leads to a pair of free radicals which are polarized identically in phase and magnitude. In the example shown they would both be emissive, but this depends upon which triplet sublevel is accessed in the original ISC step. The polarization persists long enough (for around T_1 s) in the radicals for it to be observed.

The primary essential step is the state-selective ISC. In the case of molecules containing one or more heteronuclear atoms this is driven normally by the spin-orbit coupling via which the increase in spin-angular momentum occurs with a concomitant decrease in the orbital-angular momentum while conserving their sum. In the approximation that the two electrons experience coupling with the same coupling constant ξ the Hamiltonian may be written

$$\mathscr{H}_{so} \approx \frac{1}{2}\,\xi(\underline{\ell}_1 + \underline{\ell}_2)\cdot(\underline{s}_1 + \underline{s}_2) + \frac{1}{2}\,\xi(\underline{\ell}_1 - \underline{\ell}_2)\cdot(\underline{s}_1 - \underline{s}_2) \tag{31}$$

where $\underline{\ell}$ and $\underline{s}$ are the orbital and spin angular momenta of the individual electrons. The first term lifts the degeneracy of the singlet and triplet levels while the second serves to mix spin and orbital functions of opposite parity into the wave function and is the source of the singlet-triplet ISC:

$$<T_j|\mathscr{H}_{so}|S> \neq 0 \tag{32}$$

where $|T_j>$, $j = x, y, z$ are the zero-field eigenstates in the molecule. This matrix is negligibly small unless

$$\Gamma_{so}(T_j) \supset \Gamma_o(S) \tag{33}$$

since the spin-orbit Hamiltonian is a scalar in both spin and orbital space and the spin part of the S wave function also transforms as A_1. The spin parts of the triplet representation transform as rotations about the x, y, and z Cartesian axes and it frequently happens that their direct products with the orbital part yields only a single sublevel of similar symmetry to the singlet. It is these states between which ISC occurs.

This symmetry description obviously requires knowledge, or assumption, of the orbital symmetries of the singlet and triplet states but luckily the local symmetry of the chromophore often suffices and symmetry-breaking substituents usually do not alter the overall qualitative conclusion. This is also so when vibronic coupling occurs to assist the ISC. If the sublevel populated in the reactive triplet state lies below the barycenter of the sublevel system, the triplet is absorptively polarized, and so are the radicals subsequently formed. The converse leads to emission. This provides an important test of the TM for the phase of the signals should depend upon the orbital symmetries involved. The nitrogen heterocyclic pyrazine, of D_{2h} symmetry, yields absorptively polarized radicals through the reaction of its ground $n\pi^*$ triplet state while the lower-symmetry quinoxaline (C_{2v}) molecule yields emissively polarized ones,[42] in accordance with prediction. In pyrazine ISC occurs to an upper $\pi\pi^*$ state and is followed by internal conversion to the ground state with conservation of spin alignment, while in its tetramethyl derivative the ground state is $\pi\pi^*$ and is accessed directly in the ISC step. Despite the molecules having the same overall symmetry the polarization is in the opposite phase in the resulting radicals.[43] This occurs since the sign of the zero-field coupling varies between the $n\pi^*$ and $\pi\pi^*$ states, inverting their sublevel energy order in the molecular frame. This is again direct evidence for the TM as a source of polarization.

The creation of a polarized triplet in the molecular frame of reference would not be expected, at first sight, to lead to observable polarization in the laboratory where the triplet levels are usually described as the Zeeman states, $T_{0, \pm 1}$, inside the magnetic field of the spectrometer. This follows from our knowledge of the behavior of oriented triplet states, e.g., in crystals: if the external field is shone along a given molecular axis the energy of the corresponding state is unaffected by it while the two orthogonal ones interact with it to cause one state to become the highest in energy and the other the lowest. Which T_j states do which depends upon which axis is colinear with the field. In solution the molecule tumbles freely with respect to the field direction and the state overpopulated in the ISC step fluctuates in its energy and correlates with each of the Zeeman states in turn, population transfer from the molecular to the laboratory frames being most efficient between the states closest to degeneracy. In this way we should expect the Zeeman states to become populated equally. Similarly in the high-field limit ($g\mu_B B >> |D|$, the zero-field parameter) each of the (true) Zeeman states contains equal admixtures of the zero-field ones and no polarization would be expected in the laboratory frame again.

This paradox is removed by realizing that this high-field limit is not reached in the magnetic field of the spectrometer and the spin Hamiltonian must include both the zero field and the Zeeman terms. Perturbation theory carried to second order[3] then discloses that the zero-field state populated in the ISC correlates most strongly with the intermediate-field one which is closest in energy to it and this consequently acquires a proportionate amount of its character. The triplet polarization created in the molecular frame appears, reduced, in the laboratory one. It happens that for the molecules studied, mainly heterocycles and carbonyls, $|D| \sim g\mu_B B$ at X-band. This optimizes the population transfer to the laboratory and also implies that if, for example, the zero-field state highest in energy is overpopulated then so will be the upper, T_{+1}, state in the laboratory; an emissive signal will result when this reacts to form radicals.

The magnitude of the polarization falls as the rotational correlation rate (τ_R^{-1}) of the triplet increases. This is understood by realizing that this broadens the energy levels, via an uncertainty argument, and reduces the specificity of the state correlations accordingly.

The corollary of this description is that the magnitude of TM polarization should vary with the angle the molecular axis system makes with the external field. In solution an elegant experimental test of the TM was suggested by Adrian who pointed out that (temporarily) oriented triplets could be created by producing triplet states by irradiation with plane-polarized light; only those molecules whose electric dipole transition moments are correctly oriented with respect to the electric vector of the light are excited. Calculations suggested that a 10% variation in the polarization magnitude might result if this vector was applied parallel or perpendicular to the field of the spectrometer. Despite some reports that this effect has been observed,[44-46] recent careful work shows that it is below the limits of detection.[24]

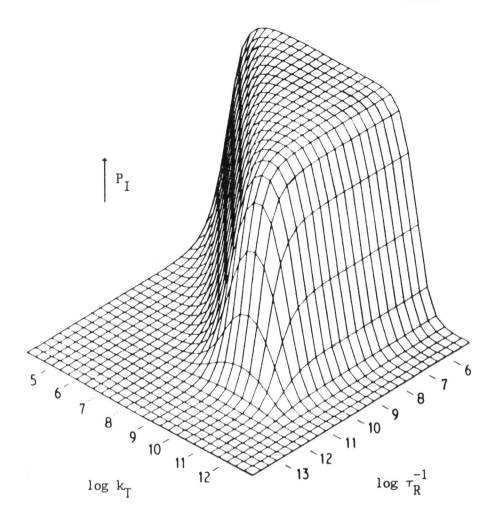

FIGURE 10. The dependence of the electron polarization due to the TM on the rotational correlation time of the triplet, τ_R, and its rate of reaction, k_T, assuming a perfectly selective ISC stage, $|D| = 1000$ G, a cylindrically symmetric triplet and a Zeeman splitting of 6×10^{10} s^{-1}. The magnitude of P_I is discussed in the text. It is seen that P_I is low if the rotational motion is too fast, or if k_T is either too fast or too slow, but it happens that it has a finite value for typical values of these quantities in chemistry.

The final stage of the production of spin-polarized radicals is the reaction of the spin-polarized triplet with a suitable reagent. In the laboratory the polarization decays via very efficient spin-lattice relaxation processes on the nanosecond time scale in solutions of normal viscosity and is transferred to the radicals produced on reaction only if the reaction rate competes with this. If however the rate constant for chemical quenching k_T is too high the laboratory triplet levels are uncertainty broadened in their turn and the efficiency of the molecular to laboratory frame polarization transfer is again impaired. The joint effects of molecular rotation and chemical reaction of the triplet on the observed polarization in the radicals are summarized in Figure 10, calculated from the detailed theory of Atkins and Evans[39] and of Freed and Pedersen.[40] These two sets of authors provided quite similar dynamical analyses of the overall generation of polarization using density matrix methods.

From the figure it can be deduced that TM polarization is expected to be observed only in electron, proton-, and some atom-transfer reactions occurring at or near diffusion-controlled rates. The observed initial polarization is related to the polarization of the triplet P^T by the relation

$$P_I = \frac{P^T \cdot k_T + P_{eq}^T \cdot {}^3T_1^{-1}}{k_T + {}^3T_1^{-1}} \qquad (34)$$

where ${}^3T_1$ is the spin-lattice relaxation time of the triplet. The rate constant k_T is normally a pseudo-first-order one and can be written $k_2[M]$, where k_2 is the second-order rate constant and [M] is the concentration of the reagent. This equation has been used, together with a consideration of P_{eq}^T, to yield an expression which allows the value of ${}^3T_1$ to be determined in solution.[41]

In conclusion, although TM polarization may arise on reaction of a triplet molecule, it is by no means inevitable and it depends upon a number of different stages any of which can diminish its magnitude. Substitution of reasonable physical values into the theoretical expressions, and assuming that state selection in the ISC stage is complete, predicts P_I values of around 250 P_{eq}, although in practice values of 10 to 90 P_{eq} are observed.

A notable feature of the TM is that the spin-orbit interaction which creates the primary polarization in the triplet is independent of nuclear spin angular momentum, as are the subsequent processes which lead to polarization appearing in the radicals. In consequence the relative intensities of the hyperfine lines in the ESR spectra of the radicals are as in equilibrated radicals although their absolute intensities are affected. Calculation of these relative intensities is precisely as for normal ESR and depends upon the degeneracies of the hyperfine states.

C. THE RADICAL PAIR MECHANISM[29-31,47]

Whenever free radicals are created photochemically or thermally in solution they are produced in pairs, for example, according to Equation 30. Since chemical reactions normally proceed with conservation of electron spin alignment, at the moment of their birth the overall electron spin angular momentum is conserved and they are formed with specific correlations between the electron spins on the two radicals. If formed from reaction of a triplet state, for example, the spins on the two radicals are parallel and, although the species might eventually react with zero activation energy, this inhibits immediate reaction (bond formation normally involves electrons with antiparallel spins). The radicals consequently diffuse apart but in their random diffusion in solution they may reencounter later. At this time the initial pure triplet state of the spin-correlated radical pair may have evolved to give it some singlet character, and then reaction can occur. The radical pair is consequently an essential intermediate in the radical reaction and occurs between the time that the excited precursor molecule exists and the time that free radicals appear. It is the spin evolution which happens within it that determines the future course of the reaction and it is largely this same evolution which, together with the electron exchange interaction, causes the radicals which escape reaction in this geminate cage process to be observed in an electron spin-polarized state (Figure 11).

The generation of CIDEP in this way is entirely a quantum phenomenon but its origins can be understood using a simple argument.

Using the Dirac method, the time dependence of the wave function of the radical pair may be written in terms of time-dependent coefficients in a linear combination of the singlet and triplet wave functions $|S>$ and $|T>$:

$$|\Psi(t)> = c_S(t)|S> + c_T(t)|T> \qquad (35)$$

It is stressed that we are considering the spin states here of the spin-correlated radical pair and not of the precursor molecule as in the TM. In the magnetic field of the spectrometer the degeneracy of the triplet state is removed and we need to know whether or not all three substates, $T_{0,\pm 1}$, can mix with S. Mixing can occur only if the energies of the states are near degenerate and our first question must be what their energies are. An insight is gained

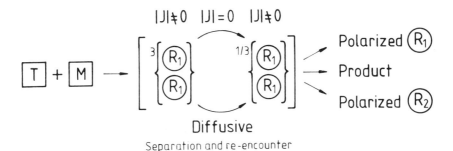

FIGURE 11. When a triplet molecule, T, reacts with a suitable molecule, M, a pair of radicals, R_1 and R_2, are produced with a triplet-spin correlation as a result of the conservation of electron-spin alignment in the reaction. The radicals cannot react but diffuse apart. A small fraction subsequently reencounter later, by which time the radical pair has attained some singlet character so that its component radicals may react. If reaction does not occur the radicals which diffuse away from the reencounter are spin polarized. In the basic model there is a close analogy between how geminate radical reactions occur and the generation of ST_0 RPM electron spin polarization. Note that during the diffusion the radicals move from a region where the exchange interaction J is large, to one where it is zero and, subsequently, they reexperience a large J interaction at the reencounter. This nonzero value is essential for polarization development.

by considering the simplest radical pair of all, a pair of hydrogen atoms, and this is most obvious if we time-reverse the diffusive separation described above; we are then asking what happens to the energies of the atoms as they are brought together to short distances. This is the basic problem of bond formation in chemistry and leads to creation of the 1sσ bonding and 1sσ* antibonding orbitals whose potential energy curves are well known. By applying the Pauli principle, which requires that for spin-$1/2$ particles the total wave function (orbital × spin) must be antisymmetric with respect to their exchange, we discover the 1sσ to be the singlet state and the 1sσ* to be the triplet one. These two states are separated by the electron exchange interaction, normally defined as 2J(r), which is distance dependent and varies between being completely dominant at low separations to near zero at high ones. The short-range nature of the interaction has important implications (see below). In the magnetic field of the spectrometer the triplet state is split into its full Zeeman components at interradical separations at which J(r) is negligible (Figure 12).

From the figure it is apparent that at all interradical separations the T_{+1} state is well removed from the S one in energy and the two do not mix. At short distances the T_{-1} level, for a negative J(r), crosses the S one and it might be expected that they would mix; indeed the hyperfine interaction to the nuclei may cause this. However, this crossover occurs at short interradical separations and the system traverses this region in the first few diffusive steps; the interaction time is too short for effective mixing. This situation can change if hyperfine couplings are large, as in radicals in which the electron is centered on a magnetic nucleus, or if diffusion is restricted in high-viscosity solutions or in micelles. In normal solutions it is only the S and T_0 levels which attain near degeneracy for long periods and are mixed at large interradical separations. Their electronic wave functions are

$$|S> \ = \ 2^{-1/2}(|\alpha\beta> \ - \ |\beta\alpha>) \quad \text{and} \quad |T_0> \ = \ 2^{-1/2}(|\alpha\beta> \ + \ |\beta\alpha>) \quad (36)$$

both of which contain equal amounts of α and β spin on the two radicals. Spin mixing cannot alter this and we obtain the important result that ST_0 mixing cannot produce true electron spin polarization.

These states are mixed under the influence of the spin Hamiltonian of the radical pair, conveniently written as

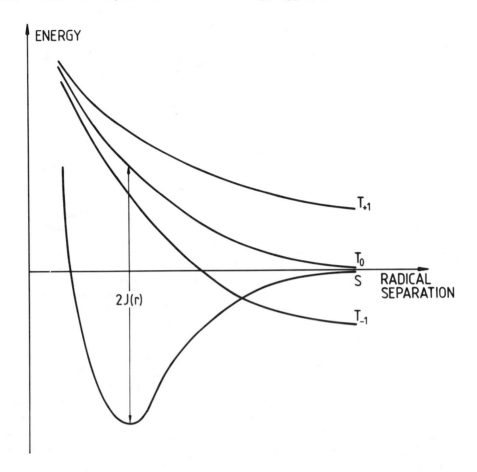

FIGURE 12.   An indication of how the energies of the S and $T_{0,\pm 1}$ states of a spin-correlated radical pair varies with inter-radical separation. The size of the Zeeman splitting between the T states is vastly exaggerated relative to the J(r) one at short distances. Only the S and T_0 states develop the near degeneracy necessary for their mixing as the radicals diffuse apart for long enough for the process to be significant. ST_{-1} mixing can be promoted by restricting the diffusion, by slowing the radical separation, or by providing an efficient means of causing it (a large hyperfine interaction).

$$\mathcal{H}_{RP} = \mathcal{H}_J + \mathcal{H}_M \qquad (37)$$

where $\mathcal{H}_J$ is the operator describing the exchange interaction and $\mathcal{H}_M$ the magnetic operator:

$$\mathcal{H}_J = -J(r)(2S_1 \cdot S_2 + \tfrac{1}{2}) \qquad (38)$$

and $\qquad \mathcal{H}_M = \mu_B(g_1\underline{S}_1 + g_2\underline{S}_2) \cdot B + \sum_n a_{1n} \cdot \underline{I}_{1n} \cdot \underline{S}_1 + \sum_m a_{2m} \cdot \underline{I}_{2m} \cdot \underline{S}_2 \quad (39)$

Here, for example, $\underline{I}_{1n}$ is the nuclear-spin angular momentum of the nth nucleus on radical 1 and $\underline{s}_1$ is the electron-spin angular momentum of that radical. The development of RPM polarization depends upon the action of $\mathcal{H}_{RP}$ on the radical pair states and in our simple model we make use of the fact that J(r) is short-ranged. When the radicals are close together we shall assume that $\mathcal{H}_J$ alone is important while at longer distances only $\mathcal{H}_M$ will be. This is an approximation which is unnecessary to the analysis of the system but is useful in building a physical model of polarization development. We note too that only $\mathcal{H}_M$ can cause spin mixing, the effect of $\mathcal{H}_J$ being to remove the degeneracy of the S and T_0 states at close encounter distances. Both terms in Equation 37 are necessary for polarization development.

The radicals are created together in a pure spin state which can evolve only after separation has caused J(r) to fall, in this model to zero. Thereafter, $\mathscr{H}_M$ operates to mix in the other state to yield the wave function given in Equation 35. Inserting the wave functions from Equation 36 and collecting up gives

$$|\Psi(t)> = \{[c_S(t) + c_{T_0}(t)]|\alpha\beta> + [c_{T_0}(t) - c_S(t)]|\beta\alpha>\} \tag{40}$$

This allows us to calculate the populations of the α and β states for each radical. For example, for radical 1,

$$n_\alpha = [c_S(t) + c_{T_0}(t)][c_S(t) + c_{T_0}(t)]^* \tag{41}$$

and substituting for n_α and n_β in Equation 2 gives the polarization,

$$P_1 = c_S(t) \cdot c_{T_0}^*(t) + c_S^*(t) \cdot c_{T_0}(t) \tag{42}$$

Throughout the asterisk denotes a complex conjugate.

It remains to obtain the values of the coefficients by solution of the time-dependent Schrödinger equation

$$\mathscr{H}_M|\Psi(t)> = i\hbar\left[\frac{d}{dt}|\Psi(t)>\right] \tag{43}$$

These are given by

$$c_S(t) = c_S(o)\cos Qt - ic_{T_0}(o)\sin Qt \tag{44}$$

and

$$c_{T_0}(t) = c_{T_0}(o)\cos Qt - ic_S(o)\sin Qt \tag{45}$$

where

$$Q = <S|\mathscr{H}_M|T_0>$$

$$= \frac{1}{2}(g_1 - g_2)\mu_B B + \frac{1}{2}\sum_n a_{1n}m_{1n}^{(a)} - \frac{1}{2}\sum_m a_{2m}m_{2m}^{(b)} \tag{46}$$

and m_{in}^a is the magnetic quantum number of the nth nucleus of radical 1 which exists in the overall nuclear spin state (a). This shows that the rate of mixing depends upon the different total (Zeeman plus hyperfine) magnetic fields at the two electrons. Insertion of Equations 44 and 45 into Equation 42 gives

$$P_1 = c_S(o)c_T^*(o) + c_S^*(o) \cdot c_{T_0}(o) \tag{47}$$

where the polarization is now expressed in terms of the known coefficients at the instant of creation. If the pair is formed in a triplet state $c_S(0)$ is zero, as is $c_T(0)$ if the initial state is singlet. In either case P_1 is zero and this shows that spin mixing under the influence of $\mathscr{H}_M$ does not on its own produce spin polarization.

We now return to our model of the geminate reencounter process and allow the radicals to diffuse back together and reexperience the exchange interaction. This situation differs fundamentally from their initial separation in that now the wave function is a mixed one. We shall consider the reencounter to occur at time τ and to persist for a time $\Delta\tau$ during which J(r) is taken as constant (J) and dominant. This delta function description is of course a further approximation, J(r) normally being taken as exponential even in solution (although little evidence exists for this). At the reencounter the wave function evolves further under the influence of $\mathcal{H}_J$ according to the equation

$$\mathcal{H}_J|\Psi(\tau)> = i\hbar\,\frac{d}{dt}\,|\Psi(\tau)> \tag{48}$$

Solution of this for the mixed wave function at τ gives the values of the coefficients as the radicals separate once more:

$$c_S(\tau + \Delta\tau) = c_S(\tau)\,.\,\exp(-iJ\Delta\tau) \tag{49}$$

and $\qquad c_{T_0}(\tau + \Delta\tau) = c_{T_0}(\tau)\,.\,\exp(iJ\Delta\tau) \tag{50}$

Insertion of these values into Equation 44 then gives

$$P_1 = [c_S(o)\,.\,c_S^*(o) - c_{T_0}(o)\,.\,c_{T_0}^*(o)]\sin(2Q\tau)\,.\,\sin(2J\Delta\tau) \tag{51}$$

The effect of the exchange interaction has been to remove the product of mixed coefficients of Equation 47 and we see that if the pair is created in the singlet state ($c_s(0) = 1$) the first term has the value $+1$, whereas from a triplet state the value is -1. In either case a nonzero value of the polarization has been obtained, as has the essential result, which is that fundamental to the use of CIDEP in photochemistry, that the sense of the polarization differs between singlet and triplet precursors.

During random diffusion in solution there is no single time τ at which radicals which were created together reencounter later, but rather a distribution which is given at its simplest by a random-walk model. Ignoring fluctuations in J(r) due to random diffusion we may write

$$P_1 = [c_S(o)c_S^*(o) - c_{T_0}(o)c_{T_0}^*(o)]\sin(2J\Delta\tau)\,.\,\int_0^\infty f(\tau)\sin(2Q\tau)d\tau \tag{52}$$

where $f(\tau)$ is the probability of reencounter at τ. Inserting the Noyes formula for this[48] and integrating yields[49]

$$P_1 \approx 0.85[c_S(o)c_S^*(o) - c_{T_0}(o)c_{T_0}^*(o)]\,.\,\sin(2J\Delta\tau)[Q\tau_D]^{1/2} \tag{53}$$

where τ_D is the mean time between diffusive steps. This equation gives the dependence of P_1 on $Q^{1/2}$, and hence on the Zeeman and hyperfine interactions, essential for the calculation of spectra. It is interesting to note that the square-root dependence arises from the nature of the diffusive motion. For typical values of $Q \sim 10^8$ rad $\cdot$ s^{-1}, $\tau_D \sim 10^{-11}$ s, and with $\sin(2J\Delta\tau)$ estimated as ~ 0.1,[49] $P_1 \sim 10\,P_{eq}$.

An alternative and elegant approach to obtaining Equation 53 is through a vector analysis applied to the same impulse model.[47] This model has been useful not only in obtaining the essential polarization results in Equation 53 but also preserving a close correspondence between RPM polarization development and the related geminate reaction process. It is of course an approximation and a fuller treatment is to allow the spin system to evolve under the simultaneous action of the exchange and magnetic terms of the Hamiltonian with a more realistic description of J(r) and with the molecular diffusive dynamics considered simultaneously. This is possible through the Stochastic Liouville equation. With

$$J(r) = J_0 exp(-\lambda r) \tag{54}$$

where λ is a simple parameter, Monchick and Adrian[47] obtained the asymptotic solution in the slow mixing condition ($Q\tau_D << 1$),

$$P_1 = \pi \frac{|QJ_0|}{QJ_0} [c_S(o)c_S^*(o) - c_{T_0}(o)c_{T_0}^*(o)] \cdot \frac{\sqrt{3}}{\lambda d} (Q\tau_D)^{1/2} \tag{55}$$

where d is the distance of closest approach. This condition can be countervened in practice by radicals with large hyperfine coupling constants (e.g., alkyl radicals). This problem is overcome in the more general theory of Freed and Pedersen,[30,31] whose equation, under the conditions experienced experimentally, can be contracted to

$$P_1 \propto [Q^{1/2} - aQ] \tag{56}$$

where the coefficient (a) is usually determined empirically.[37] This coefficient is significant only for radicals with high Q values, in practice those such as alkyl radicals with a large spectral spread; it is usually neglected. We have seen above that the generation of spin polarization requires the action of both $\mathcal{H}_J$ and $\mathcal{H}_M$, and there exists a brief period as the radicals initially separate when the wave function is slightly mixed and the pair also experience the exchange interaction. In a given pair this produces little polarization but all radicals undergo this process whereas only a few reencounter; in the ensemble this inefficient polarization is weighted strongly relative to the reencounter polarization described above, and it depends directly upon Q.[50]

One final necessity in calculating the spectrum expected for a radical escaping from a geminate pair is to realize that the matrix element Q was defined in Equation 46 with relation to a pair of radicals in the specific overall nuclear spin states a and b. In the ensemble the polarization of a single line in the spectrum of one of the radicals is the sum of the polarizations generated by it forming pairs with the counterradicals in all of their possible hyperfine states:

$$P_{1,a} = \frac{1}{x_2} \sum_n P_{1,ab_n} \tag{57}$$

where b_n is the nth nuclear spin and Zeeman state of radical 2 and x_2 is the total number of such states.

1. The Patterns of Geminate RPM Polarization

The basic features of RPM-polarized spectra can be deduced from Equations 46 and 55. We recall that the parameter Q depends upon the magnetic quantum numbers, m, of the nuclei, and these also determine the positions of the lines in the normal ESR spectrum (Equation 1). In a symmetric radical pair ($g_1 = g_2$) the lines to either side of the center correspond to opposite signs of m and consequently, from Equation 55, to opposite signs of the polarization. This is the source of the other notable feature of RPM ST_0 polarization, that one half of the spectrum appears in emission and the other half in absorption, but the overall spin-polarized intensity is zero. When the radicals are dissimilar the position of the phase crossover deviates from the center of the spectrum since Q now depends upon the g-factor difference (Δg) too, but the general result remains.

If Q is dominated by this difference (which is extremely rare) all the lines in the spectrum of one radical appear in emission and those from the other in absorption, a so-called net effect. The sign of polarization is then given by the rule

$$\Gamma_n = \mu \cdot J \cdot \Delta g, +A, -E \tag{58}$$

where J is negative if the S state of the radical pair lies below the T_0 one in energy, μ is positive for T-pairs and negative for S ones and Δg is positive for the radical with the higher g-value.

At the other extreme, $\Delta g = 0$, which is often observed experimentally, a pure multiplet effect is observed with the sign of the polarization given by

$$\Gamma_m = \mu \cdot J, +A/E, -E/A \qquad (59)$$

The most common observation is of a pair of dissimilar radicals in which the hyperfine interaction dominates a nonzero Zeeman one, and this leads to, for example, the E*/A and E/A* patterns mentioned previously. Note that the sum of the polarization in the two radicals is the same (zero) and that the sign rule Equation 59 still holds to allow the spin multiplicity of the precursor to be deduced.

The stick spectra calculated for some simple cases of radicals in each of which coupling is to a single proton are shown in Figure 13. Besides illustrating the sign rules these accentuate a further aspect of ST_0 polarization that, due to the dependence on $Q^{1/2}$, the lines in the center of the spectrum have the lowest intensity (at exact center, zero) and the outer ones are exaggerated in intensity. The effect of state degeneracies, which dominates the appearance of normal ESR spectra, is largely lost. An example of an ST_0 RPM polarized spectrum of a pair of radicals created by reaction of a triplet precursor is given in Figure 14.

2. Polarization in F Pairs

The expectation for polarization in radicals escaping from F-pair encounters is that they should behave qualitatively as they would had they been created as triplet geminate pairs, as argued above. They would be expected therefore to obey the sign rule Equation 59, also with μ positive. There is however some dispute as to the phases of the signals. In experiments designed to accentuate polarization in F-pairs, but with observations soon after radical formation, the expected E/A pattern has been reported.[51] However, in experiments performed specifically to discriminate in favor of polarization produced in bimolecular encounters, A/E patterns have sometimes resulted.[52] Furthermore, in all cases where E/A patterns have been observed from geminately polarized radicals and which have been observed for sufficiently long periods thereafter, the phase invariably inverts to A/E in time.[50,53,54] Some of the observations are now in doubt. The situation is confused further by opposite polarizations having been observed from the same radical pairs in different experiments all of which were believed to be observing the effects of F-pair encounters. These discrepancies have recently been removed by the discovery that the observed behavior is concentration dependent.[55] At present the data is consistent with two entirely different interpretations. One is that true F-pair polarizations are A/E in character, in contradiction to the theory, and that the concentration dependence results from the second-order lifetimes of the radicals. The other is that some radicals, e.g., alkyl ones, give rise to A/E polarizations while others, e.g., ketyl ones, yield E/A ones. This implies that the sign of J differs between them, possibly as a result of spin polarization of an intervening solvent molecule in the former case.[53] If this is so, the later inversion of E/A signals in time must arise from an unrecognized new mechanism of polarization or from population redistribution by cross-relaxation processes. Although the latter can in theory cause this, behavior experiments to test this appeared to show that this does not happen;[56] these results may not be genuine.

The theoretical treatments of the RPM ST_0 polarization mentioned above suggest that the intensities of lines in F-pair-polarized radicals should also depend upon $Q^{1/2}$; whatever their true phase the appearance of the spectra is similar to those from geminate polarization processes.

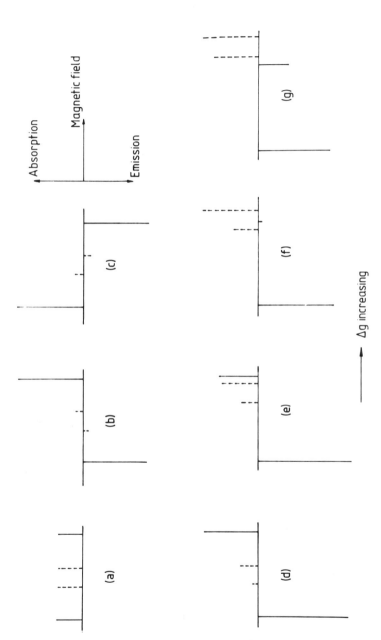

FIGURE 13.   Radicals produced by the photolysis of acetone in cyclohexanol solution. The E/A polarization pattern discloses that they were formed by reaction of the excited triplet state of the acetone molecule. The sharp lines originate from the $Me_2\dot{C}OH$ radical and the broader ones from the cyclohexyl-1-ol one (Figures 1 and 2). Inspection shows that the g-factors of the two radicals are almost exactly identical, so that ST_0 RPM polarization would be expected to yield the central line in near-zero intensity (although lines just off center would be of low intensity with opposite phases, Figure 1). Further examination of the whole spectrum discloses that the total intensity in absorption exceeds that in emission. Both of these observations demonstrate that there is an absorptive TM contribution, of small magnitude, to this spectrum.

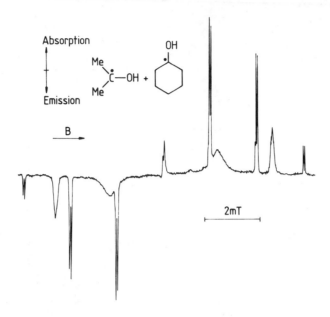

FIGURE 14.   Calculated spectra from two radicals in each of which coupling is to one proton, with a different hyperfine coupling in each. Their transitions are distinguished by full and hatched lines. (a) The unpolarized, or absorptively TM polarized, relative intensities, (b) and (c) the ST_0 RPM patterns from triplet and singlet precursors, respectively, showing an E/A phase pattern in the first and an A/E one in the second. Notable also is the large difference in intensities between the transitions from the two radicals, the line intensity depending upon $Q^{1/2}$ (see text) and hence on the distance of a line from the spectral center, (d), (e), (f), and (g) show that as the value of Δg $(= g_1 - g_2)$ is increased from 0.001 to 0.002 to 0.003 and eventually to 0.004 the pure multiplet effect depicted in (b) and (c) becomes increasingly distorted by a net effect which eventually causes the spectra of the two radicals to appear in opposite phases. Note also that in some of these cases some transition intensities become so small that the lines might not be detected above the noise level of an experiment. (From Buckley, C. D., Grant, A. I., McLauchlan, K. A., and Ritchie, A. J. D., *Faraday Discuss. Chem. Soc.*, 78, 257, 1984. With permission.)

3. ST_{-1} RPM Polarization[37,57]

Under those conditions where it can occur (see above) ST_{-1} mixing causes entirely different polarization behavior from the ST_0 form. In particular the exchange interaction is no longer necessary to produce it, nor is the separation and reencounter process within the geminate cage. Unlike the ST_0 effect, ST_{-1} interaction produces true spin polarization. It occurs through the hyperfine term in the spin Hamiltonian which we rewrite in terms of the raising and lowering operators, $I_{\pm}$ and $S_{\pm}$:

$$a \cdot \underline{I} \cdot \underline{S} = aI_z \cdot S_z + \frac{1}{2} a(I^+S_- + I_-S_+) \qquad (60)$$

Action of the I_-S_+ term on an initial T_{-1} state of the radical pair causes mixing with the S state by simultaneous flipping of the nuclear and electron spins. This can happen only if one of the members of the radical pair contains a nucleus in its α spin state and so the process is clearly hyperfine dependent; the interaction connects the T_{-1} and S states which differ in m by one:

$$<S, I, (m - 1)|I_-S_+|T_{-1}, I, m> = 8^{-1/2}a[I(I + 1) - m(m + 1)]^{1/2} \qquad (61)$$

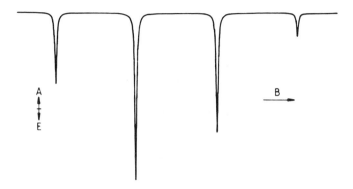

FIGURE 15. The ST_{-1} RPM polarization pattern calculated for a pair
of identical radicals, with three equivalent protons which would cause the
radicals at equilibrium to exhibit a 1:3:3:1 intensity pattern, formed from
a triplet precursor. The total polarization is the sum of hyperfine-dependent
and -independent terms, the former making zero contribution to the highest-
field line if the hyperfine coupling constant is positive. If the radicals were
created from a singlet precursor the pattern would be in absorption and
the relative intensity pattern of the lines would be reversed.

Here I is now the spin of a particular nucleus, with hyperfine coupling constant a in
one member of the radical pair.

The hyperfine-dependent contribution to the polarization is proportional to the square
of the matrix element in Equation 61 but there exists also an important hyperfine-independent
one. This arises since the mixing occurs in a period when the overall electron spin of the
radical pair may be coupled to any nucleus in the system. What eventually becomes each
separate radical may, at this time, form a radical pair with the counterradical in any of its
possible hyperfine states. As a result it acquires a hyperfine-independent polarization given
by the degeneracy-weighted average of the ST_{-1} polarizations from all the nuclear spin
states of the counterradical, calculated as above.[37] The relative magnitudes of the hyperfine-
dependent and hyperfine-independent contributions depends upon the specific radicals involved and vary greatly from system to system.

From a triplet pair the result of these contributions is to give an overall hyperfine-
dependent emission, in contrast to the hyperfine-independent emission which may result
from the TM. The polarization is least for the highest-field line in the spectrum (if a is
positive) and becomes zero if the counterradical has no significant couplings. A calculated
pure ST_{-1}-polarized spectrum for a radical with three equivalent positive hyperfine couplings
and formed in an identical radical pair from a triplet precursor is shown in Figure 15. The
highest-field line has zero contribution from the hyperfine-dependent part of the polarization,
as follows from Equation 61. A change in the sign of a would reverse this to the lowest-
field one. Although no case has been reported yet in practice, a similar reversal would occur
with positive a if the radical pair was created in a singlet state.

It must be expected that the systems which exhibit ST_{-1} polarization may also be subject
to independent polarization from ST_0 and TM processes. This has been the invariable
experience to date. To distinguish the contributions it is necessary to calculate the possible
contributions due to each and to vary their proportions empirically in a summed spectrum
until agreement is reached with experiment. The test for an ST_{-1} contribution in a radical
produced from a triplet precursor is that the overall spectrum should exhibit an excess of
emission and that this emission be hyperfine dependent. In practice the conditions of the
experimental system should suggest its possible contribution, that is, that it may appear if
hyperfine couplings are large (although usually not at room temperature in a C-centered
radical), if solutions become viscous (as at low temperatures) or if experiments are performed
in micelles.

4. Polarization When Motion is Restricted

Besides the enhancement of ST_{-1} effects by causing the S and T_{-1} crossing to be traversed more slowly in systems where diffusion is restricted, an entirely new form of polarization may arise. So far we have dealt with the spectra of the radicals which, although they have been created and spin-polarized in a geminate pair, have escaped from it before they are observed. In micelles and in viscous solutions it is however possible to prolong the lifetime of the radical pair by inhibiting the speed and/or extent of its dissociation into separate radicals. It then becomes possible to observe the ESR spectrum of the radical pair itself. This is the source of some extraordinary CIDEP spectra in which apparently each ESR line exhibits E/A polarization (from a triplet precursor) in spectra observed during the first microsecond of the existence of the radical.[34,58]

The explanation lies in the fact that the lines do not arise in individual transitions but in two separate ones of opposite polarization. These lines are due to an extra splitting caused directly by the average value of the exchange interaction with the pair.[32,33] Neglecting any hyperfine coupling, the spectrum of the radical pair is simply that of two electrons coupled via J, and is completely analogous to the NMR spectrum of two protons coupled by their spin-spin coupling constant: an AB pattern results. The four states involved are the T_{+1} and T_{-1} ones, which are eigenstates and have highest and lowest energy, respectively, and two mixed S and T_0 ones of intermediate energy. In the pair-creation step the $T_{0,\pm1}$ levels are populated equally, leading to equal populations in the T_{+1} and T_{-1} radical-pair states and populations of x(<1) and (1 − x) in the other two. Calculation of the intensities of the allowed transitions for each of the two electrons then shows that each should yield a doublet with equally intense lines in opposite phase. This causes each of the lines in a hyperfine-coupled spectrum to split into two of the opposite phase.

This spectrum persists only for as long as the radical pair exists and at later times the spectrum attains its normal spin-polarized form as the separated radicals are observed.

D. SECONDARY POLARIZATION

As mentioned above, reaction of spin-polarized radicals within their spin-lattice relaxation times may yield polarized secondary radicals. The conditions for this have been discussed theoretically and it has been shown that from such studies the polarization in the primary species may be deduced as can its rate constant for reaction to form the secondary.[15] The reaction occurs with conservation of spin alignment so that TM polarization is transferred without change in phase, but the situation is less clear when RPM polarization occurs.

If the primary has some net polarization (e.g., an E*/A pattern) and if the electron is transferred from a radical in a specific hyperfine state onto a molecule in a random hyperfine state without the nuclear hyperfine coupling in the two being correlated, it is only the net contribution which is conserved.[26,59] In this case the secondary would exhibit pure single-phase polarization without hyperfine distortion and it might be confused for a primary species polarized in a TM process. However, since RPM polarization is normally dominated in C-centered radicals by hyperfine effects, the net polarization transferred is small and only weak signals usually result. A different situation occurs if the hyperfine coupling is correlated between the two radicals. For example, a spin-polarized primary radical might add to a double bond to form a secondary in which the electron remains coupled to the same proton as in the primary; alternatively, a radical rearrangement might occur. In this circumstance intensity alternations could occur in the spectrum of the secondary species.[60] This has been observed in the reaction of P-centered radicals with olefins.[61] Here the sense of the polarization can alter between the primary and secondary radicals if the coupling constant changes in sign between the two.

A further situation, almost uninvestigated experimentally, is that secondary radical formation might occur within the geminate separation and reencounter process. In this case the pair of radicals which reencounter differs from that originally formed. Such a pair substitution would be expected to modify the magnitude of the polarization and the appearance of the spectrum, but not to vary the, for example, E/A phase characteristic.[18]

E. MEASUREMENT OF POLARIZATION RATIOS

The detailed theory of CIDEP of both the TM and RPM varieties reveals it as a potential source for investigating molecular motion, the properties of triplet states and their reactions, the details of radical reaction probabilities, etc. To extract this information requires accurate measurements of the magnitude of the polarization ratio (P_t/P_{eq}), and this remains difficult experimentally although some tests of RPM theory have been made.[53,62-64] Several different methods have been attempted including the analysis of individual decay curves obtained at a fixed magnetic field value, [17] data manipulation of TIS spectra obtained with two different time integration periods,[19] and one based upon the MISTI technique.[19] The most accurate methods, capable of comparing polarization ratios within 0.5%, are based upon the line shape analyses described above.[24] These line shapes depend however on the values of T_1, T_2, $\Delta\omega$, and ω_1 besides the polarization ratio, and to extract an accurate value of the latter requires accurate knowledge of the former. In practice this causes at least 15% errors in its measurement. A critical account of early attempts to verify theoretical predictions for the TM and the RPM has been given[3] but little recent work has been performed on this topic.

V. CALCULATION OF CIDEP SPECTRA[26]

As described above, several independent polarization contributions to a CIDEP spectrum may occur and as time increases the equilibrium contribution to the signal (Equation 12) also increases. To calculate a spectrum for comparison with experimental ones we first calculate the lines positions from a knowledge of the g-factors and hyperfine couplings of the radicals. These also allow the TM and RPM contributions to the line intensities to be obtained using the theory above, and the various contributions are weighted empirically and added until, together with the absorptive single-phase contribution from the P_{eq} term of Equation 12, they reproduce the observed spectrum. It then remains to convolute these delta-function intensities with the line shapes calculated previously (Equations 12 and 17 to 27), remembering that these change in time. This implies that the appearance of a CIDEP spectrum is time dependent both in the intensities and the shapes of the lines. In this convolution it must also be remembered that the polarized and unpolarized contributions to the signal have different line shapes and have to be calculated separately before adding the two together to synthesize the final spectrum. The overall process is summarized in Figure 16.

If required the signals observed in experimental spectra may be subject to data-handling transformations to increase either their resolution or their S/N ratio.[22]

VI. CIDEP IN PRACTICE — THE PHOTOLYSIS OF MALEIMIDE SOLUTIONS

CIDEP has been used in conventionally irradiated systems to investigate the precursors to polymerization present in maleic anhydride[65] and maleimide (MI) solutions.[66] Here we give some flash-photolysis results for the latter system.[67]

The experimental and calculated spectra observed in solution in propan-2-ol are shown in Figure 17. Two species are observed, the MI radical with sharp ESR lines and an adduct radical, with broader lines due to unresolved couplings, formed by addition of the Me$_2$ĊOH radical created in the geminate process to a normal MI molecule:

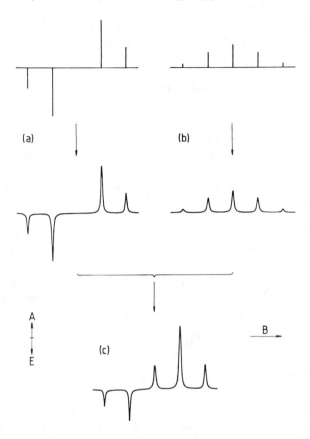

FIGURE 16.   The calculation of the spectrum of a radical, formed in an identical pair, in which the polarization is of pure triplet ST_0 RPM origin but which is observed at a time when an equilibrium contribution to the overall spectrum has become appreciable. The polarized and unpolarized spectra have different line intensities and different line shapes, (a) shows the calculated ST_0 RPM stick spectrum, with the central line characteristically missing, which is then convoluted with its line shape (from $G_y(t)$, Equation 12), (b) shows the stick spectrum of the same radical at equilibrium, with normal relative intensities between the lines, which is then convoluted with its different line shape (from $G_y(t)$). The two are added together to produce the calculated spectrum at the given time, (c). If a spectrum of the same general appearance (more intensity in one phase than in the other) was observed early on in the lifetime of the radical, it would be expected that it arose from a sum of TM and RPM contributions rather than equilibrium and RPM ones. In this case the stick spectrum in (b) would be the same but the line shape would be that appropriate to a polarized system. In general both RPM and TM effects contribute to the polarized spectrum, (a).

$$^3\left[\begin{array}{c}\mathord{\underset{\displaystyle O}{\overset{\displaystyle O}{\bigcirc}}}NH\end{array}\right] + Me_2CHOH \longrightarrow \mathord{\overset{\displaystyle O}{\underset{\displaystyle O}{\bigcirc}}}NH + Me_2\overset{\bullet}{C}OH \overset{MI}{\longrightarrow} \mathord{\overset{\displaystyle O}{\underset{\displaystyle O}{\bigcirc}}}NH \qquad (62)$$

It is immediately obvious that the spectrum exhibits more intensity in one phase (here emission) than in the other. Under the room-temperature and low-viscosity conditions of the experiment, where ST_{-1} effects would not be expected, this immediately implies the operation of both TM and RPM effects and this is what causes the spectrum to have its

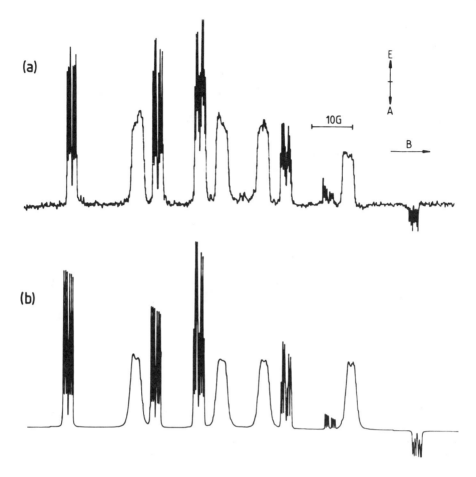

FIGURE 17.   (a) Observed and (b) calculated spectra from radicals observed on photolysis of a 0.1 *M* solution of maleimide in propan-2-ol with time integration from 1.9 to 7.2 μs after the photolysis flash. Two radicals are present, a neutral one derived from maleimide, with sharp lines, and an adduct with broad lines (Equation 62). The spectrum shows an excess of emissive phase, due to a TM process, together with an E/A distortion due to an ST_0 RPM process. Both identify the triplet state as the radical precursor.

lopsided appearance. The next problem is to identify the source of the RPM contribution, i.e., the nature of the radical pair which causes it. There are obviously three candidates and which two are relevant depends upon how fast the secondary radical is formed. Here inspection of the spectrum again helps; if the secondary was formed outside of the geminate cage by reaction with a primarily polarized $Me_2\dot{C}OH$ radical, its spectrum would not show a hyperfine-dependent polarization (see above; here the correlated couplings are too small to yield a noticeable effect). We conclude that since the intensities are hyperfine dependent it must be formed very rapidly and be, to some extent at least, the polarization partner of the MI radical. The theoretical spectrum shown is consequently obtained by calculating the RPM pattern expected for this pair and adding sufficient TM contribution to reproduce the observed spectrum. We note that the E/A nature of the hyperfine distortion, together with the TM contribution, both confirm that the molecule reacts through its excited triplet state.

The analysis reproduces the spectrum of the MI radical well but fails to reproduce the hyperfine dependence of the adduct spectrum completely. This implies that a further contribution to its polarization occurs, either from reaction of the polarized $Me_2\dot{C}OH$ or from a different radical pair than that considered. More work is required on this problem.

If triethylamine is added to the solution the adduct is completely suppressed and the spectrum of the radical anion formed by acid-base equilibration of the MI neutral radical is alone observed (Figure 18). This figure shows pure single-phase polarization, entirely due to the TM. This is the common experience with radical anions and it is surprising since the radicals are always formed in pairs and RPM polarization would be expected. In fact RPM polarization does occur but its effects are removed even at very low exchange rates by electron-transfer processes, usually involving the parent molecule.[68]

VII. THE EFFECTS OF EXCHANGE IN SPIN-POLARIZED RADICALS

As mentioned above, electron exchange is ubiquitous in solutions of radical anions and it produces notable effects on the spectra of these single-phase polarized systems, as does proton exchange. Effects due to exchange can also be observed in RPM-polarized radicals, typically those undergoing site exchange in radicals interconverting between degenerate conformers. In either situation the problem can be treated theoretically using the McConnell extension to the Bloch equations, in which terms are added to Equations 6 and 9 to account for magnetization transfer to or from a given radical.

The effect of electron transfer between degenerate states is to transfer the electron from a radical which is in a particular hyperfine state to a molecule whose hyperfine state is random:

$$A_i^{\cdot-} + A \rightleftarrows A + A_{i,j}^{\cdot-} \tag{63}$$

This causes the electron in general to experience a different coupling in each molecule so that it experiences a fluctuating magnetic field which may affect its relaxation. The net rate of electron transfer depends upon the degeneracy of the initial state, D_i, and the overall degeneracy of the radical D. With certain assumptions[15] the Bloch equations can be written

$$\underline{M}_i(t) = L\underline{M}_i(t) + T_1^{-1}M_{eq,i} + \frac{1}{D\tau}\sum_j D_j \cdot \underline{M}_j(t) - \frac{1}{\tau}\underline{M}_i(t) \tag{64}$$

where τ is the mean lifetime of the radical between electron jumps. In the case of fast and intermediate rates of exchange, that is, at frequencies greater than those of the line separations, this equation can be solved only numerically. Provided observations are made at times well before phase crossover of an initially emissively polarized signal, the spectra exhibit the features normally observed in the magnetic resonance of exchanging systems. Exchange first causes the lines to broaden progressively as the rate increases until they overlap and collapse into a single broad line. Thereafter further increase in the exchange rate causes exchange narrowing and at very fast rates a single line is observed with a line width similar to that observed in the nonexchanged system.[69] This means that exchange rate constants can be obtained by studies of highly transient radicals, thus extending the range of exchange studies.

In the slow exchange region, defined as that exchange regime in which the rate is less than the line separation, Equation 64 can be solved analytically and a new phenomenon is predicted.[70] This is that the effective relaxation rate of the radicals depends upon the degeneracy of the specific hyperfine state that the radical is in; spectral lines of the highest degeneracy in an emissively polarized radical relax through zero first, followed by the others in the order of their degeneracy. As the phase change is approached the lines display the distinctive line shape discussed at phase crossover in turn, as shown in Figure 19. This behavior is often observed in low-concentration solutions of radical anions[70,71] and the

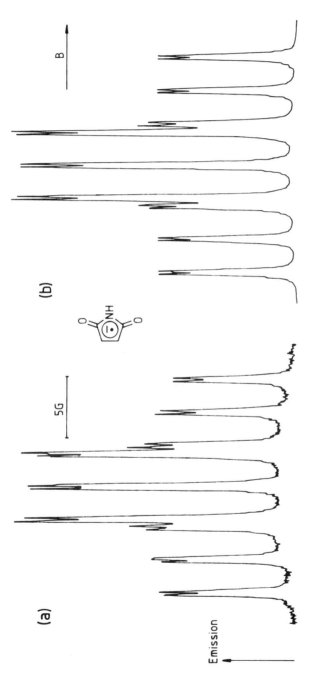

FIGURE 18.   (a) Observed and (b) calculated spectra of the maleimide radical anion, produced by photolysis of a 0.075 *M* solution of maleimide in a 4:1 propan-2-ol/triethylamine mixture, observed 1.4 to 6.2 μs after the flash. This exhibits pure emissive TM polarization, any RPM contribution having been averaged out by electron exchange with the parent molecule.

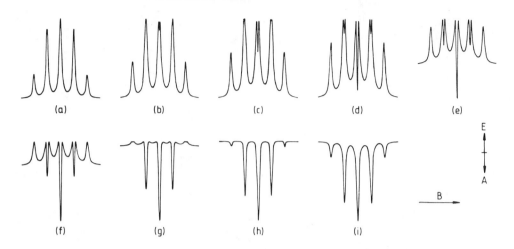

FIGURE 19.   The predicted variation in time of the TIS spectrum of a pure emissively TM-polarized radical containing four equivalent protons, under the conditions of slow electron exchange. The calculations used the following values: $T_1 = 5$ μs, $T_2 = 1$ μs, exchange lifetime = 1 μs, $\omega_1 = 1$ rad MHz, and $(P_1/P_{eq}) = -10$. All the spectra are normalized to the same total amplitude and so give no indication as to how the absolute intensity falls. The integration period was 0.5 μs throughout and the spectra were calculated with this starting at a series of times after radical creation: (a) 5 μs, (b) 8 μs, (c) 10 μs, (d) 12 μs, (3) 14 μs, (f) 16 μs, (g) 20 μs, (h) 23 μs, and (i) 30 μs. As time progresses each line in turn, starting with the strongest, relaxes though zero intensity in the order of its degeneracy. As it does the line exhibits the characteristics shown in Figure 7a. Eventually all the lines have inverted and the signal observed is the equilibrium one.

qualitative effect is observed even at intermediate exchange rates, if the individual lines still can be resolved.[69]

The physical origin of this effect can be seen from Figure 20. In the slow exchange situation a single transition is excited at a given field position, and we can think of all the radicals in other hyperfine states as constituting a large reservoir connected to the level involved in this single transition by exchange. For a radical initially overpopulated by CIDEP in the upper-energy states the population of the upper level of the single transition is reduced in time by relaxation and by transitions driven by the resonant microwave field; in the slow exchange limit this latter process does not occur for the off-resonant other lines in the system. As the population of the level falls it is refilled by net exchange from the radicals in the reservoir. Assuming that exchange occurs in the absence of a nuclear spin correlation and the process is random, the rate of refilling depends upon the difference in the degeneracy of the individual line, which determines the rate at which electrons leave the level by exchange, and that of the total system, which determines the refilling rate. Overall the rate of depopulation, i.e., the effective relaxation rate, depends on all three processes affecting the level population and is greatest for the line of highest degeneracy. Through the dependence on the microwave field strength the conditions may be adjusted to observe the effect during the lifetime of the radical as a polarized species.

At still lower rates of exchange the effects on relaxation times of this process become too small to detect in this way but they are apparent directly in the line widths. This is always the case for radical anions at low concentration. It is a characteristic of the CW flash-photolysis experiment that line shapes observed in the first few microseconds after radical creation are dominated by T_2. In the presence of slow exchange its effective value for a given hyperfine line is[15]

$$(1/T_{2i})^{eff} = (1/T_{2i}) + (D - D_i)/D\tau \qquad (65)$$

Proton exchange involving single-phase polarized spectra can be treated in a similar

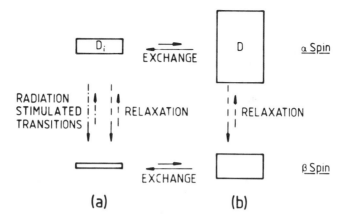

FIGURE 20. The origin of hyperfine-dependent effective relaxation in an emissively-polarized system undergoing slow electron exchange. In (a) are designated the energy levels and degeneracies of the radical in the specific hyperfine state whose transition is being excited. All other transitions are off-resonance and are treated as an ensemble, with much greater overall degeneracy, in (b). Three competing processes, relaxation, microwave pumping, and exchange, determine the subsequent population of the upper state in (a). The net rate of population transfer between the upper states in (a) and in the ensemble due to exchange depends upon the different degeneracies in the two and is greatest for the individual transition of highest degeneracy; this consequently has the shortest effective relaxation time.

way to electron exchange, now by amending the Bloch equations to include magnetization transfer between two radicals.[69] As with electron exchange in radicals observed soon after their creation, proton exchange causes the same phenomena as in equilibrated radicals, here an alternating line width effect.

Site exchange, as between two interconverting forms of the cyclohex-1-olyl radical, raises new possibilities for now the radicals are RPM-polarized and the exchange may connect lines of either the same or different phases. Now the observed signal is proportional to the y-component of the total magnetization given by the sum of the contributions $\underline{M}_A(t)$ and $\underline{M}_B(t)$ from the different sites A and B:[69]

$$\underline{M}(t) = \underline{M}_A(t) + \underline{M}_B(t) \tag{66}$$

where, for example, $\underline{M}_A(t)$ is obtained from the equation

$$\dot{\underline{M}}_A(t) = L\underline{M}_A(t) - k\underline{M}_A(t) + k\underline{M}_B(t) + T_1^{-1}\underline{M}_{A,eq} \tag{67}$$

This is similar to the equation for proton exchange.[69] Using the nomenclature of earlier sections, for a two-site exchange process six coupled differential equations must be solved, given in matrix form by

$$\begin{bmatrix} \dot{\underline{M}}_A \\ \dot{\underline{M}}_B \end{bmatrix} = \begin{bmatrix} (L_A - k\mathbf{1}) & k\mathbf{1} \\ k\mathbf{1} & (L_B - k\mathbf{1}) \end{bmatrix} \begin{bmatrix} \underline{M}_A \\ \underline{M}_B \end{bmatrix} + \begin{bmatrix} \underline{M}_{A,eq} \\ \underline{M}_{B,eq} \end{bmatrix} T_1^{-1} \tag{68}$$

At high interconversion rates the equations predict line-broadening effects, as for electron exchange above, but at low rates another new feature emerges: the effect of exchange is to transport population between the two sites so that the connected transitions tend towards the same resultant intensity.

If the two lines occur on the same side, low field or high field, of the spectrum of a radical which exhibits symmetric ST_0 polarization, this simply varies the intensities of the lines as calculated from the basic polarization theory. However, for such a system exchange between sites whose transitions fall on opposite sides of the center averages the intensity of each to zero. In general however spectra are not symmetrical in their polarization characteristics due to dissimilar radicals constituting the pair, a TM contribution, or a ST_{-1} polarization contribution. In this case very slow exchange causes lines to show the average of the nonexchanged intensities in the spectrum and this may cause, for example, a line expected to be in absorption to appear in emission. A case has been reported for the cyclohex-1-olyl radical at low temperature when ST_{-1} mixing occurs; at 223 K the spectrum observed between 1 to 2 μs after the flash exhibits successive lines in E/A/E/A phases.[68] Such behavior is impossible without exchange.

VIII. CONCLUDING REMARKS

In this by no means exhaustive account of flash-photolysis ESR and CIDEP we have attempted to introduce the principles of how spectra are obtained and analyzed with respect to both intensities and line shapes. It cannot be stressed too strongly that the spectra must be subject to complete analysis for safe conclusions as to radical natures and the multiplicities of their precursors to be made. In particular spin-polarized spectra differ from those from equilibrated radicals in that the relative intensities of the lines vary with time after radical creation, due to the operation of relaxation and of different polarization effects, as do their line shapes. Nevertheless the experiments do give information difficult to obtain in any other way in a remarkably direct manner, as, for example, in the demonstration of singlet-state reactivity in some systems.[72] No other method is capable of identifying the multiplicity of the specific precursor which leads to a specific radical; the photophysics and photochemistry of the system are directly linked.

Besides this, perhaps the greatest aspect of CIDEP, there has arisen recently the exciting possibility of studying radical pairs directly in systems of restricted diffusion. Here polymers and polymer solutions are sure to play their part. Furthermore, we have shown that all manner of exchange processes can be studied at rates varying from the very slow to the very fast.

All of these studies depend upon the understanding of spin polarization and the patterns it produces in all of its many manifestations. A large part of the article has consequently gone in attempting to explain its origins and effects. Here we have attempted to provide a readable and simple account, after Adrian,[29] rather than to use the language of density matrices and stochastic Liouville equations much of the original theory is expressed in. It must be realized however that the simple account is distilled from this, and if the reader wishes greater understanding and an appreciation of the details of the physical phenomena underlying the magnitude of polarization it is to this theory he must turn. It is elegant work which will repay his attention.

Finally, this article has drawn freely on the published and unpublished work of five ex-members of my research group, P. J. Hore, S. Basu, C. D. Buckley, A. J. D. Ritchie, and D. G. Stevens. I am grateful to them.

REFERENCES

1. **Fessenden, R. W. and Schuler, R. H.,** Electron spin resonance studies of transient alkyl radicals, *J. Chem. Phys.,* 39, 2147, 1963.
2. **Muus, L., Atkins, P. W., McLauchlan, K. A., and Pedersen, J. B.,** *Chemically Induced Magnetic Polarization,* D. Reidel, Dordrecht, Netherlands, 1977.
3. **Hore, P. J., Joslin, C. G., and McLauchlan, K. A.,** Chemically induced dynamic electron polarization, *J. Chem. Soc. Spec. Period. Rep. ESR,* 5, 1, 1978.
4. **McLauchlan, K. A. and Stevens, D. G.,** Flash photolysis electron spin resonance, *Acc. Chem. Res.,* 21, 54, 1988.
5. **Atkins, P. W., McLauchlan, K. A., and Simpson, A. F.,** A flash-correlated 1 μs response electron spin resonance spectrometer for flash photolysis studies, *J. Phys. E,* 3, 547, 1970.
6. **Smaller, B., Remko, J. R., and Avery, E. C.,** Electron paramagnetic resonance studies of transient free radicals produced by pulse radiolysis, *J. Chem. Phys.,* 48, 5174, 1968.
7. **McLauchlan, K. A. and Stevens, D. G.,** Field-time two-dimensional transient electron spin resonance spectroscopy, MISTI methods extended to complete spectra, and a comparison of existing time-resolved methods, *Mol. Phys.,* 57, 223, 1986.
8. **Basu, S., McLauchlan, K. A., and Sealy, G. R.,** A novel time-resolved electron spin resonance spectrometer, *J. Phys. E,* 16, 767, 1983.
9. **Trifunac, A. D. and Thurnauer, M. C.,** Chemically induced dynamic electron polarization. Pulse radiolysis of aqueous solutions of alcohols, *J. Chem. Phys.,* 62, 4889, 1975.
10. **Trifunac, A. D., Thurnauer, M. C., and Norris, J. R.,** Submicrosecond time-resolved epr in laser photolysis, *Chem. Phys. Lett.,* 57, 471, 1978.
11. **Torrey, H. C.,** Transient nutations in nuclear magnetic resonance, *Phys. Rev.,* 76, 1059, 1949.
12. **Trifunac, A. D., Norris, J. R., and Lawler, R. G.,** Nanosecond time-resolved epr in pulse radiolysis via the spin-echo method, *J. Chem. Phys.,* 71, 4380, 1979.
13. **Pedersen, J. B.,** Theory of transient effects in time resolved esr spectroscopy, *J. Chem. Phys.,* 59, 2656, 1973.
14. **Verma, N. C. and Fessenden, R. W.,** Time-resolved esr spectroscopy, IV. Detailed measurement and analysis of the esr time profile, *J. Chem. Phys.,* 65, 2139, 1976.
15. **Hore, P. J. and McLauchlan, K. A.,** The time dependence of esr intensities of spin-polarized (CIDEP) transient radicals produced by flash-photolysis, *Mol. Phys.,* 42, 533, 1981.
16. **Stevens, D. G.,** Studies of Electron Spin Polarization in Transient Radicals, Ph.D. thesis, University Oxford, 1987.
17. **Hore, P. J. and McLauchlan, K. A.,** CIDEP and radical relaxation times in solution, *Rev. Chem. Intermed.,* 3, 89, 1979.
18. **Basu, S.,** Flash Photolysis Electron Spin Resonance and Electron Spin Polarization, Ph.D. thesis, University of Oxford, 1983.
19. **McLauchlan, K. A. and Sealy, G. R.,** A microwave-switched time-integration (MISTI) method in the study of spin-polarized radicals. A two-dimensional electron spin resonance experiment, *Mol. Phys.,* 52, 783, 1984.
20. **Furrer, R., Fujara, F., Lange, C., Stehlik, D., Viethand, H. M., and Vollmann, W.,** Transient esr nutation signals in excited aromatic triplet states, *Chem. Phys. Lett.,* 75, 332, 1980.
21. **Hore, P. J. and McLauchlan, K. A.,** CIDEP and spin-relaxation measurements by flash-photolysis epr methods, *J. Magn. Reson.,* 36, 129, 1979.
22. **Hore, P. J., McLauchlan, K. A., Muus, L. T., and Frydkjaer, S.,** Structure in time-resolved esr spectra, *Chem. Phys. Lett.,* 77, 127, 1981.
23. **Basu, S., McLauchlan, K. A., and Sealy, G. R.,** The continuous wave flash photolysis electron spin resonance spectra of spin-polarized (CIDEP) radicals using time-integration spectroscopy, *Mol. Phys.,* 42, 431, 1984.
24. **McLauchlan, K. A. and Stevens, D. G.,** Chemically induced dynamic electron polarization (CIDEP) in radicals produced by photolysis with plane-polarized light, *Mol. Phys.,* 60, 1159, 1987.
25. **Fessenden, R. W.,** Chemically induced electron polarization of radiolytically-produced radicals, in *Chemically Induced Magnetic Polarization,* D. Reidel, Dordrecht, Netherlands, 1977, 119.
26. **Buckley, C. D., Grant, A. I., McLauchlan, K. A., and Ritchie, A. J. D.,** Flash-photolysis electron spin resonance in solution studies of free radicals, *Faraday Discuss. Chem. Soc.,* 78, 257, 1984.
27. **Atkins, P. W. and McLauchlan, K. A.,** Electron spin polarization, in *Chemically Induced Magnetic Polarization,* Lepley, A. R. and Closs, G. L., Eds., Wiley-Interscience, New York, 1973, 44.
28. **Wan, J. K. S., Wong, S. K., and Hutchinson, D. A.,** Chemically induced dynamic electron polarization. II. A general theory for radicals produced by photochemical reactions of excited triplet carbonyl compounds, *J. Chem. Phys.,* 58, 985, 1973.

29. **Adrian, F. J.,** Theory of anomalous electron spin resonance spectra of free radicals in solution. Role of diffusion-controlled separation and re-encounter of radical pairs, *J. Chem. Phys.,* 54, 3918, 1971.
30. **Pedersen, J. B. and Freed, J. H.,** Theory of chemically-induced dynamic electron polarization. I, *J. Chem. Phys.,* 58, 2746, 1973.
31. **Pedersen, J. B. and Freed, J. H.,** Theory of chemically-induced dynamic electron polarization. II, *J. Chem. Phys.,* 59, 2869, 1973.
32. **Buckley, C. D., Hunter, D. A., Hore, P. J., and McLauchlan, K. A.,** Electron spin resonance of spin-correlated radical pairs, *Chem. Phys. Lett.,* 135, 307, 1987.
33. **Closs, G. L., Forbes, M. D. E., and Norris, J. R.,** Spin-polarized electron paramagnetic resonance spectra of radical pairs in micelles. Observation of electron spin-spin interactions, *J. Phys. Chem.,* 91, 3592, 1987.
34. **Sakaguchi, Y., Hayashi, H., Murai, H., and I'Haya, Y. J.,** CIDEP study of the photochemical reactions of carbonyl compounds showing the external magnetic field effect in a micelle, *Chem. Phys. Lett.,* 110, 275, 1984.
35. **Trifunac, A. D., Nelson, D. J., and Mottley, C.,** Chemically induced dynamic electron polarization. Examples of $ST_{\pm 1}$ polarization, *J. Magn. Reson.,* 30, 263, 1978.
36. **Trifunac, A. D.,** Chemically induced dynamic electron polarization in pulse radiolysis. Viscous solutions, *Chem. Phys. Lett.,* 49, 457, 1977.
37. **Buckley, C. D. and McLauchlan, K. A.,** The influence of ST_{-1} mixing in spectra which exhibit electron spin polarization (CIDEP) from the radical pair mechanism, *Chem. Phys. Lett.,* 137, 86, 1987.
38. **McLauchlan, K. A.,** Flash photolysis electron spin resonance, in *Chemically Induced Magnetic Polarization,* D. Reidel, Dordrecht, Netherlands, 1977, 151.
39. **Atkins, P. W. and Evans, G. T.,** Electron spin polarization in a rotating triplet, *Mol. Phys.,* 27, 1633, 1974.
40. **Pedersen, J. B. and Freed, J. H.,** Theory of chemically induced dynamic electron polarization. III. Initial triplet polarizations, *J. Chem. Phys.,* 62, 1706, 1975.
41. **Atkins, P. W., Dobbs, A. J., and McLauchlan, K. A.,** Measurement of spin-lattice relaxation times of triplet states in fluid solution, *Chem. Phys. Lett.,* 29, 616, 1974.
42. **Basu, S., McLauchlan, K. A., and Sealy, G. R.,** Chemically induced dynamic electron polarization (CIDEP) in nitrogen heterocyclic radicals; a stringent test of polarization theory, *Chem. Phys. Lett.,* 88, 84, 1982.
43. **Buckley, C. D. and McLauchlan, K. A.,** Flash photolysis electron spin resonance and CIDEP studies of radicals derived from nitrogen heterocyclics. II. The photophysics and photochemistry of methyl pyrazines, *Chem. Phys.,* 86, 323, 1984.
44. **Dobbs, A. J. and McLauchlan, K. A.,** CIDEP from photolysis with plane-polarized light, *Chem. Phys. Lett.,* 39, 257, 1975.
45. **Adeleke, B. B., Choo, K. Y., and Wan, J. K. S.,** Chemically induced electron polarization. VI. Dependence of spin polarization of 1,3-benzosemiquinone and 9,10-anthrasemiquinone radicals on the orientation of the exciting polarized light, *J. Chem. Phys.,* 62, 3822, 1975.
46. **Yamauchi, S., Tominaga, K., and Hirota, N.,** Existence of the triplet mechanism in the CIDEP spectrum of acetone, *J. Phys. Chem.,* 90, 2367, 1986.
47. **Monchik, L. and Adrian, F. J.,** On the theory of chemically induced electron polarization (CIDEP): vector model and an asymptotic solution, *J. Chem. Phys.,* 68, 4376, 1978.
48. **Noyes, R. M.,** A treatment of chemical kinetics with special applicability to diffusion controlled reactions, *J. Chem. Phys.,* 22, 1349, 1954.
49. **Adrian, F. J.,** Radical pair mechanism of chemically induced magnetic polarization, in *Chemically Induced Magnetic Polarization,* D. Reidel, Dordrecht, Netherlands, 1977, 77.
50. **McLauchlan, K. A. and Stevens, D. G.,** The electron spin polarized (CIDEP) spectra of isopropyl radicals. A test of radical pair mechanism theory, *J. Magn. Reson.,* 63, 473, 1985.
51. **Thurnauer, M. C., Chiu, T.-M., and Trifunac, A. D.,** The sign of electron exchange in random encounter pairs, *Chem. Phys. Lett.,* 116, 543, 1985.
52. **Carmichael, I. and Paul, H.,** CIDEP during the photolysis of di-*tert*-butyl ketone, *Chem. Phys. Lett.,* 67, 519, 1979.
53. **Basu, S., Grant, A. I., and McLauchlan, K. A.,** The remarkable spin-polarization behavior of radicals derived from aliphatic ketones, *Chem. Phys. Lett.,* 94, 517, 1983.
54. **McLauchlan, K. A. and Stevens, D. G.,** Two-dimensional transient electron spin resonance spectroscopy, *J. Chem. Soc. Faraday Trans. 1,* 83, 29, 1987.
55. **Jent, F., Paul, H., McLauchlan, K. A., and Stevens, D. G.,** The concentration-dependence of the phase of radical pair mechanism chemically induced dynamic electron polarization (CIDEP), *Chem. Phys. Lett.,* 141, 443, 1987.
56. **Valyaev, V. I., Molin, Yu. N., Sagdeev, R. Z., Hore, P. J., McLauchlan, K. A., and Simpson, N. J. K.,** *Mol. Phys.,* 63, 891, 1988.

57. **Burkey, T. J., Lusztyk, J., Ingold, K. U., Wan, J. K. S., and Adrian, F. J.,** Chemically induced dynamic electron polarization of the diethoxyphosphonyl radical: a case of mixed ST_0 and ST_{-1} radical pair polarization, *J. Phys. Chem.,* 89, 4286, 1985.
58. **Murai, H., Sakaguchi,Y., Hayashi, H., and I'Haya, Y. J.,** An anomalous phase effect in the individual hyperfine lines of the CIDEP spectra observed in the photochemical reactions of benzophenone in micelles, *J. Phys. Chem.,* 90, 113, 1986.
59. **Pedersen, J. B.,** Determination of the primary reactions of photosynthesis from transient esr signals, *FEBS Lett.,* 97, 305, 1979.
60. **McLauchlan, K. A. and Stevens, D. G.,** On alternating phase effects in individual spin multiplets in electron spin-polarized (CIDEP) free-radical spectra, *Chem. Phys. Lett.,* 115, 108, 1985.
61. **McLauchlan, K. A. and Simpson, N. J. K.,** The transfer of electron spin polarization (CIDEP) to secondary radicals, *Chem. Phys. Lett.,* 154, 550, 1989.
62. **Pedersen, J. B., Hansen, C. E. M., Parbo, H., and Muus, L. T.,** A CIDEP study of *p*-benzosemiquinone, *J. Chem. Phys.,* 63, 2398, 1975.
63. **Fessenden, R. W.,** Time-resolved esr spectroscopy. I. A kinetic treatment of signal enhancements, *J. Chem. Phys.,* 58, 2489, 1973.
64. **Paul, H.,** Second order rate constants and CIDEP enhancements of transient radicals in solution by modulation esr spectroscopy, *Chem. Phys.,* 15, 115. 1976.
65. **Roth, H. K., Hoernig, S., and Wuensche, P.,** CIDEP and kinetic studies of photolytically generated transient radicals of maleic anhydride by time-domain esr, *Polym. Photochem.,* 4, 409, 1984.
66. **Ayscough, P. B., Elliot, A. J., English, T. H., and Lambert, G.,** Electron spin resonance studies of elementary processes in radiation- and photo-chemistry. XIV. Photolysis of solutions containing maleimides, *J. Chem. Soc. Faraday Trans. 1,* 73, 1302, 1977.
67. **Ritchie, A. J. D.,** Chemically induced dynamic electron polarization of transient free radicals in solution, Ph.D. thesis, University of Oxford, 1985.
68. **McLauchlan, K. A. and Stevens, D. G.,** The effects of site exchange on the esr spectra of free radicals with radical pair mechanism chemically-induced dynamic electron polarization (CIDEP), *J. Chem. Phys.,* 87, 4399, 1987.
69. **McLauchlan, K. A. and Ritchie, A. J. D.,** A flash-photolysis electron spin resonance study of radicals derived from anhydrides of carboxylic acids; spin-polarized (CIDEP) spectra under conditions of fast electron exchange, *Mol. Phys.,* 56, 1357, 1985.
70. **McLauchlan, K. A. and Ritchie, A. J. D.,** A flash-photolysis electron spin resonance study of radicals formed from carboxylic acids; exchange effects in spin-polarized radicals, *Mol. Phys.,* 56, 141, 1985.
71. **Basu, S., McLauchlan, K. A., and Ritchie, A. J. D.,** Time resolved electron spin resonance with electron spin polarization (CIDEP) as a sensitive probe of degenerate electron exchange reactions, *Chem. Phys. Lett.,* 105, 447, 1984.
72. **Buckley, C. D. and McLauchlan, K. A.,** ESR studies of excited states in solution, *Mol. Phys.,* 54, 1, 1985.

INDEX

A

Absorption coefficient, 5
Absorption spectra, monitoring of, 129
Accelerated particle beam pumping, 8
ACENA, see Acenaphthylene copolymers
Acenaphthylene (ACENA) copolymers, 71—72
Additives, 34—36
A/E polarizations, 288—290, see also E/A polariza-
    tions
Alexandrite laser, 15
Alumina crystal, 9—12
Alumina tube, 26
Aluminum-yttrium crystals, 10
Ammonium dihydrogen phosphate (ADP), 29
Amorphous polymer blends, 148—149
Amplification, 5—7
  coefficient, 6
  parametric, 31
  of picosecond dye laser pulses, 42
  pulsed dye, 42—43
Amplified spontaneous emission, 43
Amplifying band, width of, 8
Anisotropy, see also Fluorescence anisotropy
  data analysis of, 217
  fluorescence, 66, 112, 115—118
  measurement, 94
  time-dependent, 139—140
Anisotropy decays, 63—66, see also Fluorescence
    anisotropy decay
Anthracene polymers
  energy migration and transfer in, 73—74
  sorption experiments with, 200—201
  time-dependent anisotropy for, 225
A polymers, 96
Autocorrelation function, 223
Axial flow laser, 19
Azines, 32

B

Barium emission, 24
Bessel function, 222, 224
Bimolecular dissipation processes, 58
Binary interaction parameters, temperature-
    dependent, 191
Birefringent filters, 36
Birks' kinetic scheme, 74, 77
Black body, 5—6
Bloch equations, 261, 268, 272, 296, 299
Bohr magneton, 260
Boltzmann's equation, 7
Boltzmann's law, 3
Boxcar methods, 266, 274
B polymers, 95—96
Broadband light amplification, 42
BVN homopolymer, 72—73

C

Carbon dioxide lasers, 2, 16—18
  axial flow, 19
  hybrid TEA, 18—19
  metal vapor, 24—26
  pulsed TEA, 18
  sealed, 19—21
  ultraviolet excimer, 21—24
Carrier generation layer, 81
Carrier transport layer, 81
CCF, see Conformation correlation function
Cesium-dihydrogen arsenate, 29
Characterization techniques, 162
Charge transfer band, 57
Chemically induced dynamic electron polarization
    (CIDEP), 261, 268, 276—278
  advantages of, 300
  calculation of spectra of, 293
  measurement of polarization ratios in, 293
  phenomenon of, 263
  in practice, 293—296
  radical pair mechanism of, 282—292
  secondary polarization in, 292—293
  triplet mechanism of, 278—282
Chemical pumping, 8
Chlorophyll P700, 130—134
Chromium ions, 10
Chromophores, 65
  aromatic, 164
  binding of, 75
  classification of in polymers, 95—97
  density of, 157
  excitation transfer between, 62
  face-to-face arrangement of, 74
  fluorescence of, 106—107
  monomeric, 151
  naphthalene, 156
  naphthyl, 72, 153
  nonhomogeneous distribution of, 55—56
  orientation effect of, 56
  photoselection of, 136
  along polymer chain, 64
  separation of, 165, 186
  transfer rates between, 65
CIDEP, see Chemically induced dynamic electron
    polarization
CLPs, see Core-labeled particles
CLP-Phe, 207—208
Colloidal polymer particles, 198—199
Colloidal stability, 198
Combinatorial entropy contribution, 179
Conformation correlation function (CCF), 222—224
Conformation jumps, 222—223
Continuous wave dye lasers, 36, 40—42
Continuous wave lasers, 30, 103, 105, 261, 267
Continuous wave microwave radiation, 266

Continuous wave mode-locked laser sources, 28—29
Convolution, 99—101
Copper vapor laser relaxation, 7
Core-labeled particles (CLPs), 199, 207
Core-shell structures, 198, 209
Core-stabilizer interface, 207—208
Coulombic interaction, 81
Coulombic repulsion, 80—81
Coumarins, 32
Crystals, 10, see also specific materials
CVL emissions, 24
CVL tube, 24—26
Czochralski's method, 14

D

Deconvolution procedure, 100
Density matrices, 300
Dexter transfer, see Energy transfer
DFDL, see Distributed feedback dye laser
Dichloromethane solutions, 248—249
Differential scanning calorimetry, 162
Diode pumping, 15
Diphenylhexatriene probe, 230
Dipole-coupling mechanism, 201—202
Dipole-dipole electronic energy transfer, 62, 242—244, 247
Dipole-dipole interactions, 58, 61
Dirac method, 282
Distributed feedback dye laser (DFDL), 39
DODCI solution, 38, 41
Donor-acceptor separation, 247
Donor-donor analysis, 166
Donor-donor energy migration, 247
Doped particles, fractal analysis of, 201—202
Durbin-Watson parameter, 102—103
Dye amplifiers, pulsed, 42—43
Dye laser pumping, 13
Dye lasers, 31—32, 136—137
    continuous wave, 36
    excimer-pumped, 141
    flashlamp-pumped, 37—38
    laser-pumped picosecond, 38—40
    photophysical properties of, 32—34
    picosecond and femtosecond continuous wave, 40—42
    picosecond tunable, 39—40
    pulsed, 36—37
    pulsed amplifiers for, 42—43
    solvent effects and additive influence in, 34—36
    stimulated emission of, 34
    tunable, 37, 39—40
Dyes, see Laser dyes
Dynamic mechanical spectroscopy, 162

E

E/A polarization, 292—293, see also A/E polarization
EDA, see Electron donor acceptor complex
EET, see Electronic excitation transport
Einstein coefficient A, 33

Einstein coefficient B, 4, 34
Electromagnetic waves, 54
Electron donor acceptor (EDA) complex, 59
Electron exchange, 61, 244, 298—300
Electronic energy migration, see Energy migration
Electronic energy transfer, see also Energy transfer
    kinetic models of, 244—246
    laser studies of in polymers, 238—255
    mechanisms of, 242—244
    theory of, 241—247
Electronic excitation transfer
    medium interaction of, 62
    strong interaction of, 61
    theory of, 61—64
    weak interaction of, 62—64
Electronic excitation transport (EET), 163—166
Electronic transitions, 213
Electronic wave function, 57
Electron spin resonance (ESR), 158, 260
    intensities of, 261
    spectra of, 277
    spectrometers for, 261—262
    transient, 263
Electron spin spectroscopy, transient, 265—267
Electron transfer processes, 60, 80—83
EM, see Energy migration
Emission spectra, time-resolved, 113—115
Emission wavelength, 126, 128
Energy, 3, 6
Energy migration (EM), 58—59, 181, 246—247
    efficiency of, 178
    experimental observation of in polymers, 67—74
    fluorescence decay and, 112—113
    one-dimensional, 70
    in polymers, 60—64
    three-dimensional, 71
Energy states, 3, 4
Energy transfer, 58—59, see also Electronic energy transfer
    by dipole-coupling mechanism, 201—202
    dipole-dipole, 242
    experimental observation of in polymers, 67—74
    nonradiative, 242—244
    in polymers, 60—64, 238—255
    radiative, 242
    single step, 246—247
    studies of in polystyrenes, 247—255
Epitaxy, 15—16
ESR, see Electron spin resonance
Excimer dissociation rate, 204
Excimer emission, 202—207
Excimer fluorescence
    delayed, 153, 156
    as molecular level probe, 191
    for polymer-polymer interactions, 172
    for predicting blend immiscibility, 174
    as quantitative photophysical tool, 162—163
    for temperature dependence of blend miscibility, 178
Excimer formation, 75, 107
Excimer forming sites, 163

concentration of, 171
  intermolecular, 164, 170—171
  intramolecular, 164
  in pure polystyrene, 247
  sampling of, 166, 170
  types of, 164
Excimer laser excitation, 141
Excimer lasers, 21—24
Excimer molecular probe, 166—167, 170
Excimer phosphorescence, 153—155
Excimers, 59
  binding energy of, 75
  emission of from labeled stabilizer, 202—206
  fluorescence decays of, 108—112
  formation of, 68, 158—159
    in aromatic polymers, 107—108
    in polymers, 74—78, 163—164
    in polystyrene, 68—71
    due to rotational sampling, 180
  high energy, 75—76
  polystyrene, 68—71, 239—241
  reaction mechanisms of, 21—22
  second, 75
  tacticity effect on, 76
Exciplexes, 59, 78—80
Excitation hopping, 64—65
Excitation radiation, scattered, 99—100
Excitation transfer, in molecular aggregates, 64—
    66
Excited states, 92—95, 97
Exciton, 157, 164, 166, 170

F

Fabry-Perot cavity, 26
Fabry-Perot etalons, 8—9, 36
Fabry-Perot interferometer, 37
FAD, see Fluorescence anisotropy decay
FELs, see Free electron lasers
Femtosecond continuous wave dye lasers, 40—42
Fitting functions, 106—113
Fitting routine, 123—124
Flashlamp-pumped dye lasers, 37—38
Flashlamp systems, 37, 104
Flash photolysis, 129
  of maleimide solutions, 293—296
  methods of, 260
  picosecond diffuse reflectance laser, 135
  principles of, 130
Flash photolysis electron spin resonance, 260—267
Flash photolysis electron spin resonance, magnetic
    resonance and, 268—276
Flash pumping, 14
Flash spectroscopy, 130, 260
Flory-Huggins configurational entropy, 188
Flory-Huggins lattice treatment, 174, 175, 178
Flory-Huggins theory, 167, 172, 179, 191
Fluorescence
  atoms, 3
  intensity of, 94
  polarization of, 117

for polymer blends, 162
  quantum yield of, 92—93
Fluorescence anisotropy, 115—118, 163, 232
Fluorescence anisotropy decay (FAD), 115—117,
    212—213
  alternative methods of measuring, 118—129
  of anthracene label, 229
  apparatus for, 213—215
  chain motion studied with, 230
  comparison of with other spectroscopic methods,
    232—233
  with energy migration, 112—113
  heterogeneity and, 106—112
  with motion, 112
  multiexponential, 126—127
  in OACF study, 224—226
  ratio of to total decay, 163
  with relaxation, 113
  single exponential analysis of, 109—110
  triple exponential analyses of, 111
Fluorescence decay time, 93, 94
Fluorescence depolarization, 165
Fluorescence polarization, 68
Fluorescence probe method, 55
Fluorescence probes, 218, 219, 226—230
Fluorescence quenching, 81
Fluorescent decay, form of, 99
Fluorometer, variable-frequency phase modulation,
    124
Fluorometers, phase-modulation, 122—123
Fluorophores, 115, 124—126
Förster critical radius, 165, 246
Förster dipole-dipole kinetics, 246
Förster kinetic equation, 65, 245—246
Förster's mechanism, 61
Förster's original theory, 201
Förster transfer, 164, 186, 254
Fourier transform, 26, 101
F pairs, polarization in, 288—290
Fractal analysis, 201—202
Franck-Condon excited state, 57
Franck-Condon transition, 33
Free electron lasers (FELs), 2, 43—45
Free radicals, 260—261
Frequency conversion, 29—31
Frequency domain phase-modulation fluorimetry,
    119—129
Frequency tripling, 31
Frequency upconversion, 31

G

Gain characteristics, 42—43
Gain media, 27
Galilean telescope, 27
Gas circulation system, for high repetition rate TEA
    laser, 19
Gaussian beam, 10
GDL model, 224, 226, 229
Geminate radical pair mechanism polarization, 287—
    288

Geminate reencounter process, 285—286
Glasses, 10, see also specific materials
Glass transition temperatures, 162
Graft copolymers, 198, 209
Grating-pair technique, 41
Green function, 66

H

Hall-Helfand model, 222—226, 229
Hamamatsu photodiodies, 134
He-NE mode locking, 27
Heterogeneity, fluorescence decay with, 106—112
Heterojunctions, 16
Hole drift mobility, 81—82
Holographic transient polarization grating (HTPG),
    136—139, 212, 215, 224, 233
Homojunctions, 15
HTPG, see Holographic transient polarization grating
Hyperchromism, 57
Hyperfine coupling, 260, 286, 290—292
Hyperfine coupling constants, 287
Hyperfine-dependent polarization, 291
Hyperfine states, 296
Hypochromism, 57

I

IC, see Internal conversion
Impulse response function, see Instrument response
    function
Impurity fluorescence, 100
Impurity phosphorescence, 153
Instrument response function, 99, 124
Internal conversion (IC), 92
Intersystem crossing (ISC), 92
IR, see Iterative reconvolution methods
ISC, see Intersystem crossing
Iterative reconvolution (IR) methods, 217

J

Jones-Stockmayer (JS) expression, 227—228
JS, see Jones-Stockmayer expression

K

KDP, see Potasium dihydrogen phosphate
Kerr cell, 119
Kerr effect, 213
Kinetic models, 244—247
Kinetic spectrometry, 130, 131
KTP, see Potassium titanyl phosphate

L

Labeling, 218—221
Laplace transforms, 66, 101, 217, 270
LASER acronym, 2

Laser cavity properties, 8—9
Laser diodes, 15
Laser dyes, 31
    absorption spectrum of, 33
    photophysical properties of, 32—34
    principal classes of, 32
    radiative lifetime of, 33—34
    refractive index of, 34—35
    solubility of, 34—36
Laser emission spectrum, conditions of, 8—9
Laser photoexcitation, 152—153
Laser photolysis, 131, see also Flash photolysis
Laser pulse-probe methods, 157—158
Laser-pumped picosecond dye lasers, 38—40
Lasers
    carbon dioxide, 16—26
    dye, 31—43
    free electron, 2, 43—45
    high spatial coherence of, 2
    metal vapor, 3
    for photochemistry, 2—45
    pulsed, 92
    solid state, 10—16
    time-correlated single-photon counting and, 103—
        106
    types of, 2
Least square fitting, 101—103, 123
Leuko Crystal Violet, 83—84
Light sources, 216—217
Lippert-Mataga analysis, 79
Longitudinal cavity modes, 8—9
Longitudinal flow carbon dioxide laser, 20
Longitudinal multimode, 9
Luminescence, 148—151
Luminescence depolarization, 232—233
    local dynamics studies and, 221—231
    probes and labels for, 217—220
    techniques for study of, 212—217
Luminescence quenching, 60, 157
Luminescence signals, decays of, 159
Luminescence spectroscopy, time-resolved delayed,
    155—156
Luminescent site, extrinsic, 217—218

M

Magnetic wiggler, 44
Main-chain labeling, 218—220
Maleimide radical anion, 297
Maleimide solutions, photolysis of, 293—296
MeHPB copolymers, 250—251
Merocyanines, 32
Metal vapor lasers, 3, 24—26
Methanol, 36
Methods of moments, 101
Methyl acrylate copolymers, 71—72
Methyl methacrylate copolymers, 71—72
Michelson-type interferometer, 36
Microwave switched time-integration spectroscopy
    (MISTIS), 267

Microwave-switched time-integration (MISTI) technique, 271—272, 293
MISTI, see Microwave-switched time-integration technique
MISTIS, see Microwave switched time-integration spectroscopy
MOCVD, 15—16
Mode-locked continuous wave lasers, 28—29, 103, 105
Mode-locked lasers, 27, 30, 40, 95
Mode-locked picosecond fluorescence measurement, 120
Mode locking, 26—27
    active, 28
    colliding-pulse, 41
    of dye lasers, 39
    passive, 27, 29, 38, 41, 42
Mode-locking dye, 28
Modulating function, 101
Molecular aggregates, 60, 64—66
Molecular triplet states, 148
Molecular weight
    effect of on miscible polymer blends, 180—184
    effects of on spinodal decomposition, 190—191
    of polymer blend host, 172—177
Molecules, quantized levels of, 3
Monochromatic radiation, 6
Monochromator-photomultiplier system, 150
Monochromators, 130, 149
Multipole-multipole energy transfer, 61, 246

N

Nanosecond domain, 95
Nanosecond kinetic spectrometry, 131
Nanosecond measurements, 129
Naphthalene polymers, energy migration and transfer in, 71—73
Nd-glass lasers, 28
Nd:YAG lasers, 2
    active ion in, 12
    continuous mode-locked, 30, 40
    design of, 10
    diode pumping of, 15
    electric yield of, 14—15
    medium used in, 10
    picosecond, 28
    pulsed, 8
    resonant cavity of, 13—14
    stimulated emission in, 12
    time-correlated single-photon counting and, 103, 105
    use of slags in, 14
    wavelength in, 13
Neodymium ions, 10
Neodymium YAG laser, see Nd:YAG laser
Neutral density filters, 134
NMR, see Nuclear magnetic resonance
Nondedicated synchrotron, 216—217
Nonradiative energy transfer, 242—244

Nonradiative unimolecular decay paths, 57
Noyes formula, 286
Nuclear magnetic resonance (NMR) spectroscopy, 262, 268—276

O

OACF, see Orientation autocorrelation function
Observed decay, 99
Oligomer probes, 220, 232
Optical absorption, of transient triplets, 150, 157—158
Optical gating, 119
Optical multichannel analyzer, 119
Optical parametric oscillator, 31
Optical pulse compressor, 30
Orbital symmetries, 279—280
Organic photoconductors, 55
Orientation autocorrelation, analytic models for, 223
Orientation autocorrelation function (OACF), 212, 218
    continuous sampling of, 232
    measurement of, 221, 226—230
    models-of-motion studies of, 224—226
    sampling of, 217
    study of, 215
Oriented triplet states, 280
Oxygen quenching, 207—208

P

P1VN polymers, 166
P2VN
    concentration effect in, 168—172
    excimer to monomer intensity ratio for, 168
    excitation of, 163
    low concentration, 191
    molecular weight of, 172—177
    study of, 166
P2VN/PMMA blends, 174, 177
P2VN/PS blend, 174—177
PAE, 73—74
PAMMA, 73—74
Pauli master equation, 65
Pauli matrices, 223
PCVD, see Photochemical vapor deposition technology
PEHMA copolymer, 202, 206—208
PEHMA/PVAc particles, 199, 204, 208
PEPPA, 74
Perrin kinetic model, 245
Perturbation theory, 280
Perylene, fluorescence anisotropy of, 117—118
Phase matching, 29
Phase-plant plot, 101
Phase separation kinetics
    molecular weight effects on spinodal decomposition in, 190—191
    phase diagram for, 187—190
    two-phase model of, 186—187

Phase separation mechanism, 180
Phenanthrene (Phe) group, 73, 199—200
Phosphates, 10
Phosphorescence, 57, 148, 151—155
Phosphorescence spectroscopy, time-resolved, 154
Phosphor screen, 119
Photoabsorption, 56—57
Photochemical applications, 10—11, 19
Photochemical hole burning, 54
Photochemical reactions, products of, 94—95
Photochemical smog, 54
Photochemical vapor deposition (PCVD) technology, 54
Photochemistry
    applications of for Nd:YAG lasers, 13
    lasers for, 2—45
Photochromic materials, 54
Photoconductors, 54
    carrier transport layers of, 81
    organic, 55
    photoinduced electron transfer in solid polymers and, 81—83
    two-layered system of, 81
Photodegradation, 55, 153
Photodimerization, 59
Photoenergy absorption, 54
Photoexcitation pulse, 148
Photolysis, 142, 262—263, see also Flash photolysis
Photomemory, 54
Photomultiplier, 100, 213
Photon energy, 56
Photon-harvesting polymers, 73—74
Photooxidation, 130—134
Photophysical/chemical processes, 54, 56—60
Photophysical studies
    of miscible and immiscible amorphous polymer blends, 162—191
    of solid polymers, 148—159
Photopolymerization, 54
Photoresponsive materials, 54
Photosensitive materials, 54—55
Photostabilizers, 250
Photostationary-state kinetics, 163
Photoswitching, 54
Physical principles, 3—4
PIB/PMMA particles, 198—199, 202, 203
Picosecond continuous wave dye lasers, 40—42
Picosecond diffuse reflectance laser flash photolysis, 135
Picosecond laser transient fluorescence anisotropy, 163
Picosecond measurements, 129—130
Picosecond pulse generation, 39
Picosecond rare-gas halide lasers, 29
Picosecond ruby laser pulses, 27—28
Picosecond transient absorption spectrometer, 132
π-π transition, 57
PMMA, 172—174, 178—179, 200, see also PIB/ PMMA particles
PnBMA host, 178—179
PNCz, 75—76
Poisson distribution, 217

Polarization, 277
    decays, 281
    F-pair, 278, 288—290
    measurement of ratios of, 293
    radical pair mechanism, 282—292
    with restricted motion, 292
    secondary, 292—293
    ST_{-1} RPM, 290—291
    theory of, 261, 300
    triplet mechanism, 278—282
Polyacetylene, 158
Poly(alkyl methacrylate) host matrices, 167—172
Polydiacetylene, 158
Polyesters, structures of, 76
Poly(ethylene), 95
Polyisoprene main-chain dynamics, 224
Poly-L-proline, 62, 63
Polymer blends
    binary interaction parameters of, 174—175, 179
    binodal compositions of, 175—177
    concentration effects in, 168—172
    early studies of, 166
    immiscibility of, 165
    mechanical properties of, 162
    molecular weight effect in, 172—177
    morphology of
        close to immiscibility, 166—180
        miscible, 180—186
        phase separation kinetics in, 186—191
    photophysical studies of, 162—191
    solubility parameter differences in, 166—168
    temperature effects in, 177—179
    thermodynamic equilibrium in, 175, 178, 191
    thermodynamic stability of, 174
    uses of, 162
Polymer chains, 56
    aryl vinyl, 165
    cyclization properties of, 80
    flexibility of, 76
    models-of-motion of, 224—226
    theoretical models of motions of, 221—224
    zipping effect of, 78
Polymer effects, 56
Polymer photophysics, 163—166
Polymers, see also Polymer blends; Polymer chains
    antenna effect in, 73
    with aromatic chromophores, laser studies of energy transfer in, 238—255
    chromophore classification in, 95—97
    depolarization of luminescence of, 212—233
    electronic excitation transport in, 164—166
    energy transfer and energy migration processes in, 60—74
    excimer formation in, 74—78, 163—164
    exciplex formation in, 78—80
    heterogeneity of, 106—112
    interchromophore interactions in, 56
    interfaces, 198
    laser-induced production of triplet states in, 158
    photoinduced electron transfer processes in, 80—84
    photophysical and photochemical primary processes in, 54—84

solid
    delayed luminescence of, 149—151
    phosphorescence emission from, 151—155
    photophysical studies of triplet exciton processes
        in, 148—159
    triplet exciton migration in, 156—157
    specificity of in photophysics/chemistry, 55—56
    synthetic, 95—97, 107
    time-resolved emission spectra of, 113—115
    time-resolved fluorescence measurement of colloid
        morphology of, 198—209
Polymethine cyanines, 32
Poly(propylene), 95
Polypropylene oxide compounds, 230—231
Polystyrene
    atactic, 95
    conformational energy calculations of chains in,
        241
    energy migration and transfer in, 68—71
    energy transfer studies in, 247—255
    excimers and structure of, 239—241
    head-to-head, 70, 71
    isotactic, 96
    molecular weight of, 172—177
    pure, 247—248
    room temperature emission spectra of, 239—240
    spectra of, 238—240
    structure of, 239
    study of, 166
    syndiotactic, 95
Polystyrene-butadiene block copolymers, 248
Polystyrene copolymers, 248—255
Population inversion, kinetic equations for, 7—8
Potassium dihydrogen phosphate (KDP), 29, 31
Potassium titanyl phosphate (KTP), 31
PPA, 74
PPO, 248—249
Probe molecules, 97
Probes, 226—231, see also Fluorescence probes
Proton exchange, 297—300
PS/PVME blends, 163
    binary interaction parameter of, 188
    concentration in, 186
    critical solution temperature of, 183
    energy migration dimensionality in, 184
    high viscosity of, 181
    phase separation of, 187, 189—191
    temperature dependence in, 181—182
Pulse-counting technique, 129
Pulsed dye amplifiers, 42—43
Pulsed dye lasers, 36—37
Pulsed lasers, 92, 95
Pulsed ultraviolet sources, performance of, 216
Pulse fluorometry techniques, 212
Pulse pile-up errors, 100—101
Pumping
    diode, 15
    flash, 14
    four levels of, 7—8
    synchronous, 40
    for UV excimer lasers, 22
Pumping yield, 8

PVAC, 156—158
PVAc particle component, 207
PVAc/PEHMA particles, see PEHMA/PVAc particles
PVCz, 76
PVME host matrix, 183
Py groups, 202, 204, 206
Py pairs, 206

Q

Q factor, 11, 262
Q-switched solid-state lasers, 27—28, 95
Q-switching technique, 11, 12, 40
Quantum numbers, 3
Quartz plate beam splitter, 150
Quenchers, 58
Quenching, 81, 178—179
Q values, 287

R

Radiation intensity, 6
Radiative energy transfer, 242
Radiative transition probability, 58
Radical decay, 272
Radical pair mechanism, 277, 282—287
    effects of, 294—296
    in F pairs, 288—290
    patterns of germinate polarization in, 287—288
    polarization with restricted motion in, 292
    ST_{-1} polarization, 290—291
Radio frequency excitation, 21
Radiofrequency interference, 100
Raman effect, 31
Raman scattering, 31, 29, 139—141
Random mixing effects, 184—186
Random walk, 164—165
    one-dimensional, 180—181, 183, 191
    three-dimensional, 180, 183, 184
Rare gas atoms, 21
Rare gas dimers, energy potential curves of, 21—22
Redox photochemistry, 60
Red visible spectrum, 3
Reencounter polarization, 286—287
Refractive indices, 29, 34—35
Resistance-capacitance circuits, 155
Resonance Raman spectra, 95
Rhodamine, 32, 38, 41
Rotational diffusion coefficients, 117
Rouse matrix, 223
Ruby lasers, 10—12

S

Saturable absorbers, 27, 30, 42
Scattered excitation radiation, 99—100
Schrödinger equation, time-dependent, 285
Sealed carbon dioxide lasers, 19—21
Second harmonic generator, 30, 54—55
Self-phase modulation (SPM), 28
Self-Q-switching behavior, 38—39
Semiconductor circuits, high implantation density, 2

Semiconductor lasers, 3, 10, 14—16, 43
Shift routine, 100
Side-chain labeling, 218, 220—221
Silicates, 10
Singlet-singlet absorption measurement, 94
Singlet-singlet energy transfer phenomena, 62
Slabs, 14
SLP-Phe, 207—208
SLPs, see Stabilizer-labeled particles
Smoluchowski-Einstein equation, 68
Solid-state lasers, 10—11
    Q-switched and mode locked, 27—28
    specific types of, 11—16
    tunable, 15
Solvatochromism, 34
Solvent relaxation, 113
Solvents, 34—36
    polarity of, 78—80
    refractive index of, 37
Sorption experiments, 200—201
Spectrometers, 149—151, 263, 266
Spectroscopic methods, 212, 232—233, see also
        specific techniques
Spin-correlated radical pairs, 282—284
Spin-echo spectroscopy, 266—267
Spin-forbidden process, 57
Spin Hamiltonian, 280, 283—284, 290
Spin-lattice relaxation time, 282
Spinodal decomposition, 190—191
Spin-orbit Hamiltonian, 279
Spin polarization, 277, 278, 285, 300
Spin-polarized radicals, 281, 296—300
Spin wave function, 57
SPM, see Self-phase modulation
ST_{-1} polarization contribution, 290—291, 300
Stabilizer-core-material interface, 209
Stabilizer-labeled particles (SLPs), 199, 202, 205—
        206
Stab-Phe, 207
Stab-Py, 202, 204, 205
Stern-Volmer kinetics, 207—208, 244—246
Stimulated emission, 2, 4—7
    in carbon dioxide lasers, 17
    coefficient, 5
    of dye lasers, 34
    in ruby laser, 11
Stochastic Liouville equations, 286, 300
Straight line fitting, 101
Streak camera, 119, 122
Strickler-Berg equation, 33
Subpicosecond pulse generation, 42, 43
Superconducting cavity linear accelerators, 45
Surfactants, 36
Synchrotron radiation, 44, 216
Synchrotrons, 216—217, 233

T

TAC, see Time-to-amplitude converter
TCSPC, see Time-correlated single-photon
        counting
TEA lasers, 18—19

Temperature effects, 180—181
Thermal equilibrium, 5, 10
Thermodynamic equilibrium, 3
Thickeners, 36
Thiourea, 36
Third harmonic generator, 54
Time-correlated single-photon counting (TCSPC),
        97—99
    fitting functions in, 106—113
    lasers and, 103—106
Time-integrated signal, 274—276
Time-integration spectroscopy (TIS), 262—266, 273,
        275
    purpose of, 274
    spectra of, 268, 276, 293, 298
Time-resolved delayed luminescence spectroscopy,
        155—156
Time-resolved emission spectra, 113—115, 126
Time-resolved fluorescence techniques, 97—118
    data analysis in, 101—103
    in polymer colloid morphology measurement,
        198—209
Time-resolved fluorescence up-conversion, 121
Time-resolved Raman methods, 139—140
Time-to-amplitude converter (TAC), 98, 104, 106,
        213—214
TIS, see Time-integration spectroscopy
Torrey wiggles, 271
Transient absorption decay, microcrystal, 137
Transient absorption methods, 129—136
Transient absorption spectra, 133, 139—140
Transient signals, time-integration of, 274—276
Transient species, alternative methods for studying,
        129—142
Transient triplets, optical absorption of, 157—158
Transition energy, 4, 6
Translational diffusion, 112
Transverse modes, 9—10
Trap concentration, 181, 184
Triplet excitons
    migration of, 156—157
    photophysical studies of, 148—159
Triplet mechanism, 277—283, 292, 294—296
Triplet quenchers, 36
Triplet spectroscopy, signal-to-noise ratios of, 149
Triplet states, 148, 158
Triplet-triplet absorptions, 151, 158—159
Triplet-triplet annihilation, 156
True decay, 99

U

Ultrashort laser pulses, 34
    continuous wave mode-locked laser sources of,
        28—29
    mode locking in, 26—27
    in picosecond rare-gas halide lasers, 29
    in Q-switched and mode locked solid-state lasers,
        27—28
Ultraviolet absorption, 212
Ultraviolet excimer lasers, 2, 21—24
Ultraviolet light sources, 216—217

Unimolecular radiative process, 57
Up conversion, 119, 121
Uranium, 3, 36
Uranium-235 atoms, 3

V

VHPB copolymers, 250, 252—254
Vibrational modes, 58
Vibrational relaxation, 92, 113
Vibrational wave function, 57
Vidicon, 130
Vinyl aromatic polymers, fluorescence decay kinetics
    of, 107—108
1-Vinylnaphthalene copolymers, 71—72
Viologen groups, 82—93
Voltage discriminators, 114
Voltz mechanism, 59

W

Wavelength dependence, see Photomultiplier

Wavelengths, 3, 23

X

XeCl exciplex laser, 21, 23, 29
XeCl formation, 21

Y

YAG crystals, 10, 14
Yield
    definition of, 2—3
    of excimer lasers, 23—24
    of metal vapor laser, 24
YLF rod, 15
Yttrium crystals, 10, 14

Z

Zeeman interaction, 286
Zeeman states, 280, 287

JDR